Technical Change and Social Conflict in Agriculture

Westview Replica Editions

The concept of Westview Replica Editions is a response to the continuing crisis in academic and informational publishing. Library budgets for books have been severely curtailed. Ever larger portions of general library budgets are being diverted from the purchase of books and used for data banks, computers, micromedia, and other methods of information retrieval. Interlibrary loan structures further reduce the edition sizes required to satisfy the needs of the scholarly community. Economic pressures on the university presses and the few private scholarly publishing companies have severely limited the capacity of the industry to properly serve the academic and research communities. As a result, many manuscripts dealing with important subjects, often representing the highest level of scholarship, are no longer economically viable publishing projects—or, if accepted for publication, are typically subject to lead times ranging from one to three years.

Westview Replica Editions are our practical solution to the problem. We accept a manuscript in camera-ready form, typed according to our specifications, and move it immediately into the production process. As always, the selection criteria include the importance of the subject, the work's contribution to scholarship, and its insight, originality of thought, and excellence of exposition. The responsibility for editing and proofreading lies with the author or sponsoring institution. We prepare chapter headings and display pages, file for copyright, and obtain Library of Congress Cataloging in Publication Data. A detailed manual contains simple instructions for preparing the final typescript, and our editorial staff is always available to answer questions.

The end result is a book printed on acid-free paper and bound in sturdy library-quality soft covers. We manufacture these books ourselves using equipment that does not require a lengthy make-ready process and that allows us to publish first editions of 300 to 600 copies and to reprint even smaller quantities as needed. Thus, we can produce Replica Editions quickly and can keep even very specialized books in print as long as there is a demand for them.

About the Book and Editors

*Technical Change and Social Conflict in Agriculture:
Latin American Perspectives*
edited by Martín Piñeiro and Eduardo Trigo

Incorporating case studies of technological change in six Latin American countries, this book presents the results of a large cooperative research project (PROTAAL) that has led to a new interpretation of the process of technical change in agricultural development. The contributors contrast the perspective emerging from PROTAAL with two other views of technical change in agriculture: the theory of induced innovation and the political economy approach. They then describe the methodology developed by PROTAAL, which is highlighted in their analysis of the case studies. In the concluding chapters, the authors address important issues concerning the organization of agricultural research activities at the national and international levels and consider theoretical and policy implications for the analysis of technical change in Latin American agriculture.

Martín Piñeiro, formerly coordinator of the Cooperative Research Project on Agricultural Technology in Latin America (PROTAAL) at the Inter-American Institute for Cooperation on Agriculture (IICA), is a researcher at the Centro de Investigaciones Sociales sobre el Estado y la Administración in Argentina. Eduardo Trigo is co-coordinator of PROTAAL and coordinator of the Institutional Task Force on Technology Transfer and Adoption at IICA.

Technical Change and Social Conflict in Agriculture

Latin American Perspectives

edited by Martín Piñeiro
and Eduardo Trigo

Westview Press / Boulder, Colorado

A similar English version of Chapter 3 was published in *Food Policy* (England), Vol. 4, No. 3:169–177, 1979.

A similar Spanish version of Chapter 4 was published in *Comercio Exterior* (Mexico), Vol. 31, No. 3:303–318, 1981.

A similar Spanish version of Chapter 6 was accepted for publication in *Desarrollo Económico* (Argentina).

A similar English version of Chapter 7 was accepted for publication in *Food Policy* (England) in 1983. A Spanish version of the same chapter was published in *Desarrollo Económico* (Argentina), Vol. 21, No. 84:435–468, 1982.

A similar Spanish version of Chapter 8 was accepted for publication in *Comercio Exterior* (México).

A Westview Replica Edition

All rights reserved. No part of this publication may be reproduced or transmitted in any form or by any means, electronic or mechanical, including photocopy, recording, or any information storage and retrieval system, without permission in writing from the publisher.

Copyright © 1983 by Westview Press, Inc.

Published in 1983 in the United States of America by
  Westview Press, Inc.
  5500 Central Avenue
  Boulder, Colorado 80301
  Frederick A. Praeger, President and Publisher

Library of Congress Catalog Card Number: 83-50881
ISBN 0-86531-802-6

Printed and bound in the United States of America.

10 9 8 7 6 5 4 3 2

CONTENTS

List of Tables	ix
List of Figures	xiii
Preface	xv
Acknowledgements	xvi
About the Authors	xvii

PART I: TECHNOLOGICAL PERSPECTIVES

An Induced Innovation Interpretation of Technical Change in Agriculture in Developed Countries .... *Vernon W. Ruttan* 1

Aspects of the Political Economy of Technical Change in Developed Economies ... *Phillip LeVeen and Alain de Janvry* 25

Technical Change in Latin American Agriculture: A Conceptual Framework for its Interpretation .... *Martín Piñeiro, Eduardo Trigo and Raúl Fiorentino* 37

PART II: A LATIN AMERICAN PERSPECTIVE: SOME EMPIRICAL EVIDENCE

Introduction: A Few Words on Methodology 45

Social Relations of Production, Conflict and Technical Change: The Case of Sugar Production in Colombia .... *Martín Piñeiro, Raúl Fiorentino, Eduardo Trigo, Alvaro Balcázar and Astrid Martínez* 47

Agriculture in the Argentine Pampas: Technology Adoption in Corn Cultivation from 1950 to 1978 .... *Jorge Federico Sábato* 71

Technology as a Social Issue: Agricultural Research Organization in Latin America . ... *Eduardo Trigo, Martín Piñeiro and Jorge Sábato* 125

Social Articulation and Technical Change .... *Martín Piñeiro and Eduardo Trigo* 139

PART III: AGRICULTURAL RESEARCH ORGANIZATION IN LATIN AMERICA: ISSUES FOR THE FUTURE

Introduction 163

Foundations of a Science and Technology Policy for Latin American Agriculture ... *Eduardo Trigo and Martín Piñeiro* 165

International Technology: The International Agricultural Research Centers .... *John K. Coulter* 175

Legal Systems and Private Sector Incentives for the Invention of Agricultural Technology in Latin America .... *Donald D. Evenson and Robert E. Evenson* 189

Appendix 1 217

Conclusions: Towards an Interpretation of Technical Change in Latin American Agriculture .... *Martín Piñeiro and Eduardo Trigo* 235

References 241

LIST OF TABLES

An Induced Innovation Interpretation of Technical Change in Agriculture in Developed Countries:

1. Agricultural Output, Factor Productivity, Factor Endowments and Factor Price Ratios in Six Countries, 1880-1970 9

2. Annual Rates of Change in Agricultural Output, Factor Productivity, and Factor Endowments in Six Countries, 1880-1970 10

3. Relationships Between Fertilizer Use per Hectare and Relative Factor Prices in Six Countries 11

4. Relationship Between Use of Feed Concentrates per Hectare and Factor Prices. 12

5. Relationships Between Land per Worker and Relative Factor Prices in Six Countries 13

6. Relationship Between Power per Worker and Relative Factor Prices in Six Countries 14

7. Necessary Elasticity of Substitution to Explain Differences in Land-Labor Ratio by Price Ratio Differences: The Cross–Section Test 15

8. Necessary Elasticity of Substitution to Explain Differences in Land-Labor Ratio by Price Effect: the Time-Series Test 16

9. Price-Corrected Shares, Actual Factor Shares and Factor Prices: United States Agriculture, 1912-68 18

Aspects of the Political Economy of Technical Change in Developed Economies:

1. Characteristics of Agrarian Structure: U.S. and California 26

Social Relations of Production, Conflict and Technical Change: The Case of Sugar Production in Colombia:

1. Colombia: Area Cultivated and Harvested, Agricultural and Industrial Yields and Production of Sugar, White Sugar and Molasses, 1960-1977 (Total Country) 49

2. Colombia: Area Harvested and Production of Sugar Cane in Eight Surveyed Mills According to Source of Supply 50

3. Colombia: Percent Participation in Sugar Production of Mills, by Size: 1961-1974 51

4. Colombia: Chronological Order of Technological Innovations Adopted by Mills 57

5. Colombia: Indices and Ratios for Factor Prices in Sugar Production 1960-1978 . 60

6. Colombia: Distribution of the Income Generated by the Sugar Producing Sector Among the Social Groups that Participate in the Pre-Harvest Activities (Constant Prices Expressed in Thousands of 1970 Pesos) 61

Agriculture in the Argentine Pampas: Technology Adoption in Corn Cultivation from 1950 to 1978:

1. Indices of Physical Volume for Production in the Pampas (Five-Year Average, 1935-1939 = 0) 73

2. Value of Exports (FOB), and Agricultural Exports as a Percent of the Total (Five-Year Averages in Millions of 1950 Dollars) 73

3. Argentine Foreign Sector, by Broad Categories 1950-1977 (Millions of Constant Dollars) 74

4. Grain Harvests of Over 20 Million Tons per Year (Wheat, Corn, Grain Sorghum, Sunflower, Soy, Flax, Oats, Barley and Rye) 76

5. Annual and Five-Year Yields of Wheat, Corn, Grain Sorghum, and Soy, in Kilograms per Hectare Harvested (1950-1977) 76

6. Five-Year Price Averages (1960-100) 77

7. Indices of Total Physical Volume of Production for the Eight Selected Items (Index of Five-Year Averages, 1960-1961 to 1964-1965 = 100) 80

8. The Data in Table 7 Broken Down by Departments Within and Outside the Corn Belt 80

9. Constitución and Pergamino: Land Area Used for Farming and for Livestock (Thousands of Hectares, Five-Year Averages) 81

10. Constitución and Pergamino: Trends in Total Production Value by Hectare (Millions of 1960 Pesos, Five-Year Averages) 82

11. Constitución and Pergamino: Area Planted to Corn, Wheat and Soy (Thousands of Hectares, Five-Year Averages) 83

12. Trends in Intermediate-Sized Operations (from 101 to 1000 Hectares) in the Selected Departments; 1947 and 1969 Census Data 86

13. Price Variability: Wheat Compared with Corn (Average Prices for Each Period, in Dollars per Quintal) 87

14. Wheat—Income Risks, Production Risks and Market Risks 90

15. Corn: Income Risks, Production Risks and Market Risks 90

16. Beef: Income Risks, Production Risks, and Market Risks 90

17. Five-Year Fluctuations in Individual Income from Each Major Farm Crop 92

18. Fluctuation in Income from Farm Production and Overall Agricultural Production 93

19. Annual Variations in Land Planted to Each Commodity, and Total Value of Farm Production (Projected vs. that Obtained the Preceding Year). Total for the Eight Departments from 1950 to 1978 94

20. Wheat and Alfalfa (Thousands of Hectares Under Cultivation) 109

21. Province of Buenos Aires: Cattle Herds, Percent of Creole Stock, Total Value and Value per Head 109

22. Changes in the Land Surface Used for Livestock and for Farming in the Pampas (in Millions of Hectares), 1920-1954 112

23. Labor Requirements for Corn Production 117

24. Tractors: Subsidy on Mechanization 118

25. Subsidy Implicit in Credit as a Percentage of Slaughter Value 119

Technology as a Social Issue: Agricultural Research Organization in Latin America:

1. Budgetary Resources Allocated to Agricultural Research in Latin America and the Caribbean, between 1960 and 1980, Selected Year (Constant Value of 1975; Official Money Exchange Rate: National Currency/US Dollars, per Year Selected) 128

2. Human Resources (Professional Personnel) in Agricultural Research in Latin America and the Caribbean, from 1960 to 1980 (Selected Years) 129

Social Articulation and Technical Change:

1. Annual Rates of Increase in Agricultural Production and Productivity by Continent and by Groups of Products (1958-1978) 140

2. Production and Yields of Cotton, Beans and Corn in Northeastern Brazil 147

3. Rates of Increase for Annual Production and Yields for Products Under Study, 1958-1978 149

4. Relationships Among the Structural Dimensions and the Nature of Social Articulation: Eight Case Studies 154

International Technology: The International Agricultural Research Centers:

1. Research Priorities by Commodities... 183

2. Research Priorities by Production System 184

3. Research Priorities by Production Factors 185

Legal Systems and Private Sector Incentives for the Invention of Agricultural Technology in Latin America:

1. Comparison of "LAARC" Research Results and Availability of Patent/Variety Protection for Various Countries 194

2. Invention Patents Granted by Country: Selected Years 197

3. Regional Summary 199

4. Patenting in Foreign Countries: Selected Origin Countries - 1976 201

5. Distribution of Patents Granted by Class According to the International Patent Classification: Average for the Period 1970-74 202

6. Utility Models (Petty Patents) Granted 1967 204

7. Plant Patents Granted 1975, 1980 205

8. Private Sector Agricultural Research Activity in Selected Countries 206

9. Patenting Activity Related to Agricultural Production, Selected Countries 207

10. International Agricultural Patenting Activity by Per Capita Income Group, 1965-72 208

11. Origin of Agricultural Patents Granted by Selected Countries and Ratio of Total Patents Granted Domestically to Total Patents Granted to Nationals Domestically and Abroad, 1972-77 209

LIST OF FIGURES

An Induced Innovation Interpretation of Technical Change in Agriculture in Developed Countries:

1. Historical growth paths of agricultural productivity in the U.S., Japan, Germany, Denmark, France and the United Kingdom, 1880-1970 3

2. Input-output and land-labor ratios in agriculture, 1965 4

3. Relation between fertilizer input per hectare of arable land and the fertilizer arable land price ratio 5

4. Relation between tractor horsepower per male worker and the price of machinery relative to labor, 1970 6

5. Indices of bias on technical change 17

6. Machinery bias, actual machinery share, and price of machinery inputs in relation to cost of all agricultural inputs 19

A. Factor prices and induced technical change 23

Social Relations of Production, Conflict and Technical Change: The Case of Sugar Production in Colombia:

1. Colombia: flow of goods and services in sugar cane production in the Cauca Valley 52

2. Colombia. Ratio between export sugar prices and official mill price 55

3. Colombia: Use of fixed capital per ton of cane produced, excluding harvest. Average for the mills, the suppliers and for the total sugar-growing sector 58

4. Colombia: Land use per ton of cane produced. Average for mills, suppliers and for the total sugar-producing sector. 59

5. Colombia: Factor use ratios in the production of one ton of cane, excluding harvest, for the eight mills analyzed 65

6. Colombia: Factor use ratios for average independent suppliers 66

Agriculture in the Argentine Pampas: Technology Adoption in Corn Cultivation from 1950 to 1978:

1. Wheat, corn, grain sorghum and soy, aggregate production and land surface under cultivation (indexed with 1960 = 100):1950–1978 78

2. Constitución and Pergamino: Land surface planted to major commodities (1950/1951 – 1978/1979). Thousands of hectares 83

3. Annual index of pampas farm prices (wheat, corn and flax) and beef prices, 1923–1965 (Base 1935–1939 = 100) 88

4. Corn and beef prices (Liniers exchange price, on the hoof) 89

5. Effect of broad commodity price fluctuations on the adoption of technologies requiring increased fixed capital endowments 96

6. Effect of broad commodity price fluctuations on the adoption of technologies with different variable cost structures 98

7. Price and yield fluctuations: How they influence the adoption of technologies with different variable cost structures 100

8. Production alternatives, opportunity cost of capital, and technology adoption . 102

9. Limits on technical progress in the pampas region 104

10. Effects of implied subsidies in credit for the adoption of technology requiring increased fixed capital endowments 116

11. The new phase of technical progress in the pampas 120

International Technology: The International Agricultural Research Centers:

1. International Agricultural Research Centers. Annual core expenditures and core operating expenditures 1960-80. (In terms of constant 1981 dollars) 177

Legal Systems and Private Sector Incentives for the Invention of Agricultural Technology in Latin America:

1. Cost Curves for Monopoly and Competitive Firms........................... 212

PREFACE

This book presents the intellectual production of the first phase of the Cooperative Research Project on Agricultural Technology in Latin America (PROTAAL) and the most relevant papers presented by invitees at a meeting held in San José, Costa Rica in September 1981.

PROTAAL was initiated in early 1977, in an attempt to interpret the puzzling role of technology in Latin American agriculture, especially as it concerned public institutions like the National Research Institutes. From an organizational point of view, PROTAAL is a relatively extensive research network composed of independent teams, funded by a group of several international and national institutions, and a coordinating team with the principal executing agency.

The preocupation with the technological performance of Latin American agriculture arose as a natural outcome of our experience in the economic department of the Argentine National Research Institute at Castelar and the perceptions that we developed there about the overriding importance of public policy in agricultural development in general and technological modernization in particular. First, we attempted to explain the erratic behavior of Argentina's agricultural sector and the attitudes of political interest groups toward public policy with the traditional theoretical tools of economics. When these attempts were frustrated, we were stimulated to develop an alternative conceptual framework. These ideas were the basis for the analytical model used in PROTAAL.

When events that took place in Argentina during 1976 forced us to leave, first the institution and ultimately the country, some of us decided to continue together and build a small research operation that would allow us to pursue some of the ideas that had interested us previously. In the end, circumstances dictated that only two of us would see the project through, a process that took six more years.

As it is common in these cases, the objectives we set ourselves were impressive. We would try to characterize technical change, linking it to the performance of the public sector, and providing an explanation for its intensity and qualitative characteristics on the basis of economic and social structure and the conflicts that were generated from it. Furthermore, we would explain the effects of technical change on agricultural development, after which we would make normative suggestions as regards public policy and the organization of research activities.

In retrospect, it must be recognized that we did not reach these overly ambitious goals. However, a large number of papers and four books in Spanish are testimony to the productivity of the network which developed through the years and the potential of this style of research operation. Furthermore, we think that the perceptions developed by PROTAAL have had an influence in the development of new ideas regarding an explanation of technical change and have helped to set in motion in a number of Latin American countries social processes that will lead to better and more effective research institutions.

The analytical framework followed by the project attempts to mesh the basic propositions of the two leading theories of technical change, that is, the theory of induced innovation and the political economy of technical change. The project owes a considerable intellectual debt to Vernon Ruttan and Alain de Janvry for their work on this subject.

The main proposition of the theory of induced innovation is that the qualitative characteristics of technical change (its biases) are determined by market forces. More specifically, the theory suggests that microeconomic decisions regarding the adoption of innovation and the allocation of resources by research institutions respond to the relative scarcity of factors of production as reflected by relative market prices. In this way, technical change has eased relative factor scarcity and has efficiently contributed to a balanced economic growth.

Opposing this view, the political economy of technical change emphasizes the imperfection of markets, especially in developing economies, and argues that the principal determinant of the intensity and biases of technical change is the relative power of interest groups. In addition, political economy places a considerable importance on the international dimension and the subordination of agriculture to international industrial and financial capital.

The analytical framework proposed by PROTAAL borrows from both perspectives, integrating the historical dimensions of political economy with the more detailed and precise economic reasoning of neoclassical microeconomics. In addition, it gives center stage to the concepts of conflict and power and their role in the definition of public policy.

The difficulties of integrating different theoretical perspectives into one interpretative framework proved to be considerable. Some of them came up during the development of the empirical studies; others presented themselves during the final comparative analysis and while formulating normative prescriptions for public policy. Because of these problems, it was decided to have a seminar where the PROTAAL results could be analyzed in a comparative manner with the work of some of the leading exponents of the theories from which we had drawn.

This volume is a selection of the work done in PROTAAL and of the papers presented in the seminar. In this selection we have attempted to showcase the PROTAAL effort, but in a way that it can be evaluated in relation to other perspectives.

The first part of the volume presents papers by Ruttan, as a main architect of the application of induced innovation theory to agriculture, and by de Janvry and LeVeen as exponents of the political economy perspective. It also includes a paper by Piñeiro, Trigo and Fiorentino written at the very beginning of the PROTAAL project, as a basic analytical framework.

The second part develops the PROTAAL work. The first two papers, by Piñeiro and others, and Sábato, are summary presentations of empirical studies that illustrate clear and definite cases of innovative process that are systematized and explained in the following two papers by Trigo, Piñeiro and Sábato, and Piñeiro and Trigo. They make a comparative analysis of the eight empirical studies developed within the project, giving the reader an insight into the determinants of technical change in Latin American agriculture.

The third part of the volume begins with a chapter by Trigo and Piñero which presents the main policy implications derived from the PROTAAL research. The two following papers by Coulter, and Evenson and Evenson expand two policy issues touched upon in the previous paper: the role of the International Centers and the growing importance of the private sector in the innovative process.

ACKNOWLEDGEMENTS

PROTAAL owes a great deal to many people. We would like to make special mention of some who, through their efforts, made the project and this book possible. Reed Hertford, Dick Dye, Jim Himes and Peter Hakim from the Ford Foundation had the vision and courage to help us make the initial steps in difficult times. Ubaldo García convinced us to establish ourselves in IICA and helped us to do so. José Emilio G. Araujo had the courage to accept a proposition that implied considerable risk.

The funding institutions and the persons that at different times were responsible for their support include: K. Rao, Ruth Zagorin, David Steadman, Elizabeth Fox and very especially Anthony D. Tillett from IDRC; Mario Valderrama, Jan Flora, William Saint and Michael Redcliff from the Ford Foundation; Michael Gucovsky, Emma Torres, Angel Herrera, and Eduardo Gutiérrez from UNDP; Ralph Davidson from the Rockefeller Foundation; Diego Londoño, at that time with the Colombian Institute for Agriculture (ICA); and Raúl do Valle from the Brazilian Government.

Our colleagues from IICA include Augusto Donoso, Hugo Cohan, Ernesto Liboreiro, Mario Kaminsky, Arnaldo Veras, Manuel Rodríguez, Malcolm H. MacDonald, and Enrique Blair, all of whom helped us at different times in many ways.

Colleagues in the project who shared with us their efforts and knowledge in problem solving are: Jorge Sábato, Celia Barbato, Otto Flores, Osvaldo Barsky, Gustavo Cosse, Miguel Murmis, Oscar Marulanda, Humberto Rojas, Raúl Fiorentino, Alvaro Balcázar, Astrid Martínez, and very especially Albert Hirschman, Orlando Fals Borda, Guillermo O'Donnell, Alain de Janvry, Armando Samper, Luis Marcano, José Marull and Ubaldo García, who acted as advisers to different parts of the project. We would also like to thank all those who contributed to the seminar, especially Vernon Ruttan, who has always willingly accepted difficult challenges and has helped us to organize our thinking, even though it might be different than his own.

Finally, we would like to express our deep appreciation to Elizabeth Lewis for her excellent translation of Chapters 5 and 8 originally written in Spanish; to María Cuvi for the organization of the texts, and to Michael Snarskis for the preparation of the manuscript and his editorial work.

Martín Piñeiro and Eduardo Trigo – 1982

AUTHORS

MARTIN PIÑEIRO. Coordinator, Cooperative Research Project on Agricultural Technology in Latin America (PROTAAL), Centro de Investigación sobre el Estado y la Administración (CISEA), Buenos Aires, Argentina.

EDUARDO TRIGO. Until December, 1982 was Co-coordinator, Cooperative Research Project on Agricultural Technology in Latin America (PROTAAL). Since January, 1983, Senior Research Officer, International Service for National Agricultural Research (ISNAR), The Hague, Netherlands.

ALVARO BALCAZAR. Investigator, Oficina de Investigaciones Socio-Económicas y Legales (OFISEL), Bogotá, Colombia.

JOHN K. COULTER. World Bank, Washington, D. C.

ALAIN DE JANVRY. Professor, Department of Agricultural and Resource Economics, University of California, Berkeley.

DONALD D. EVENSON. Partner, Craig and Antonelli, Washington, D. C.

ROBERT E. EVENSON. Professor of Economics, Yale University.

RAUL FIORENTINO. Coordinator, Projecto de Desenvolvimento Rural Integrado do Nordeste do Brasil (DRIN), Recife, Brasil.

PHILLIP LEVEEN. Director, Public Interest Economics-West, University of California, Berkeley.

ASTRID MARTINEZ. Professor, Universidad Nacional, Bogotá, Colombia.

VERNON RUTTAN. Professor, Department of Agricultural and Applied Economics and Department of Economics, University of Minnesota.

JORGE FEDERICO SABATO. Investigator, Centro de Investigaciones sobre el Estado y la Administración (CISEA), Buenos Aires, Argentina.

PART I

TECHNOLOGICAL PERSPECTIVES

AN INDUCED INNOVATION INTERPRETATION OF TECHNICAL CHANGE IN AGRICULTURE IN DEVELOPED COUNTRIES

Vernon W. Ruttan

ABSTRACT

Technological change in agriculture is analyzed from the induced innovation perspective, in which such change represents a dynamic response to shifts in resource endowments and in the social and economic environments. The theory of induced innovation is tested against the history of agricultural development in the United States, Western Europe and Japan, and is presented as a useful tool in the planning of future research.

INTRODUCTION

The tools of the economist are relatively blunt instruments with which to confront the grand theme of epochal growth and decline. Until a few decades ago comparative statistics was the most powerful tool available to the economist as a guide to empirical knowledge. Even modern neoclassical growth theory has been based primarily on an application of the tools of comparative statistics to the analysis of alternative growth paths. In the simple Harrod-Domar-Mahalanobis models which dominated growth theory in the 1950s, increases in the capital/labor ratio represented the only source of increase in per capita income. Even in the more sophisticated models that have been available in more recent years the growth of output is narrowly determined by growth of the labor force and of physical capital and by technical change and improvements in the quality of human capital. Technical change continues with few exceptions to be treated as exogenous to the economic system.

On the other hand, outside of growth theory proper, substantial progress has been made in the effort to interpret the process of technical change as endogenous rather than exogenous to the economic system. In this view technical change represents a dynamic response to changes in resource endowments and in the social and economic environment. The induced innovation perspective implies a much more optimistic view of the relationship between resource endowments and the possibilities for economic growth than the view that progress in science and technology is essentially autonomous and hence unresponsive to social and economic forces. The induced innovation perspective suggests that "the fundamental significance of technical change is that it permits the substitution of knowledge for resources, or of less expensive and more abundant resources for more expensive resources, or it releases the constraints on growth implied by inelastic resource supplies" (Ruttan, 1971:708).

In this paper I review the evolution of thought on induced technical change; summarize the series of attempts to test the theory of induced technical change against the history of agricultural development in the United States, in Western Europe and in Japan;[1] and address the question of whether the induced innovation perspective can become a useful tool in research planning.

THE THEORY OF INDUCED TECHNICAL CHANGE

The theory of induced technical change represents an attempt to clarify the impact of relative resource endowments, as mediated through factor and product markets, on the rate and direction of technical change. The term "induced innovation" was first used with reference to bias in the direction of technical change by Sir John Hicks in his *Theory of Wages* (1932). Hicks argued that changes in factor prices induce biases in the direction of technical change which save the progressively more expensive factors. He did not attempt to specify the mechanism of induced innovation.

Interest by economists in the issue opened up by Hicks lagged until the 1960s. In 1960 W. E. G. Salter objected to the Hicks formulation on the grounds that there is no incentive for competitive firms to develop new knowledge designed to save a particular factor.[2] While Salter's criticism diverted attention from work on the theory of induced innovation, it did encourage the emergence of a body of work on the choice of technology (Sen, 1959, 1962).

Greater interest in the process of induced innovation emerged in the early 1960s as a result of efforts to explain the apparent stability of the shares of capital and labor, in spite of a rising capital/labor ratio, in the United States (Fellner, 1961, 1971). A second source of interest in the theory of induced innovation grew out of a concern with technology policy in the field of economic development. During the late 1960s and early 1970s it was gradually perceived that technology policies based on a choice of technology perspective represented an inadequate response to a situation where the available agricultural and industrial technologies had been developed under conditions of factor endowments and factor prices that were sharply different from those in many underdeveloped economies. Let us turn to a brief review of the several schools or traditions that have emerged in the literature on induced technical change since the mid-1960s.

Growth Theory Approaches

The most fully developed attempt to construct a theory of induced innovation involved an attempt by Kennedy (1964), Samuelson (1965), and Drandakis and Phelps (1966) to incorporate the process of technical change into modern growth theory.[3] It was primarily the implications of factor share stability which interested Kennedy and the other growth theorists. They were apparently not interested in the issue of research resource allocation.

The staggering burden of assumptions carried by the Kennedy growth model approach to induced innovation has been examined by Nordhaus (1967, 1973), Wan (1971), and in Binswanger and Ruttan (1978). First, there is the assumption of an exogenously given budget for research and development of new techniques that is unresponsive to the productivity of research investment and hence to changes in the size of the firm which occur as a result of successful technical innovation. Second, there is the assumption of a given "fundamental" trade-off or transformation function (which Kennedy termed the innovation possibility frontier-IPF) between the rate of labor augmentation (or reduction in labor requirements) and the rate of capital augmentation (or reduction in capital requirements) which is stable over time, and is therefore independent of achieved levels of labor or capital augmentation.

Although the Kennedy growth theory approach to induced innovation was developed to permit technical change to occur endogenously —as a result of the working of economic factors rather than by postulation— it attempted to achieve this result by postulating an IPF in which technical change turns out to be just as exogenous as in a growth model without an IPF. It is hard to escape the conclusion that there is no real world research and development process which is consistent with a Kennedy type IPF. This inadequate microeconomic foundation accounts for the lack of any significant empirical research on induced innovation along the lines suggested by Kennedy, Samuelson, and Drandakis and Phelps.

Microeconomic Approaches

A microeconomic approach to induced innovation can be built directly on Sir John Hicks' original observation that changes on factor prices induce biases in the direction of technical change which save the progressively more expensive factors. The first attempt to develop a microeconomic approach to induced innovation was Ahmad's seminal article in 1966.

Ahmad employed the concept of a historical innovation possibility curve (IPC) which is the envelope of all unit isoquants of the subset of potential processes which an entrepreneur might develop given the existing state of knowledge in science and technology and a given research and development budget. Each process in the set is characterized by an isoquant with a relative low elasticity of substitution between capital and labor.

The Ahmad model shares one fundamental limitation with the Kennedy model: it does not treat research and development as a resource-using activity. It is clear, however, that research and development are resource-using investments that lead to benefits which accrue over time. The rate of technical change is a function of investment in research and development and the direction of bias is a function of the mix of labor-saving or capital-saving research projects in the R & D portfolio of a firm (or an industry or a nation). From an investment perspective, factor prices (including the rate of interest), product prices and the size of the market should be included among the factors which determine the optimum R & D investment portfolio and hence the rate and direction of technical change.

An attempt to extend the Ahmad model to incorporate the allocation of research resources to achieve technical change was introduced by Hayami and Ruttan (1970, 1971). Their attention was directed to the process of induced innovation as a result of a study of historical differences in the rate of productivity growth over time and of differences in productivity levels among countries in the agricultural sector (Figs. 1 and 2). It appeared completely unreasonable to expect

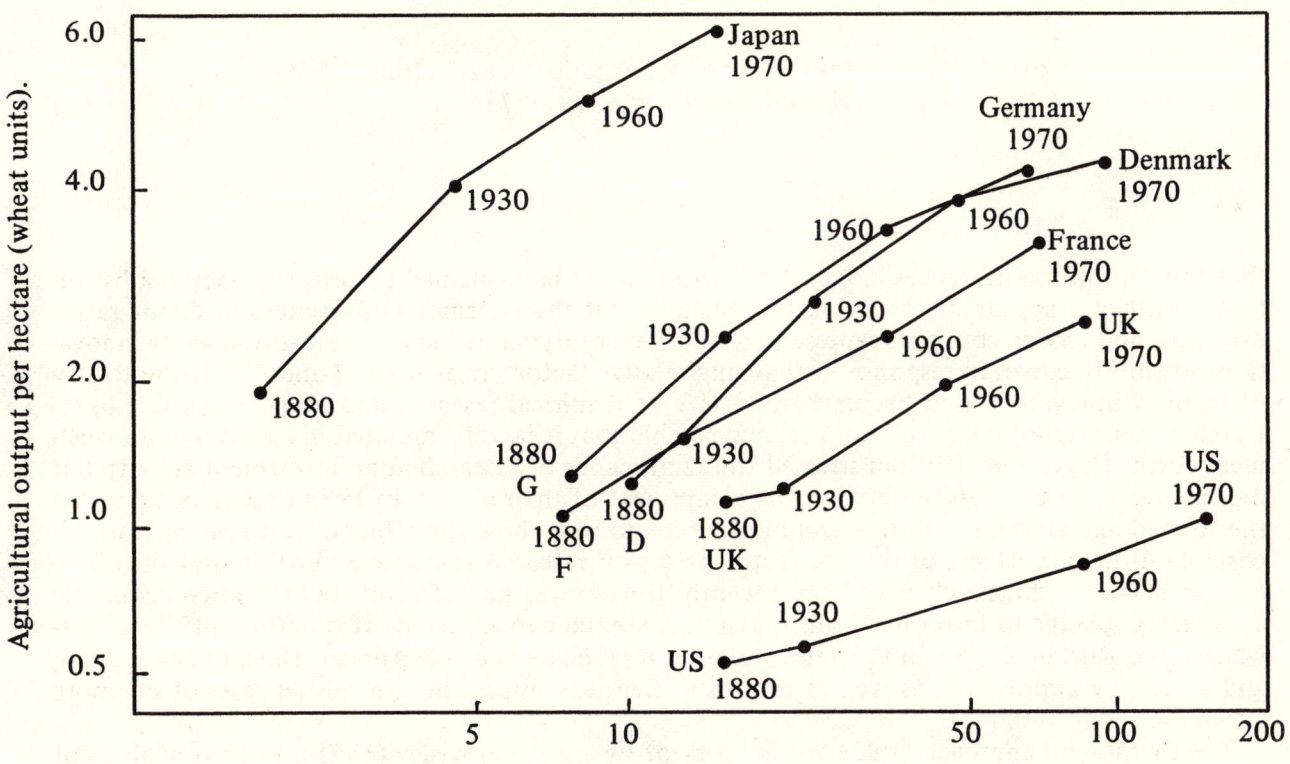

Fig. 1. Historical growth paths of agricultural productivity in the US, Japan, Germany, Denmark, France and the United Kingdom, 1880–1970 (Binswanger and Ruttan, 1978:44–87).

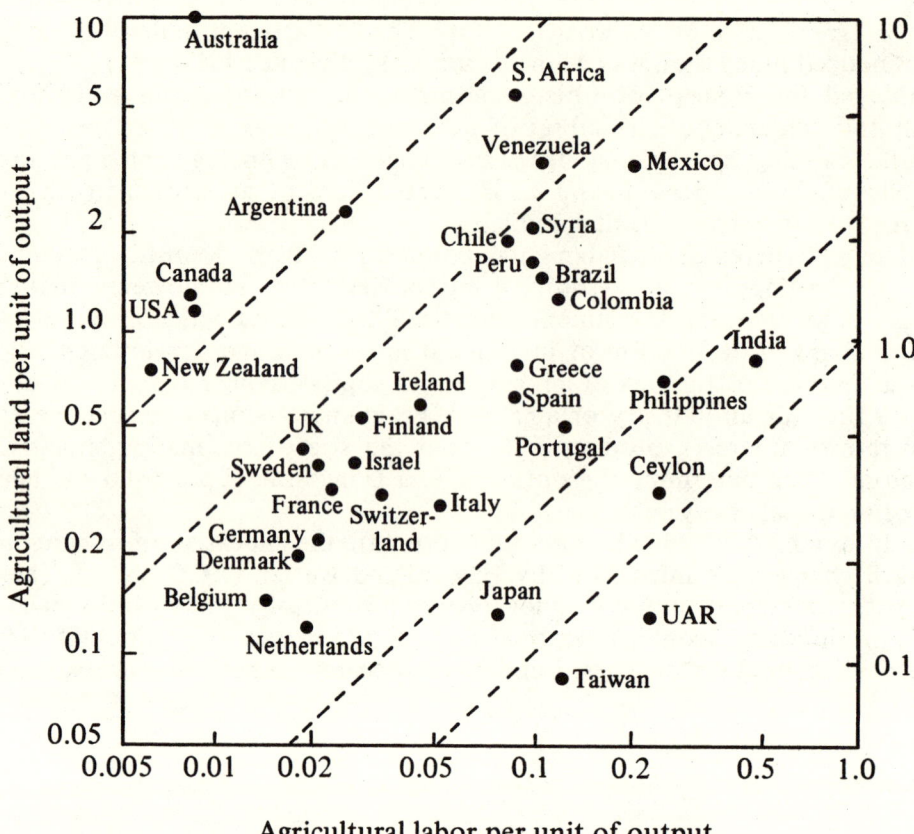

Fig. 2. Input-output and land-labor ratios in agriculture, 1965 (Hayami and Ruttan, 1971:73).

that the enormous differences in land/labor ratios could be explained by ordinary factor substitution.[4] Rather it appeared reasonable to conclude that the enormous differences in factor ratios over time and among countries represented a process of dynamic factor substitution along innovation possibility curves in response to changing relative factor prices (Figs. 3 and 4).[5] If the theory of induced innovation is to become productive of empirical research and useful as a guide to research resource allocation, it would seem desirable that it be reformulated in an explicitly investment form. Hayami and Ruttan stressed the significance of research as an investment activity but did not formally develop the investment component of their model. In 1973 de Janvry extended the Ahmad and Hayami — Ruttan graphical model to show how the effect of introducing research costs modifies the effects of relative factor prices on research resource allocation and the direction of bias in technical change. More recently Binswanger has reformulated the microeconomic approach explicity to introduce both research costs and expected payoff functions (1974a; Binswanger and Ruttan, 1978). In this reformulation it is shown that the Ahmad, Hayami and Ruttan, and de Janvry approaches, as well as the static Kennedy approach, are special cases of the more general investment model.

The investment approach builds on the work of Evenson and Kislev (1975). In their analysis of crop-breeding research they treat research as a sampling process. A probability distribution of potential yield increases is assumed in which the potential yield increases are determined by nature, the state of basic science, and plant-breeding techniques. Research is viewed as drawing successive trials from this distribution. The Evenson-Kislev view of the research process is related to the induced innovation process by identifying the research objective as shifts in the factor demand curves (per unit of output) corresponding to a given production process.[6] Research resource allocation decisions are cast in terms of deciding to pursue different lines of research which

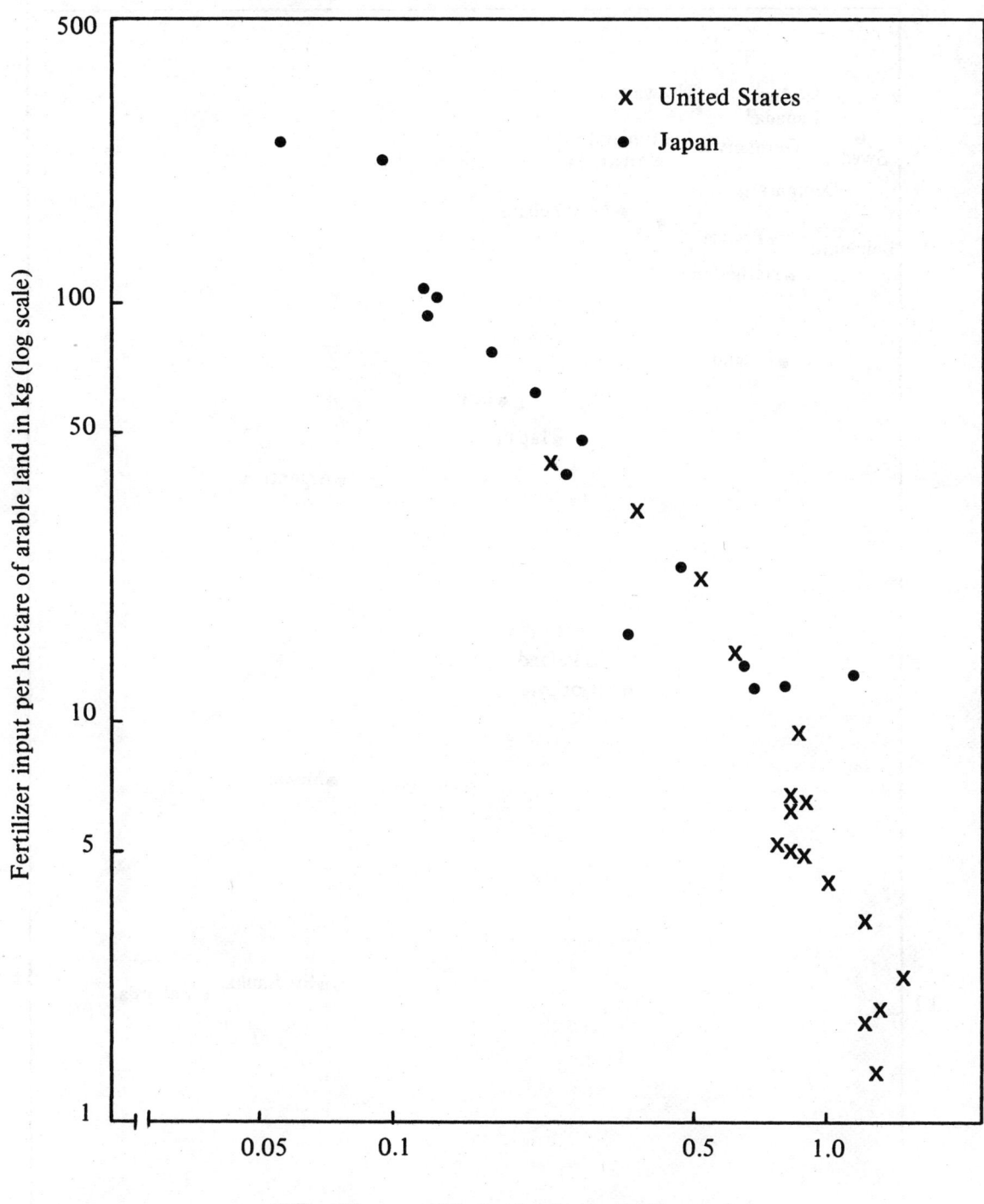

Fig. 3. Relation between fertilizer input per hectare of arable land and the fertilizer-arable land price ratio (Hayami and Ruttan, 1971:127).

result in the reduction of factor demands. Each research activity reduces labor and capital demands to a different extent, so that the research activities can be ordered according to the extent to which they move the production process in alternative factor-saving directions.[7] An investment model is then built in which the entrepreneur chooses a portfolio of research activities.

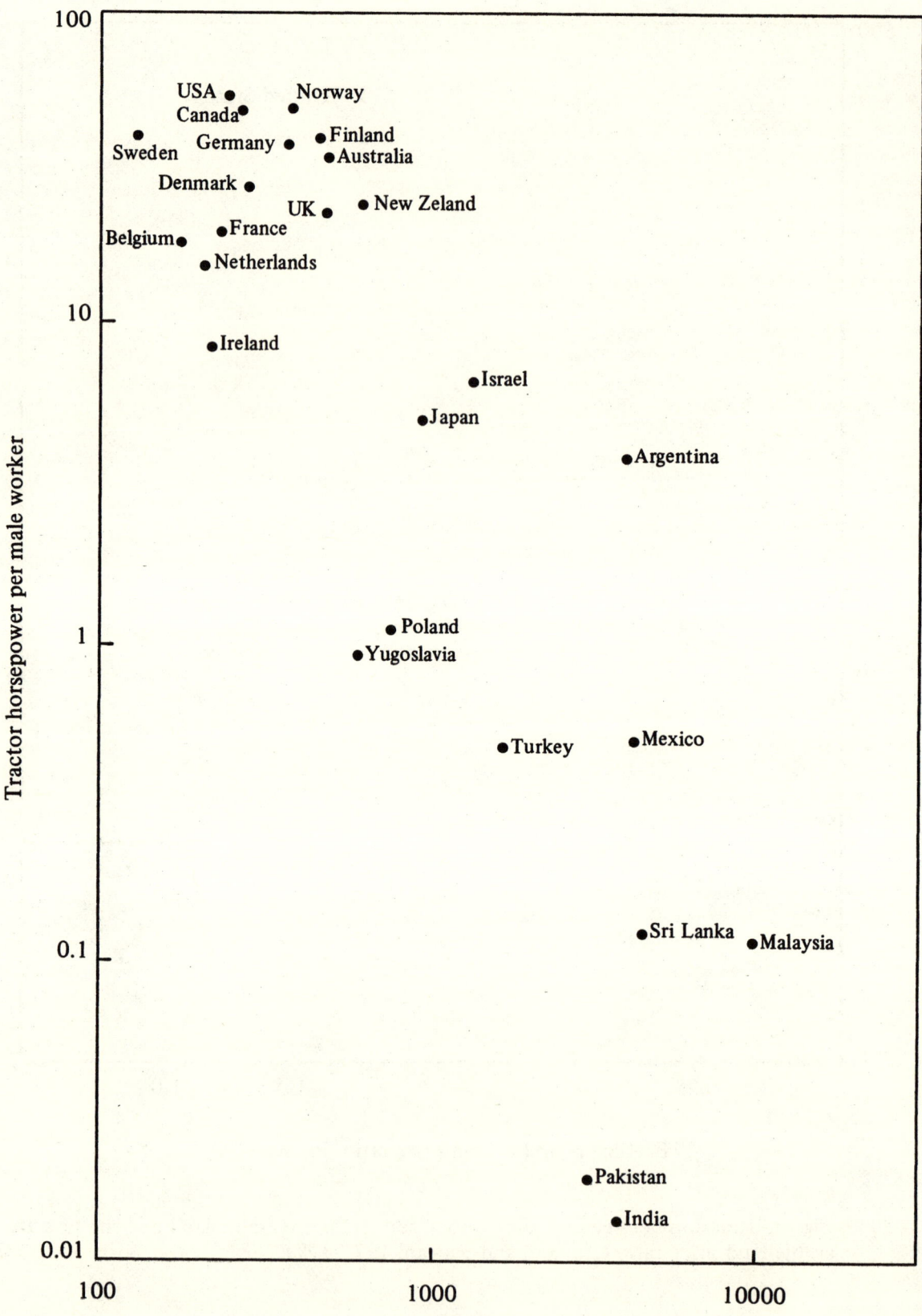

Fig. 4. Relation between tractor horsepower per male worker and the price of machinery relative to labor, 1970 (Yamada and Ruttan, 1980:540).

In the investment model, biases and rates of technical change are determined jointly by the following factors: (a) the relative productivity of alternative research lines; (b) changes in the cost of capital-saving (labor-saving) research; (c) changes in the scale of output; and (d) changes in the *present value of factor costs*. A rise in the wage rate, or the initial labor-output ratio, tends to increase labor-saving lines of research relative to other lines. In general this will lead to a more pronounced labor-saving bias.

Demand and the Rate of Technical Change

The theory of induced innovation has been directed primarily to efforts to understand the role of resource endowments and factor prices on the direction of technical change. At the microeconomic level, empirical work on demand-induced innovation began with the work of Griliches (1957) on hybrid corn and with the work of Schmookler on the railroad, petroleum, paper and agricultural equipment industries (1966, 1972). Griliches demonstrated the importance of demand in determining the location and diffusion of the invention of hybrid corn varieties. Schmookler concluded, from an exhaustive study of patent statistics, that the rate of return to inventive activity was of far greater importance than advances in the state of knowledge in explaining technical change in the four industries that be studied.

At the macroeconomic level, the responsiveness of technical change to final demand was first empirically tested and confirmed in a study by Lucas (1967). Ben-Zion and Ruttan (1978) show that the rate of input-saving technical change is higher in periods of growing demand than in periods of stable or declining demand in the United States economy.

Some Qualifications

Before proceeding to summarize the results of the several tests of the induced innovation hypothesis, I would like to respond to two misconceptions about the induced innovation hypothesis that have appeared in the literature. The first misconception is that an attempt to treat the process of technical change as endogenous to economic systems—as induced by relative resource endowments conveyed to public and private sector research scientists and managers through relative factor prices – implies that "research resource allocation is left to the hidden hand of market forces" (Biggs, 1981:3). My colleagues and I have repeatedly stressed that the generation of technical change cannot be left to an "invisible hand" that directs technology along an "efficient" path induced by original resource endowments or by the growth of demand (Hayami and Ruttan, 1971:56–61; Binswanger and Ruttan, 1978:327–357; Ruttan, Binswanger and Hayami, 1980). Development of capacity for the production of new knowledge leading to technical and institutional change is itself a product of institutional innovation. The way in which resource endowments and final demand express themselves in factor/factor and factor/product price ratios in strongly influenced by the efficiency of market processes, by the responsiveness of political institutions, and by the existing structure of income distribution. Neither the economic or political institutions which characterize modern industrial economies have occurred naturally. They are human artifacts.

A second misconception is that the induced innovation model is in some manner inconsistent with an interest group or structuralist model of technical change in agriculture (Beckford, 1972; Grabowski, 1979; Guttman, 1979; Biggs, 1981). But it has been emphasized from the very beginning that the dialectic interaction among farmers, research scientists and administrators, and legislative bodies is most effective when farmers are organized into local and regional associations that are capable of expressing their economic interests in the political marketplace (Hayami and Ruttan, 1971:57). When the distribution of either economic or political resources is highly unequal, the focus of scientific and technical effort will reflect the resource endowments of that part of the rural community in which economic and political resources are concentrated. If initial tests indicate a path of technical change that is inconsistent with relative resource endowments it immediately opens a series of questions about the nature of the structural constraints that are inducing an inefficient path of technical change and the institutional innovations that would lead to a more efficient allocation of research resources. These issues have been explored by de Janvry

(1973, 1978) and by Sanders, Ruttan and Feeny (1978) within the framework of the induced innovation model.

In my judgement the best alternative to the theory of induced innovation as an explanation of the generation of technical change is the evolutionary approach developed by Richard Nelson and Sidney Winter (1973, 1974). Along with Schuette (1972), it represents an interesting attempt to reach beyond neo-classical theory to describe the process of technical change.

The Nelson-Winter simulation models do not contain a production function in the usual sense and do not make a distinction between moving along a given production function and shifts in the production function. Firms produce with fixed-proportion techniques in any given period of time. In the early versions of the model (1973, 1974), firms start to search for new techniques of production if profits fall below a certain margin. The models assume that in this search the firms draw samples from a distribution of input-output coefficients in a way similar to that described by Evenson and Kislev. Except for the profitability check, research outcomes are random in their early versions. The inducement mechanism comes about through competition, selection and growth of the successful firms, not through an elaborate maximization scheme.

In the more recent versions of the model (1975, 1977), Nelson and Winter move beyond the satisficing assumptions and explicity introduce directed search, while continuing to emphasize the importance of uncertainty surrounding the research process. This brings their models much closer to the Ahmad-Hayami-Ruttan-Binswanger microeconomic investment model. We continue to view investment models with directed search as having the greatest potential for generating fruitful empirical research.

TESTING THE INDUCED TECHNICAL CHANGE HYPOTHESIS

I now turn to a review of the efforts to establish the plausibility of the induced technical change hypothesis by testing it against historical experience of agricultural development in the United States, Western Europe and Japan. Primary interest will focus on the issue of whether changes or differences in relative factor prices are associated with bias in the direction of technical change.[8]

The Hayami-Ruttan Type Tests

Hayami and Ruttan first tested the induced innovation hypothesis against the historical experience of agricultural productivity growth in Japan and the United States for the period 1880–1960 (1970, 1971). The analysis was later extended by Wade (1973) to include the United Kingdom, France and Denmark and by Weber (1973) to include Germany. The time period has been also extended to permit the test to cover the period 1880–1970.

The tests involve an analysis of the relationships between factor prices and the pattern of factor use associated with growth in both output per hectare and output per worker in the six countries (Tables 1 and 2). The model of *biological technology* outlined earlier in this paper (Fig. 3) suggests that a decline in the price of fertilizer relative to the price of land can be expected to induce advances in crop technology, such as the development of more fertilizer-responsive crop varieties, which can be characterized by new short-run production functions (such as i_1 below and to the right of i_0). A decline in the price of fertilizer relative to the price of labor can also be expected to induce technical changes leading (a) to the substitution of fertilizers, or other chemical inputs such as herbicides and insecticides, for more labor-intensive husbandry practices and (b) to the substitution of chemical fertilizers for farm-produced fertilizers such as animal manures and green manures.

A strong negative relationship was observed between the fertilizer/land price ratio and fertilizer use per hectare in all six countries (Table 3). A positive relationship between the price of labor relative to land and fertilizer use per hectare was also observed, although the relationship appears to have emerged later in France and Germany than in the other four countries. Given the enormous differences in the institutional and physical environments in which crops are produced among the six counties the similarity in the fertilizer/land price (P_F/P_A) response coefficients is quite remarkable. The implication is not only that farmers have responded in a roughly compara-

TABLE 1 Agricultural Output, Factor Productivity, Factor Endowments and Factor Price Ratios in Six Countries, 1880–1970 (Binswanger and Ruttan, 1978: Appendix 3.2.)

	Year	Japan	Germany	Denmark	France	United Kingdom	United States
Agricultural output Index (Y)	1880	100	100	100	100	100	100
	1930	223	192	279	146	111	204
	1960	334	316	422	235	185	340
	1970	428	412	459	334	236	403
Agricultural output per male worker in wheat units (Y/L)	1880	1.89	7.9	10.6	7.4	16.2	13.0
	1930	4.60	16.0	24.1	13.2	20.1	22.5
	1960	8.41	35.4	47.5	33.4	45.3	88.8
	1970	15.77	65.4	94.4	59.9	87.6	157.4
Agricultural output per hectare of agricultural land in wheat units (Y/A)	1880	2.86	1.25	1.19	1.06	1.10	0.513
	1930	5.06	2.47	2.95	1.50	1.18	0.555
	1960	7.44	4.01	4.65	2.48	1.94	0.811
	1970	10.03	5.40	5.27	3.70	2.61	0.981
Agricultural land per male worker in hectares (A/L)	1880	0.659	6.34	8.91	6.96	14.7	25.4
	1930	0.908	6.46	8.18	8.80	17.0	40.5
	1960	1.131	8.83	10.21	13.44	23.3	109.5
	1970	1.573	12.20	17.92	16.19	33.5	160.5
Days of labor to buy one hectare of arable land (P_A/P_L)	1880	1.874	967	382	780	995	181
	1930	2.920	589	228	262	189	115
	1960	2.954	378	166	166	211	108
	1970	1.315	244	177	212	203	108

NOTES: One wheat unit is equivalent to one ton of wheat. The method of constructing output measures in terms of wheat units is described in Hayami and Ruttan (1971:308–325).
Definitions of agricultural land are not strictly comparable among countries and over time, but generally include all land in farms, including crop land used for crops, pasture and fallow plus permanent pasture.
In Denmark the land price includes the value of agricultural land and buildings.

ble manner to similar factor/factor price ratios but that farmers have been able to respond in a similar manner as a result of comparable shifts in the short-run production function. This implies a similar institutional response in the allocation of research resources to make more fertilizer-responsive crop varieties available to farmers. A second test of the model of induced innovation in biological technology was made on the relationship between feed concentrates and factor prices. In animal agriculture feed concentrates play a role analogous to fertilizer in crop agriculture. The results are shown in Table 4.

The model of *mechanical technology* outlined earlier suggests that a decline in the price of land relative to labor can be expected to induce advances in mechanical technology leading to an expansion in the area cultivated per worker. Technical change leading to a decline in the price of machinery relative to labor would also contribute to an expansion in the area cultivated per worker.

The results of the empirical tests of the induced innovation hypothesis were not as clear cut in the case of mechanical as in the case of biological technology (Tables 5 and 6). The hypothesis that land area per worker is negatively related *both* to the price of land relative to labor and to

TABLE 2 Annual Rates of Change in Agricultural Output, Factor Productivity, and Factor Endowments in Six Countries, 1880–1970 (Binswanger and Ruttan, 1978: Appendix 3.2.)

	Japan	Germany	Denmark	France	United Kingdom	United States
1880–1970						
Agricultural output (Y)	1.63	1.59	1.71	1.35	0.96	1.56
Output per worker (Y/L)	2.39	2.48	2.46	2.35	1.89	2.81
Output per hectare (Y/A)	1.40	1.64	1.67	1.40	0.96	0.72
Land per worker (A/L)	0.97	0.73	0.78	0.94	0.92	2.07
1880–1930						
Agricultural output (Y)	1.62	1.31	2.07	0.76	0.21	1.44
Output per worker (Y/L)	1.79	1.42	1.66	1.16	0.43	1.10
Output per hectare (Y/A)	1.15	1.37	1.83	0.70	0.14	0.16
Land per worker (A/L)	0.64	0.04	−0.17	0.47	0.29	0.94
1930–1970						
Agricultural output (Y)	1.64	1.93	1.25	2.09	1.91	1.72
Output per worker (Y/L)	3.13	3.81	3.47	3.85	3.74	4.98
Output per hectare (Y/A)	1.73	1.97	1.44	2.28	2.00	1.43
Land per worker (A/L)	1.38	1.60	1.98	1.54	1.71	3.50
1930–1960						
Agricultural output (Y)	1.36	1.67	1.39	1.60	1.72	1.72
Output per worker (Y/L)	2.03	2.68	2.29	3.14	2.75	4.68
Output per hectare (Y/A)	1.29	1.63	1.53	1.69	1.67	1.27
Land per worker (A/L)	0.73	1.05	0.74	1.42	1.06	3.37
1960–1970						
Agricultural output (Y)	2.51	2.69	0.84	3.58	2.45	1.71
Output per worker (Y/L)	6.49	6.35	7.11	6.02	6.82	5.89
Output per hectare (Y/A)	3.03	3.02	1.26	4.08	3.01	1.92
Land per worker (A/L)	3.35	3.29	5.79	1.88	3.69	3.90

NOTES: One wheat unit is equivalent to one ton of wheat. The method of constructing output measures in terms of wheat units is described in Hayami and Ruttan (1971:308–325).
Definition of agricultural land are not strictly comparable among countries and over time, but generally include all land in farms, including crop land used for crops, pasture and fallow plus permanent pasture.
In Denmark the land price includes the value of agricultural land and buildings.

the price of machinery relative to labor was confirmed in the historical experience of the United States, the United Kingdom and Germany after 1850. In all six countries, except Germany during 1880–1913, land area is, as hypothesized, negatively related to the price of machinery relative to labor. The hypothesis that power per worker is negatively related *both* to the price of land relative to labor and of machinery relative to labor was confirmed in all cases except for Denmark and for France before 1920. Where the test was run for both an early and a late period, the results tended to be weakest for the early period. This may simply reflect the relatively weak inducement for mechanization in an environment characterized by very low wage rates.

The tests reported in this section that were conducted within the framework of the Hayami-Ruttan model are clearly consistent with the induced innovation hypothesis, but they do not represent an adequate test of the hypothesis. The tests do not permit us to determine: (a) whether the historical changes in factor use reflect the response of farmers to the rising economic value

TABLE 3 Relationships Between Fertilizer Use per Hectare and Relative Factor Prices in Six Countries

Country and period	Coefficient of prices of		Coefficient of determination	Standard error of estimate	Degrees of freedom
	Fertilizer relative to land	Labor relative to land			
	(P_F/P_A)	(P_L/P_A)	(R^2)	(S)	
Japan (1880–1960)[a]	−1.274* (0.057)	0.729* (0.220)	0.974	0.0810	14
Germany (1880–1913)[b]	−1.806* (0.009)	0.083 (0.515)	0.943	0.289	13
(1950–1968)	−0.377* (0.098)	0.799* (0.093)	0.954	0.100	15
Denmark (1910–1965)[c]	−1.20* (0.348)	0.958* (0.430)	0.87	0.310	9
France (1870–1965)[d]	−0.950* (0.332)	−1.375*I (0.362)	0.56	0.776	17
(1920–1965)	−0.664* (0.259)	0.485 (0.733)	0.386	0.538	7
United Kingdom (1870–1965)[e]	−1.130* (0.025)	1.010* (0.080)	0.92	0.218	17
Unites States (1880–1960)[f]	−1.357* (0.102)	1.019* (0.168)	0.970	0.083	14

NOTE: Equations are linear in logarithms. The numbers inside the parenthesis are the standard errors of the estimated coefficients.
*Significant at 0.5 level (one-tail test); I: inconsistent with simple induced Innovation hypothesis.

SOURCES
[a] Hayami Ruttan (1971).
[b] Weber (1975:23).
[c] Wade (1973:128).
[d] Wade (1973:134, 136).
[e] Wade (1973:179).
[f] Hayami and Ruttan (1971:132).

of land in relation to the price of fertilizer or to the increasing cost of labor compared to the cost of machinery along an unchanging, neo-classical production function; or (b) whether the production function available to farmers in the six countries has itself shifted to the left as a result of scientific and technical efforts by scientists, engineers and inventors in response to changing factor price relationships. The magnitude of the changes in factor prices, and in factor use, strongly suggest that the induced innovation hypothesis has been involved.

In the next section I present a simple two-factor test of the induced innovation hypothesis developed by Binswanger, using the same data employed in this section.

TABLE 4 Relationship Between Use of Feed Concentrates per Hectare and Factor Prices

Country and period	Coefficient of prices of		Coefficient of determination	Standard error of estimate	Degrees of freedom
	Concentrates relative to land	Labor relative to land			
	(P_O/PA)	(P_C/PA)	(R^2)	(S)	
Germany (1880–1913)[a] (Net oil cake imports)	−3.333* (0.569)	3.974* (1.221)	0.712	0.337	31
(1950–1968)	−1.567* (0.254)	2.381* (0.255)	0.973	0.337	15
Denmark (1880–1925)[b] (All imported concentrates per hectare)	−0.680* (0.300)	0.494* (0.124)	0.590	0.030	7
United Kingdom (1870–1965)[c] (All concentrates per hectare)	−3.642* (0.331)	3.634* (0.331)	0.970	0.137	17

NOTE: Equations are linear in logarithms. The numbers inside the parentheses are the standard errors of the estimated coefficients.
*Significant at P = 0.05 (one-tail test).

SOURCES
[a] Weber (1973:23).
[b] Wade (1973:128).
[c] Wade (1973:149).

The Binswanger Two-Factor Test

Binswanger has suggested a simple two-factor test to overcome the limitations in the Hayami-Ruttan test. The basic task in designing a test for induced innovation is, as noted above, to divide any changes in the labor/land ratio over time, or any cross-sectional differences between countries at a particular time, into (a) a component that results from ordinary price substitution and (b) a component that is the result of technical change.

The Binswanger two-factor test proceeds in two steps. First, the *necessary elasticity of substitution* to explain the observed factor ratio differences by factor price ratio differences is computed. If these exceed the econometrically estimated elasticities of substitution by a sufficiently large margin, the hypothesis of neutral technical change can be rejected. The argument on which the two-factor test is based, and the methodology used to estimate the pairwise elasticities of substitution, are summarized in Binswanger and Ruttan (1978:Chapter 3).

The results of the time series and cross-section tests conducted by Binswanger, shown in Tables 7 and 8, indicate that four different paths of technological development can be distinguished: (1) In 1880 the United States was on the same production function as France, Germany and the United Kingdom. After 1880 agricultural technology in the United States was strongly labor-saving. (2) Continental Europe experienced neutral technical change, or possibly even labor-using

TABLE 5 Relationships Between Land[a] per Worker and Relative Factor Prices in Six Countries

Country and period	Coefficients of prices of		Coefficient of determination	Standard error of estimate	Degrees of freedom
	Land relative to labor	Machinery relative to labor			
	(P_A/P_L)	(P_M/P_L)	(R^2)	(S)	
Japan (1880–1960)[b]	0.159[I] (0.110)	−0.219 (0.041)	0.751	0.016	14
Germany (1880–1913)[c]	−0.264* (0.066)	0.066*[I] (0.018)	0.393	0.012	31
(1950–1968)	−0.177 (0.139)	−0.476* (0.087)	0.975	0.083	15
Denmark (1910–1965)[d]	0.148[I] (0.084)	−0.357* (0.072)	0.910	0.030	9
France (1870–1965)[e]	0.398*[I] (0.202)	−0.088 (0.141)	0.323	0.189	17
(1970–1965)	0.050[I] (0.226)	−0.498* (0.166)	0.460	0.164	7
United Kingdom (1870–1925)[f]	−0.129* (0.033)	−0.139* (0.070)	0.610	0.041	17
(1925–1965)	0.279[I] (0.159)	−0.065 (0.256)	0.440	0.110	6
United States (1880–1960)[g]	−0.451* (0.215)	−0.486* (0.120)	0.826	0.084	14

*Significant at P = 0.05 (one-tail test).
[I] Inconsistent with simple induced innovation hypothesis.

NOTES:
[a] Arable land per male worker in Japan, Denmark, France and the United Kingdom; agricultural land per male worker in Germany and the United States.
[b] Hayami and Ruttan (1971). Land per worker (W7); Power per worker (W9).
[c] Weber (1973:24). Land per worker—Regressions (6) and (7); Power per worker—Regressions (4) and (5).
[d] Wade (1973:128).
[e] Wade (1973:134, 136).
[f] Wade (1973:149).
[g] Hayami and Ruttan (1971:130). Land per worker (W1); Power per worker (W5).

technical change until the 1960s. After the 1960s France and Denmark began to experience labor-saving technical change. (3) The United Kingdom experienced neutral technical change until 1930. After 1930 it began to experience strong labor-saving technical change, but its technology remains much more labor-intensive than U.S. agricultural technology. (4) Japan started from an extremely labor-intensive position. Its path of technological development since 1880 has been either neutral or, particularly in recent years, slightly labor-saving.

The Binswanger tests are clearly consistent with the induced innovation hypothesis. Yet there are some observations, based on the two-factor test, that are not consistent with the simple version of the hypothesis. For example, the United States followed a more labor-saving path with

TABLE 6 Relationship Between Power[a] per Worker and Relative Factor Prices in Six Countries

Country and period	Coefficients of prices of		Coefficient of determination	Standard error of estimate	Degrees of freedom
	Land relative to labor	Machinery relative to labor			
	(P_A/P_L)	(P_M/P_L)	(R^2)	(S)	
Japan (1880–1960)[b]	−0.665* (0.261)	−0.299 (0.685)	0.262	0.219	14
Germany (1880–1913)[c]	−0.238* (0.070)	−0.607* (0.020)	0.978	0.069	31
(1950–1968)	−0.234 (0.329)	−1.358* (0.207)	0.979	0.213	15
Denmark (1910–1965)[d]	1.494[I] (1.010)	−3.180* (0.861)	0.830	0.370	9
France (1870–1965)[e]	1.704*[I] (0.880)	−0.705 (0.614)	0.160	0.810	17
(1920–1965)	−0.443 (0.976)	−2.460* (0.715)	0.550	0.705	7
United Kingdom (1870–1965)[f]	−1.120* (0.295)	−1.090* (0.527)	0.810	0.075	17
United States (1880–1960)[g]	−1.279* (0.475)	−0.920* (0.266)	0.827	0.187	14

*Significant at P = 0.05 (one-tail test).
[I] Inconsistent with simple induced innovation hypothesis.

NOTES:
[a] Arable land per male worker in Japan, Denmark, France and the United Kingdom; agricultural land per male worker in Germany and the United States.
[b] Hayami and Ruttan (1971). Land per worker (W7); Power per worker (W9).
[c] Weber (1973:24). Land per worker—Regressions (6) and (7); Power per worker—Regressions (4) and (5).
[d] Wade (1973:128).
[e] Wade (1973:134, 136).
[f] Wade (1973:149).
[g] Hayami and Ruttan (1971:130). Land per worker (W1); Power per worker (W5).

respect to its initial land/labor ratio than did the four European countries, despite the fact that in the United States the rise in the price of labor relative to that of land was less rapid than in Europe.

There are several factors that may account for the less than complete consistency between the implications of the induced innovation hypothesis and the observed differences in factor price and factor use ratios. One is, of course, that there are fundamental biases in innovation possibilities. A second is the impact of borrowing—of technology transfer—from countries with different factor/price ratios. This may be particularly important for the countries with extreme differences in factor/price and use ratios such as Japan and the United States. If a country starts the process

TABLE 7 Necessary Elasticity of Substitution to Explain Differences in Land-Labor Ratio by Price Ratio Differences: The Cross-Section Test (Binswanger and Ruttan, 1978:66)

Item	1880	1930	1960	1970
U.S. vs. other countries				
Japan	2.08*	1.35*	1.95*	3.12*
Great Britain	0.29	1.47*	2.50*	2.70*
France	0.87	1.96*	5.79*	4.13*
Germany	0.80	1.16	2.49*	4.00*
Japan vs. Europe				
Great Britain	7.01*	1.21*	1.24	2.04*
France	3.26*	0.92	0.79	1.39*
Germany	4.13*	1.29	1.00	1.28
Great Britain vs. Continental Europe				
France	a*	2.47*	a*	17.12*
Germany	a*	0.98*	1.71*	5.72*
Continental Europe				
France vs. Germany	0.46	0.38	0.50	2.02*

NOTE: Critical value to reject hypothesis of equal technology is 1.34, that is, twice the value of σ for equiproportional changes in P_A and P_L.

$$\sigma_N = \frac{(A/L)_i - (A/L)_j}{(P_L/P_A)_j - (P_L/P_A)_i} \times \frac{\sqrt{(P_L/P_A)_i (P_L/P_A)_j}}{(A/L)_i (A/L)_j}$$

a* Denotes cases where the country with the higher land/labor ratio also has the higher land/price ratio. Such behavior is possible only if the country with the higher land/labor ratio employs a more land intensive technology, i.e. the hypothesis of equal technology is rejected. No common isoquant maps can be constructed through points P and Q in Fig. A.
*The paths of the two different countries differ significantly in land/labor intensity.

of modernization from an extremely labor-intensive position, as Japan did in the 1870s and 1880s, or as some developing countries are doing today, the only technologies that are available to be transferred from other countries will be more labor-saving than would be induced by its own factor endowments and price ratios. Similarly, if a country starts the process of modernization from an extremely labor-extensive position, as in the United States in the post-Civil War period, it is likely that the technologies which it borrows will be more land-saving than would be induced by its own factor endowments and price ratios. We are not, at this stage, able to provide quantitative estimates of the effects of fundamental and transfer biases.

The Binswanger Many-Factor Test

In spite of its formal rigor, a basic limitation of the two-factor test, even in comparison with the less rigorous tests of the Hayami-Ruttan type, is that a many-factor production process was

TABLE 8 Necessary Elasticity of Substitution to Explain Differences in Land-Labor Ratio by Price Effects: the Time-Series Test (Binswanger and Ruttan, 1978:67)

Time period	United States	Great Britain	France	Germany	Denmark	Japan[a]	
						Land-price basis	Land-rent basis
1880–1930	1.03	0.16	0.20	0.04	b	—	—
1890–1930	—	—	—	—	—	c*	0.33
1890–1910	—	—	—	—	—	c*	0.33
1910–1930	—	—	—	—	—	1.09	0.34
1930–1960	16.5*	c*	0.90	0.70	0.70	c*	—
1960–1970	d*	9.43*	c*	0.74	c*	0.40	—

NOTE: The critical ratio to reject the hypothesis of neutral technical change is $\sigma = 1.34$.

$$\sigma_N = \frac{\text{Percentage change in land/labor ratio between two periods}}{\text{Percentage change of labor price/land price ratio}}$$

with geometric means as a basis for the two percentage changes, that is,

$$\sigma_N = \frac{(A/L)_{i+j} - (A/L)_i}{(P_L/P_A)_i - (P_L/P_A)_{i+j}} \div \frac{\sqrt{(P_L/P_A)_{i+j}(P_L/P_A)_i}}{(A/L)_{i+j}(AL)_i}$$

where $i = 1880, 1930, 1960, 1970$.

[a] Data for 1890–1930 are taken from Table 3.10 and data for 1930–1960 from Table 3.1.
[b] Land/labor ratio declined very slightly, but price declined as well.
[c] Price ratio and land/labor ratio rose, which implies labor-saving technical change. (No common isoquant map can be constructed through P and Q in Fig. A. in this case.
[d] No price change: technical change labor-saving.

*Significantly labor-saving.

treated as if it were a two-factor process. The test neglected the influence of prices other than land and labor on the land/labor ratio. In this section we present the results of a many-factor test designed by Binswanger (1974a, 1974b, and 1978:Chapter 7).[9]

The many-factor test is based on directly measured biases in the direction of technical change on the use of individual factors, rather than on factor ratios, for the United States agricultural sector from 1912 to 1968. Five factors were included in the measurement of bias: land, labor, machinery, fertilizer and other inputs. Land, however, was omitted from the test of the induced innovation hypothesis because agricultural land prices were, at least until recently, determined by factors that are largely endogenous to the agricultural sector.

The rationale for the many-factor test can be summarized as follows. Suppose that innovation possibilities are neutral and that factor prices are exogenous to the agricultural sector. Factor-saving bias would be inversely related to the direction of change in the factor price. Furthermore, turning points in factor price trends should be followed, after a lag, by corresponding changes in the direction of bias. If, on the other hand, innovation possibilities are not neutral, then it is possible that a factor-using bias may be associated with a rise in the price of the corresponding factor. Such an occurrence can be used to test for the presence of a fundamental bias in innovation possibilities. Induced innovation may either offset or reinforce such a fundamental bias. But in the case of a factor-saving shift in prices, an acceleration of the price rise should, after some years, result in a decrease in the rate of factor-using bias.

The many-factor test developed by Binswanger involves partitioning the observed changes in factor shares into a component due to ordinary factor substitution and a component due to bias in the direction of technical change. This was accomplished in two steps: (a) a translog production function was used to estimate elasticity of substitution parameters from an independent sample; (b) the parameters were used to adjust the time-series factor share changes to obtain the part that was caused by technical change alone. A formal statement of the above argument has been developed by Binswanger (1974a, 1978).

The price-corrected factor shares are presented along with the actual shares and the factor prices in Table 9. The indices of bias for each input, as computed from the price-corrected factor shares, are shown in Fig. 5. Technical change exhibited a very strong fertilizer-using bias over the entire period. There was also a strong machinery-using bias and, after 1948, a strong labor-saving bias.

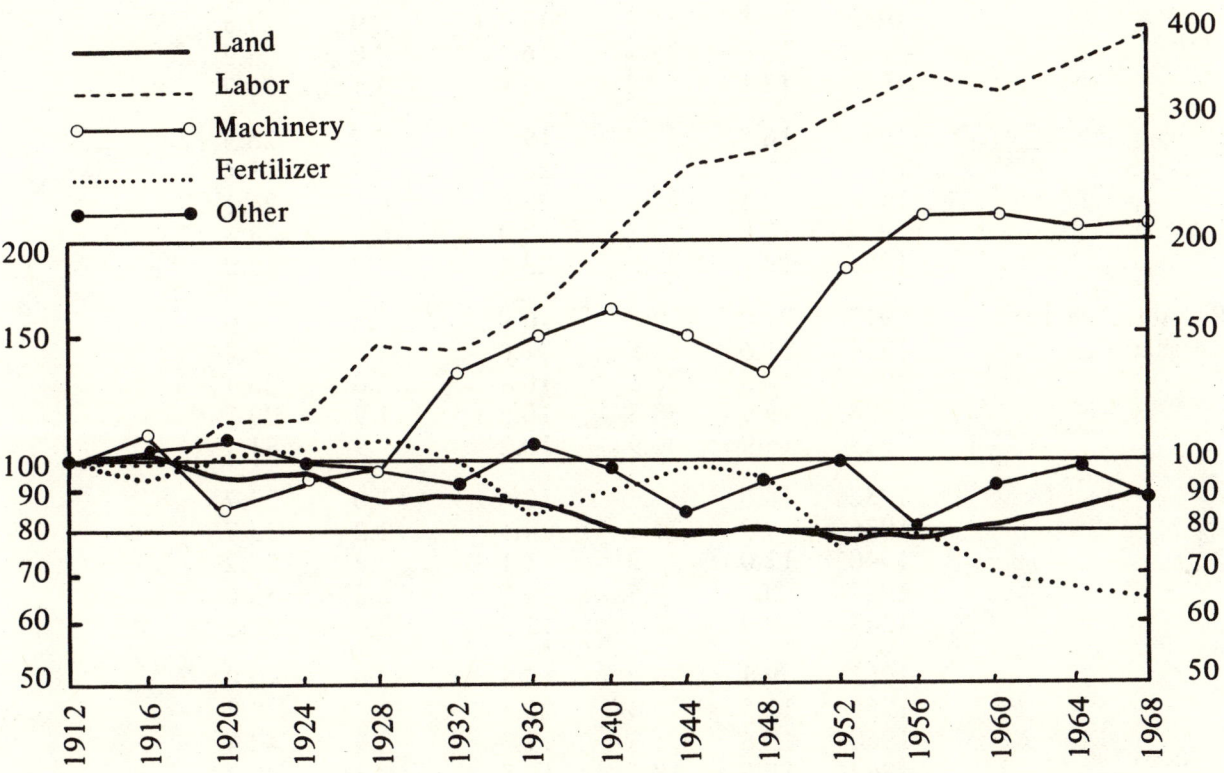

Fig. 5. Indices of bias on technical change (Binswanger and Ruttan, 1978:221).

The machinery case is particularly interesting (Fig. 6). The change in the machinery share for the period as a whole is entirely due to bias. The bias alone would have moved the machinery share from 10.9 per cent in 1912 to 23.1 per cent in 1968. The actual machinery share rose only to 19.1 per cent. A striking feature of the bias and the price movements is the sequence of turning points. From 1916 to 1920 machinery prices fell. A substantial machinery-using bias began in 1920. After 1920 the machinery price index rose slowly. By 1928 it had returned to its 1912 level. It continued to rise more rapidly until about 1940. By 1940 the bias changed from a machinery-using to a machinery-saving direction until 1948. From 1940 to 1948 machinery prices declined and eight years later another period of rapid machinery-using technical change began. The machinery price index rose again from 1948 to 1960. In 1956, eight years after prices began to rise, the machinery-using bias disappeared and technical change turned neutral.

TABLE 9 Price-Corrected Shares, Actual Factor Shares and Factor Prices: United States Agriculture, 1912–68 (Binswanger and Ruttan, 1978:222)

		Year	Land	Labor	Machinery	Fertilizer	Other	Index of input prices with respect to agricultural output prices
A.	Price-corrected factor shares S_j. Model I estimates	1912	21.0	38.3	10.9	1.9	28.0	
		1916	21.2	36.7	11.6	1.8	28.7	
		1920	19.6	39.3	9.3	2.1	29.7	
		1924	20.0	39.7	10.3	2.2	27.8	
		1928	18.1	41.4	10.4	2.7	27.4	
		1932	18.8	40.3	14.3	2.7	24.0	
		1936	18.9	32.5	16.3	3.0	29.3	
		1940	16.8	34.3	17.6	3.9	27.5	
		1944	16.5	38.4	16.1	4.8	24.2	
		1948	17.1	37.2	13.9	5.1	26.7	
		1952	16.5	29.8	19.7	5.7	28.3	
		1956	16.3	30.6	23.1	6.5	23.4	
		1960	17.1	27.2	23.4	6.1	26.1	
		1964	17.8	25.8	22.4	6.7	27.3	
		1968	19.1	25.3	23.1	7.2	25.3	
B.	Actual factor shares S_i	1912	21.0	38.3	10.9	1.9	28.0	
		1916	21.6	36.5	11.6	1.9	28.4	
		1920	17.3	40.5	10.1	2.0	30.1	
		1924	19.7	38.5	10.3	1.7	29.7	
		1928	15.9	40.9	10.2	1.9	31.1	
		1932	18.6	37.6	12.6	1.6	29.7	
		1936	14.9	34.7	14.5	2.2	33.7	
		1940	12.0	35.3	15.1	2.3	35.2	
		1944	8.5	39.5	14.0	2.3	35.6	
		1948	9.4	37.7	12.2	2.4	38.3	
		1952	9.8	29.7	17.5	3.0	40.0	
		1956	11.5	27.4	20.1	3.3	37.8	
		1960	15.6	21.3	19.8	2.9	40.4	
		1964	17.5	18.3	18.5	3.3	42.3	
		1968	20.4	15.8	19.1	3.6	41.1	
C.	Factor prices relative to agricultural aggregate input prices 1912 = 100	1912	100.0	100.0	100.0	100.0	100.0	100.0
		1916	105.2	99.2	102.1	98.1	96.4	107.7
		1920	81.1	107.1	83.5	88.0	107.8	97.4
		1924	99.2	112.1	93.1	77.6	88.8	120.0
		1928	79.7	117.2	97.7	68.4	90.4	131.5
		1932	97.8	118.4	140.7	78.2	61.7	164.5
		1936	59.5	97.2	162.1	85.4	95.0	116.7
		1940	49.6	101.7	164.1	58.8	91.0	176.0
		1944	32.2	107.4	120.7	31.1	104.7	202.3
		1948	34.0	115.5	105.6	23.5	103.9	214.5
		1952	39.6	119.0	130.6	23.2	93.1	230.6
		1956	48.2	134.8	140.0	21.8	75.9	302.7
		1960	71.6	141.6	155.0	17.7	68.0	355.1
		1964	82.9	149.6	159.9	15.5	66.4	407.9
		1968	100.8	160.7	154.2	12.2	58.8	477.2

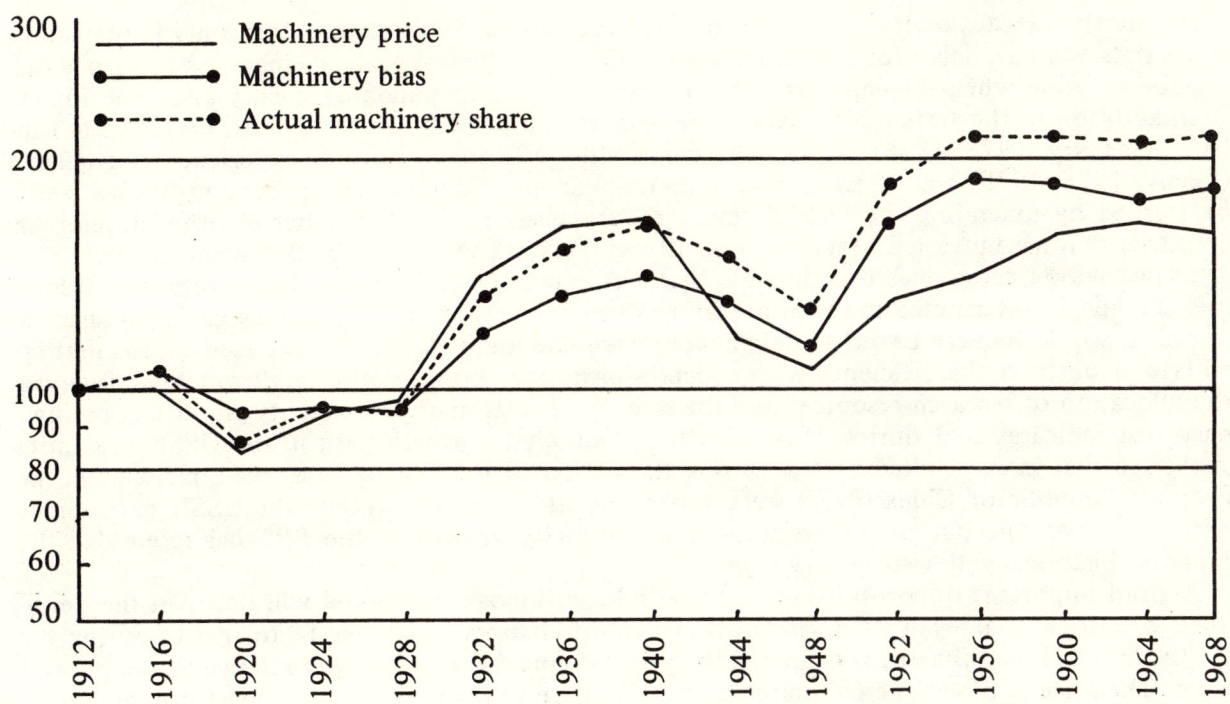

Fig. 6. Machinery bias, actual machinery share, and price of machinery inputs in relation to cost of all agricultural inputs (Binswanger and Ruttan, 1978:224).

The above sequence of turning points appears fully consistent with the induced innovation hypothesis. Indeed, the price responsiveness of the machinery biases is quite remarkable. Evenson (1968) documented a five-to eight-year lag between the initiation of a research effort and the beginning of a productivity impact from the research output in U.S. agriculture. A lag much shorter than eight years would be consistent with the choice-of-technique rather than the induced innovation hypothesis.

The results of the many-factor test support a conclusion that, in United States agriculture, the biases in technical change affecting machinery, fertilizer and labor use have been very responsive to changes in factor prices with a lag of eight years of longer. The lags have, however, tended to become shorter over time. Furthermore, in the case of machinery, fertilizer and labor, the biases in the share changes due to changes in factor prices were large and greatly exceeded the changes that could be explained by simple substitution effects. It is also clear from the machinery case that innovation possibilities have not always been neutral. But even in the case of machinery, changes in factor prices have at times partially offset and at other times have reinforced the exogenous bias in innovation possibilities.

Some Further Tests: An Agenda

The formal tests of the induced innovation hypothesis against agricultural development experience have drawn primarily on the history of presently developed economies, on the experience of economies in which the market plays an important role in the determination of factor and product prices, and from a period characterized by a long term secular decline in energy prices. It would be very useful to test the induced innovation hypothesis against a broader range of historical and institutional experience.

At the time Professor Hayami and I initiated our studies in the late 1960s, only limited time series data were available for most developing countries. The test against Japanese experience did include a period when the Japanese economy was, by present standards, clearly underdeveloped.

In addition to the tests against Argentine agricultural development experience referred to earlier (de Janvry, 1973, 1978), we now have available tests drawing on the experience of Thailand (Feeny, 1976, 1978) and of Egypt and Syria (Ahmad and Kubursi, 1977). But a great deal could be learned by extending the tests to include the experience of a number of other developing countries that have invested heavily in agricultural research, such as India, Brazil and Nigeria.

In nonmarket economies it is possible that differences or changes in relative resource endowments could be interpreted to research planners and managers through planning guides or shadow prices. It would be very useful to gain a more adequate understanding of the institutions that facilitate or obstruct the efficiency with which information about resource endowments influences the allocation of research resources and the rate and direction of productivity growth. A preliminary test indicates that during 1950–70 the productivity growth path in the USSR was quite similar to that in the United States. During the period of the first five-year plan, planners in the People's Republic of China (PRC) were apparently attempting to follow the USSR path. Since 1960, however, the pattern of agricultural productivity growth in the PRC has resembled the Japanese historical path (Ruttan, 1978).

A third important opportunity to test the induced innovation model will occur in the developed countries as a result of a series of significant changes in relative factor prices during the 1970s. In the United States for example, the price of land has risen sharply relative to the price of labor. The price of direct energy inputs has risen relative to the price of labor. Fertilizer and other chemical prices have continued to decline relative to the price of land. It is likely that the increased noise in the system, with long run trends in relative factor prices still somewhat uncertain, will result in less clear-cut results when 1970–80 data are added to the 1880–1970 data that served as a basis for the earlier tests of the induced innovation hypothesis.

INDUCED INNOVATION AND AGRICULTURAL RESEARCH POLICY

What are the implications of the induced innovation perspective for agricultural research policy? Sensitivity to the induced innovation perspective should affect agricultural research policy at two levels. One is the general economic policy and planning level. The second is the research program planning and resource allocation level.

Economic Policy and Agricultural Research

A clear implication of the induced innovation perspective is that biases in factor prices relative to each other or relative to product prices affect not only current choices of technology but also the technical choices that will become available a decade or more in the future. This implies that distortions in factor/factor or factor/product price ratios can have a much more serious effect on both the rate and direction of technical change and on the economic organization of the agricultural sector than that produced when current technology choice alone is affected by price distortions. The effects of price distortions on technology choice can, as the price distortions are eliminated, be corrected rather rapidly through the combined effect of depreciation and obsolescence. But the effect of price distortions on research resource allocation are not so easily corrected. One has only to review the history of agricultural development in Argentina over the last several decades to be convinced of the burden imposed on the generation of technology by price distortions (de Janvry, 1973, 1978). This implies that price distortions introduced to achieve income distribution objectives can be very costly in terms of the generation of growth dividends from technical change. It forces price policy back to center stage in development thought.

But are there not other options? After all, even undistorted factor/factor and factor/product price ratios can be viewed as little more than recording instruments used to register and communicate underlying resource endowments and technical relationships. It might be argued that, in a regime of severely distorted factor/factor and factor/product price relationships, policymakers and planners should be able to ask their research staffs to "discover" the underlying technical relationships and social welfare functions even though they are not revealed in the political or economic marketplaces. This implies that the planner or research manager should have available staff analytical capacity to develop a set of shadow prices for factors and products that reflects their "real" or undistorted value. Research managers could thus be directed to allocate their research resources in a manner consistent with the shadow prices rather than the distorted market prices.

But the above approach is, in my judgement, entirely unrealistic. Unless the shadow prices and the market prices can be made to converge, the output of the research effort will not be used. It is unrealistic, for example, to insist that the California Agricultural Experiment Station direct its mechanization or biological research to the needs of the 160-acre farm — unless the State of California or the federal government is prepared to support the structural policies necessary to reverse the trends toward large-scale agriculture. A research system can not be asked to produce knowledge and technology that will not be used without undermining the intellectual integrity and ultimately the scientific capacity of the system.

Yet what alternative is available in a society in which price relationships and resource ownership are highly distorted? Clearly the economic planning system and the agricultural research system must have access to the analytical capacity necessary to assess the potential magnitude and incidence of benefits and burdens in order to enter into effective dialogue with the political system about price policy and about the allocation of research resources. Economic analysis leading to a better understanding of the distortions in the economic, political and social weighting system is essential. But the objective of such research should not be to provide policy and planning agencies and officials with the weighting systems for internal research resource allocation. The objective should be to contribute to the political dialogue that will result in institutional changes that will lead to convergence of the "efficiency" and "market" weights.

If market and efficiency prices can be made to converge, research managers can be given clear signals for the allocation of research resources. When market and efficiency prices diverge, it will be almost impossible to induce research managers and scientists to allocate their effort in a manner consistent with the "shadow" prices.

Research Planning and Resource Allocation

An implication from the previous section is that research planners, managers and scientists have a very strong incentive, derived from a concern that their efforts will lead to technical innovations that will be adopted, to accept current or anticipated price relationships as a guide to research resource allocation. In a low wage economy, the generation of a technology designed to use relatively scarce and expensive inputs—a capital intensive technology in a labor surplus economy— would be less profitable and would generate lower social returns than a technology designed to absorb relatively abundant or low-cost labor inputs.[10] The research planner or manager who fails to direct research resources in a manner consistent with anticipated factor and product market relationships will find it difficult to demonstrate the productivity gains resulting from research that are necessary to induce a continued flow of research support.[11]

How can this articulation be accomplished at the research program, institute or project level? The conceptual linkage between the information generated by the market and research resource allocation decision-making has been outlined in the Evenson-Kislev investment model described previously.

During the 1960s and the 1970s, a great deal of effort was devoted to the development of more formal systems for the organization and analysis of the data needed for research decision-making (Fishel, 1971; Arndt, Dalrymple and Ruttan, 1977; Norton and others, 1981). In my forthcoming book, I have reviewed these methodological developments under the following headings: (a) scoring methods; (b) experimental approaches; and (c) benefit-cost methods.

There can be little question that a judicious selection and application of the new methodologies can represent exceedingly powerful tools in the hands of research planners and managers who have the intellectual vigor to grapple with both the substance of the research program and with the analytical tools that are available as aids in research decision-making. Without the knowledge of research impact and incidence that can be made available, the director of a national research system is in a weak position to participate in the research policy and planning dialogue with the government or with the political system.

I am, however, considerably less optimistic about the contribution of the new methodologies to research decision-making than I was a decade ago. This is because of the importance of dialectal interaction, at both the economic and political level, among research scientists, research planners and managers, research clientele and participants in the legislative process. This interaction is necessary if a political environment that can sustain agricultural research is to be maintained (Evenson, Waggoner and Ruttan, 1979; Ruttan, 1980).

Our review of the process by which agricultural research became institutionalized in Japan and the United States led Hayami and me to argue that an agricultural research system in which substantial responsibility for financial support and decision-making is decentralized to the state or provincial level has greater capacity to induce a pattern of technical change consistent with resource endowments than a highly centralized system. More recently I have argued that this is because a decentralized system tends to simulate market behavior in the allocation of research resources. What it loses in redundancy is more than made up in relevance.

This perspective has been criticized by Pastore and Alves (1977). They argue that in developing countries scientific resources are too scarce to permit the redundancy associated with a decentralized system. And they argue that cultural and political constraints often preclude participation of rural people in the dialectic interaction necessary for effective functioning of the decentralized model.

The Pastore and Alves criticism is appropriate in the small country case. Many small country research systems are no larger than a state or provincial system in a large country. The discipline of external markets can, if market linkages are not too distorted, serve appropriately to focus research effort. The Pastore and Alves criticism may also be valid in the short run in a large country. But cycles of research capacity expansion and erosion observed over the last several decades in countries such as Argentina, Colombia and Peru (Ardila, Trigo, and Piñeiro, 1981) should serve as a caution concerning the costs of failure to design agricultural research systems in a manner that encourages the mobilization of political support for their activities.

These considerations have led me to become more concerned with the issue of incentive-compatible research system design than with the methodology of research resource allocation.

NOTES

[1] Both the literature review and the empirical tests are presented in greater detail in Hayami and Ruttan (1970, 1971), Binswanger (1974a, 1974b), Binswanger and Ruttan (1978), Yamada and Ruttan (1980), and Ruttan, Binswanger and Hayami (1981).

[2] "If . . . the theory implies that dearer labor stimulated the search for new knowledge aimed specifically at saving labor, then it is open to serious objections. The entrepreneur is interested in reducing costs in total, not particular costs such as labor costs or capital costs. When labor's costs rise, any advance that reduces total costs is welcome, and whether this is achieved by saving labor or capital is irrelevant. There is no reason to assume that attention should be concentrated on labor–saving techniques, unless, because of some inherent characteristic of technology, labor-saving knowledge is easier to acquire than capital-saving knowledge" (Salter, 1960:43). Salter then stated that engineers, given existing knowledge, design machines so that they use optimal amounts of factors, given the existing factor prices. But he argued that this is not induced innovation. This amounted to eliminating induced innovation by definition.

[3] The growth theory approach to induced technical change is relatively simple. Assume that it is equally expensive to develop a new technology which would either reduce labor requirements or capital requirements by 10 per cent. If the capital share is equal to the labor share, entrepreneurs will choose indifferently between the two courses of action and technical change will be neural. If, however, the labor share is 60 per cent, entrepreneurs will choose the labor-saving alternative.

If the elasticity of substitution between capital and labor is less than one, this process will continue until labor and capital shares are equal—provided that technical change does not alter the elasticity of substitution between labor and capital. An implication of the model is that factor shares can be stable even if the capital/labor ratio changes over time.

[4] For example, in the early 1960s, the United States had a land/labor ratio of 141 hectares per worker while Japan's ratio was 1.74 hectares per worker. The U.S. ratio exceeded the Japanese ratio by a factor of 81. However, Japan's land/labor price ratio exceeded the U.S. ratio by a factor of less than 30 during the same period. To explain the difference in factor ratios by factor-price effects, the elasticity of substitution between the two factors would have to be 3 or more. Evidence from micro-level production functions suggested that an elasticity of substitution of this level is highly unrealistic.

[5] In the Hayami-Ruttan model the process of advance in *mechanical technology* is illustrated in the left-hand portion of Fig. A. I_0^* represents the land/labor isoquant of the metaproduction function (MPF) at time zero; it is the envelope of less elastic isoquants such as I_0 that correspond, for example, to different types of harvesting machinery. I_1^* is the innovation possibility curve (IPC) of time period 1. A certain technology—a reaper, for example—represented by I_0, is invented when the price ratio BB prevails for some time. When this price ratio changes from BB to CC, another technology—such as the combine—represented by I_1, is invented. The new technology represented by I_1, which permits an expansion in land area per worker, is generally associated with higher animal or mechanical power inputs per worker. This implies a complementary relationship between land and power, which may be illustrated by the line (A, M). It is hypothesized that mechanical innovation is responsive to a change in the wage rate relative to the price of land and machinery and involves the substitution of land and power for labor.

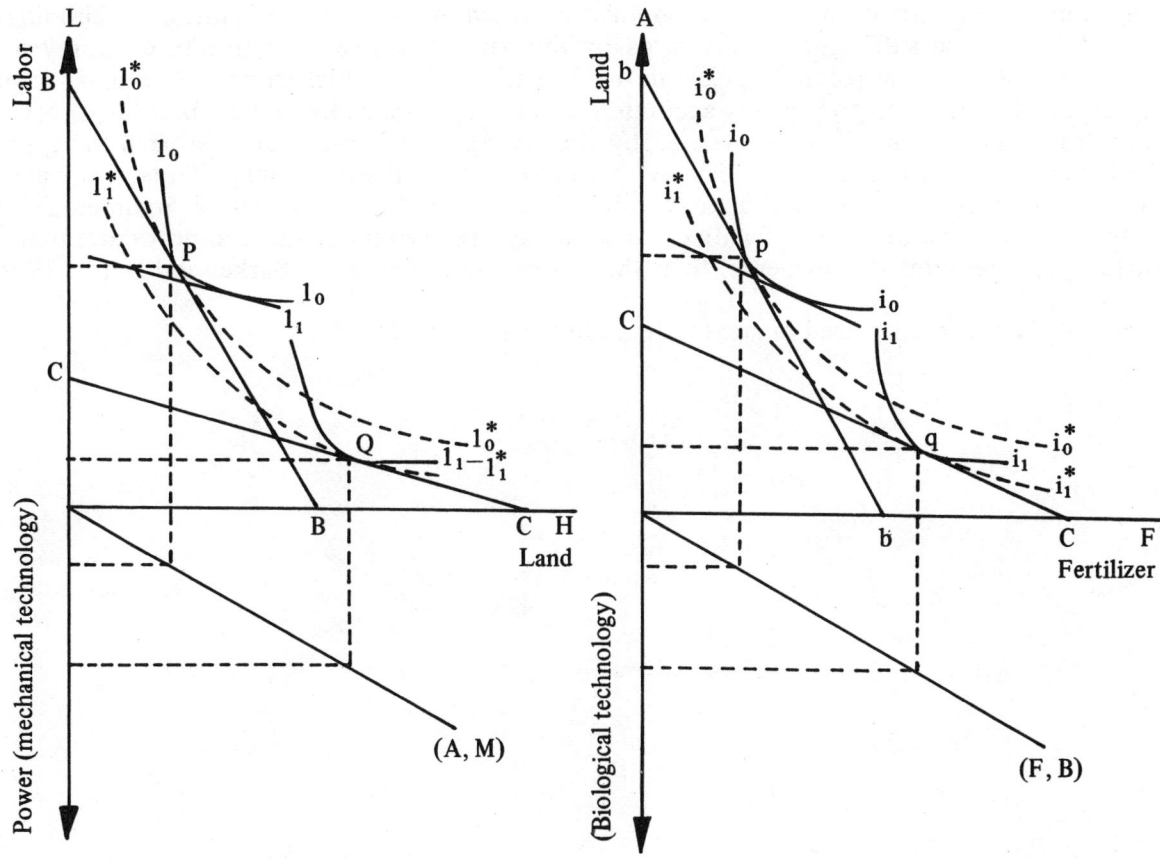

Fig. A. Factor prices and induced technical change (Hayami and Ruttan, 1971:126).

The process of advance in *biological technology* is illustrated in the right-hand portion of Fig. A. Here i_0^* represents the land/fertilizer isoquant of the metaproduction function. The metaproduction function is the envelope of less elastic isoquants (for example, i_0), that represent advances such as crop varieties characterized by different levels of fertilizer responsiveness. A decline in the price of fertilizer is seen as inducing plant breeders to develop more fertilizer-responsive crop varieties—which might be described by the isoquant I_1 along the IPC I_1^*—and as inducing farmers to adopt the new varieties as they become available.

[6] This builds on recent advances in the theory of duality between production functions and cost functions and the one-to-one correspondence between factor demand curves and cost functions (Jorgenson and Lau, 1973).

[7] For example, only by coincidence would the factor into which the new quality is embodied be the only factor whose productivity is augmented. A new rice variety will not only augment the productivity of the seed in which the new productive capacity is embodied but it will also augment the productivity of the land and labor used to produce rice. The Binswanger model is designed to cope with this problem by permitting each research activity to affect each factor augmentation coefficient.

[8] Until very recently most of the formal tests of the induced innovation hypothesis had been against historical experience from the agricultural sector. Within the last several years several tests drawing on the history of technical change in other sectors have appeared (Berndt and Khaled, 1979; Cain and Peterson, 1981; Kapp and Smith, 1981).

[9] Even the four factor test employed by Hayami and Ruttan (1970, 1971) represents a simplification of reality. In a test of the induced innovation hypothesis against Victoria (Australia) experience during 1890–1910, Ian McLean (1979) indicates that while changes in factor proportions have been consistent with the movement of factor prices the changes are more complex than suggested by the chemical-biological technology dicotomy used in the Hayami and Ruttan work. For a qualification of the Binswanger multi-factor test, see Kislev and Peterson (1981).

[10] This point is relevant to much of the populist criticism of the "green revolution" technology. It has often not been sufficiently appreciated by the critics that in a labor surplus economy simply focusing scientific and technical effort in the direction with the highest pay-off will, if factor/factor and factor/product price ratios accurately reflect resource endowments, bias the direction of technical change in the direction desired by the critics. If price ratios are distorted the appropriate focus of reform is on price policy. As hard data on the distributional effects of the green revolution has begun to replace the casual observations that fueled the earlier arguments, it is clear that most of the new "seed fertilizer" technology has been much more appropriate, that is, consistent with resource endowments, than the critics had anticipated (Barker and Herdt, 1978: 91).

[11] These two issues are discussed in greater detail in Ruttan (1982).

ASPECTS OF THE POLITICAL ECONOMY OF TECHNICAL CHANGE IN DEVELOPED ECONOMIES

Alain de Janvry and E. Phillip LeVeen

ABSTRACT

Technological change has played an important role in stimulating remarkable productivity growth in United States agriculture. In the last decade, however, there has been an ominous slow-down. This paper suggests that explanations of the U.S. experience have become excessively "economistic," and that technological change must be understood not only as a search for economic efficiency, but also as an instrument of change (or inertia) in social relations. This argument is supported by an examination of the process of technical change in California agriculture, with emphasis on the mechanized harvest of specialty crops.

INTRODUCTION

Technological change has been an important source of output and labor productivity growth in United States agriculture. During the period 1950 to 1968, for example, land productivity increased at an annual rate of 2.5 per cent while labor productivity increased at the annual rate of 6.8 per cent (President's Council of Economic Advisors, 1978). This extraordinary technological performance has, however, shown ominous signs of slow-down in the past decade. For the United States as a whole, growth in land productivity fell to an annual rate of only 0.4 per cent in 1970–75, while growth in labor productivity fell to 2.0 per cent during the same period. This observation raises the question of explaining the observed underinvestment in agricultural research as a possible cause of faltering productivity growth (Ruttan, 1978; Evenson, 1973). And this, in turn, raises the question of identifying the determinants of the inducements to the generation and diffusion of technological innovations.

In spite of the recent slow-down in productivity, the remarkable secular performance of United States (U.S.) agriculture and particularly the role played by agricultural research and extension in this past growth, continues to attract attention, especially in Third World nations hardpressed to meet the nutritional demands of rapidly growing populations and the foreign exchange requirements of late industrialization. If it is to serve as a model, however, we must clearly understand the economic and social determinants of technological progress in the U.S. system.

It is the thesis of this paper that the theories of technological change that have been proposed to explain the U.S. experience have generally fallen into excessively "economistic" arguments. Technical change must be understood not only as a quest for higher economic efficiency, but also as an instrument of change (or of resistance to change) in social relations. Hence the determinants of technical change must be sought in *both* the response to new economic conditions *and* in the struggle for the definition of social relations. The state is an essential institution through which these objective (economic) and subjective (social) forces are translated into new technologies. Any theory of technical change must incorporate a theory of the state and of how it responds through technology and other policies to both economic and political pressures.

In this paper, we support our above contentions by looking at the process of technical change in California agriculture, particularly at the mechanization of harvesting specialty crops. California agriculture provides a different perspective on the political economy of technical change than might be derived from observing, for example, Midwestern grain or dairy farming. Unlike the Midwest, California agriculture is based on social relations that are much closer to industrial capitalist enterprises, wherein the functions of ownership, management and labor are carried out, not within a family unit as in the Midwest, but by hired labor and salaried managers working for owners (many of whom are families rather than large corporations) who may or may not be directly involved in the day-to-day operation of the farm. Table 1 provides evidence supporting the above generalization, indicating (1) the greater importance of large-scale enterprise in California, as

TABLE 1 Characteristics of Agrarian Structure: U.S. and California

Characteristic	1950	1974
Number of Farms		
U.S. (million)	5.1	2.4
California (thousand)	123.1	60.0
Average Size of Farm (acres)		
U.S.	228	438
California	287	538
Percent of Farmland on Farms of More than 1000 acres (%)		
U.S.	15.7	33.7
California	40.4	55.4
Percent of Farms in Largest Census Size Class (%)		
U.S.	2.8*	7.0**
California	14.0*	17.0**
Percent of Total Farm Production on Largest Farms (%)		
U.S.	22.0	54.0
California	68.0	81.0
Share of Hired Labor in Total Farm Workforce (%)		
U.S.	23.5	31.7
California	64.9	76.1

*Refers to farms with sales of $25,000 or more.
**Refers to farms with sales of $100,000 or more.

SOURCES: U.S. Dept. of Commerce, *Census of Agriculture*, 1949 and 1974; USDA, *Agricultural Statistics*, various issues; State of California, *California Statistical Abstract*, various issues.

measured by percent of cropland held by owners with more than 1000 acres and by the concentration of production on large farms, and (2) the much greater importance of hired relative to family labor in California as compared with the rest of the U.S. The California perspective provides an opportunity to observe the inducement of technical change within a more fully developed capitalist context, and highlights social relations that are not well developed within the traditional family farming system that has been the model for most of the analysis of technical change in the U.S.

THE ROLE OF SOCIAL VERSUS ECONOMIC FORCES IN THE INDUCEMENT OF TECHNOLOGICAL INNOVATIONS

There has been a tendency, in both economics and political economy, to understand technological and institutional changes in terms of the rationality of a continuous search for higher levels of economic efficiency. Thus, changes in the ecological system, in scientific potential, or in the economic and social structure all lead to the definition of new technological and institutional opportunities that allow for potentially higher levels of productivity and profits. These opportunities create the expectation of economic pay-offs which set in motion demands on both the state apparatus and private firms towards their implementation. Public and private initiatives then generate new technologies and institutional arrangements in response to these demands. This "economistic" interpretation of technological and institutional change is particularly clear in the market-based theories of the inducement and diffusion of technological innovations.

According to the theory of induced innovations (Hayami and Ruttan, 1971), changes in relative factor scarcities are reflected in changes in relative factor prices. The new levels of factor prices create profit incentives that guide technological research towards saving on the increasingly expensive factor. As in the neoclassical theory of the firm, strict economic calculus leads to both choice of technique among actually or potentially available blueprints and of allocation of resources for a given technological design.

Similarly, the diffusion of technological change among farms is conditioned by the quest for profit, in either an inducive or coercive manner (Owen, 1966): Schumpeterian entrepreneurs adopt innovations motivated by the quest for higher profits. As the innovation diffuses, production increases, prices fall, and other farmers are forced into adopting the cost-saving innovation to cover costs. This further increases supply and depresses prices to the point where the differential rents of early innovators are cancelled, causing these adopters to again search for new opportunities for Schumpeterian profit. In both the theories of innovation and diffusion of technological change, the active social agents are individual entrepreneurs, motivated by the search for higher profits, while scientists, agribusiness entrepreneurs, and the apparatus of the state are seen as responsive to these demands. The decentralized actions of all these social agents are reconciled by the market. Institutional innovations are similarly seen to emerge from technological innovations, and hence to derive from the same economistic determinism: institutions are transformed and adapted in order to capture fully the new economic advantages provided by technological change.

This economistic interpretation of technological innovation, even if partially correct, seems to us insufficient in that it is only one element of a theory of technical change. This is because technical change is not only an instrument in the generation of an economic surplus but also an object and an instrument of social conflicts. Technical change conditions the social control of the means of production, the organization of the labor process, the social division of labor, and the social appropriation of the surplus. As such, it is a powerful instrument of social change or stasis.

A similar interpretation of the role of social conflicts versus economic determinism was advanced by Dobb (1963) in his debate with Sweezy (1978) on reasons for the breakdown of feudalism and the emergence of capitalism. Sweezy characterized the essence of feudalism as production for use (natural economy), while viewing capitalism as production for the market. As a result, the decline of feudalism occurred under the impetus of economic forces external to the system: the opening of long distance trade, the rise of towns, and the consequent gradual shift of the purpose of production from home use to production for the market. Capitalism thus developed in agriculture in response to market forces, and social relations in production were transformed through the elimination of serfdom, because the use of wage labor was more efficient in the adoption of modern techniques. Technological change was thus motivated by the quest for higher economic efficiency in response to changing market conditions.

By contrast, Dobb, and later Takahashi (1978) and Brenner (1977), defined feudalism in terms of its social relations. These authors stress the concept of serfdom, with its extra-economic coercion and imposition of tribute, especially labor rent, that characterized the social relation between feudal lords and peasants. Under feudalism, increasing the surplus captured from serfs could occur on an *absolute* basis, by raising the level of rent without a necessary concomitant increase in the productivity of labor. Under capitalism, by contrast, increasing surplus extraction must be obtained on a *relative* basis by increasing the productivity of labor.

When long distance trade emerged and feudal lords were presented with the opportunity of exchanging the surplus captured from peasants for the enticing new consumption goods brought by merchants, a strong inducement was created to increase this surplus. This could occur on an absolute or a relative basis, according to the social power of the landlords over peasants. Where peasants were weak due to high demographic pressures on the land, strong extra-economic coercive forces exercised by powerful landlords, and little access to local markets, surplus extraction was increased by raising rents and hardening feudal relations, a phenomenon which occurred in a period called the "second serfdom," particularly in Eastern Europe in the 17th and 18th centuries (Kay, 1974). Where social conflicts were strong and the power of landlords over peasants weak, increasing surplus had to be obtained on a relative basis and the feudal social relations transformed into those of capitalism. Capitalism, and its associated technological dynamics, came about not simply because it was more efficient, but because social conflicts forced a transformation of social relations and a redefinition of the organization of the labor process.

Applying this kind of analysis to the issue of induced technological change in agriculture, we argue that there are four levels at which social conflicts enter as important determinant variables of the rate and bias of technical change. In the economic theories of induced innovations, these variables are taken as exogenous or omitted. At the level of political economy, they can be endogenized, and the process of technical change understood in its full social complexity.

The first level concerns the role of social conflicts in establishing the structural and economic determinants of the inducement of innovations. The land tenure system, for instance, is nothing but a set of social relations, since land ownership is not established as a relation between people and land, but as a relationship of exclusion among individuals. The long history of the formation of the land tenure system must be understood as the struggle for appropriation of land and water with its associated dose of legal battles, bribery, violence and violation of the law.

Relative prices and terms of trade between agriculture and the rest of the economy, which are the key determinants of induced innovations in the economic theories of technological change, cannot be understood as mere reflections of scarcity, ignoring social relations. As Sraffa (1960) showed, every price system is based on a particular instance of the distribution of income between wages, profits, and rents. And this distribution is not the result of relative factor contributions to production—labor, capital, and land—but of the relative power of social classes in dividing total surplus among themselves, in particular through control of the state apparatus. This does not mean that scarcity (for example, per capita land availability in Japan versus the U.S.) does not enter as a determinant of prices, but that it so enters only in reinforcing the social power of landlords over peasants, and their consequent ability to set high rents and low wages. Prices are, in addition, directly manipulated by the state, in response to both objective and subjective forces, as the extensive scheme of farm subsidies established since the New Deal period evidences. If a theory of technical change is to be complete, it must be part of a theory of price, and a theory of price requires a theory of income distribution and hence of social classes and conflicts.

The second level at which social conflicts enter into the determination of technical change is in the operation of the research apparatus. This occurs in the delineation of the respective domains of public versus private research. It is determined not only by the nature of research—mechanical is largely private, biological is largely public—and by competition, but also by the laws that codify the possibility of private appropriation of research results. Recent changes in the patentability of biological innovations have opened the door to massive research in the breeding of new crop varieties, especially in conjunction with the manipulation of DNA. Privatization of research, in turn, implies a set of technological biases that correspond to the search for maximum profits and to attempts at gaining monopolistic control of germ plasm and seed markets.

The financing of public research is also an issue that must be understood in terms of social conflicts. This is, of course, true for the determination of public appropriations of research and extension. But it is also clear that small private grants to public research institutions can divert to specific purposes the use of large public sums. Grant-making very much reflects the relative power and financial capacity of interest groups in civil society.

A third level at which social conflicts condition technical change, particularly the mechanization of agriculture, is at the level of the organization of the labor process and of the control over this process that technology affords farm owners and managers. Technological options condition powerfully the labor process: the bunching of labor demands in particular periods of the year, the division of labor among skills and sexes, the pace of work, and the degree of socialization of the work process (Braverman, 1974; Gintis, 1978). For nonfamily farms that rely heavily on hired labor, there are two key objectives in the organization of the labor process, and hence in the choice of techniques: one is for the owner to establish control over the labor process and retain (reproduce) this control over time; the other is to maximize profit under the constraint of the establishment and reproduction of control. Both objectives are mediated by class conflicts.

One way of understanding how technology conditions control of the labor process is to conceptualize the production process as one where inputs include conditions of production (land and water), physical means of production (working capital), labor, and a state of consciousness of the workers involved. The outputs are a product and by-products—pollution and other external effects—and a certain state of modified consciousness. Reproduction of control implies reproduction of a state of consciousness such that the production process will be perpetuated over time. Technology is important for that purpose since it allows the segmentation and specialization of productive tasks and the increased division of labor that serve to alienate workers from control over the production process, as well as the "de-skilling" of many tasks for which the demand for labor may be high. It also allows an increase in the productivity of labor and hence material rewards from the work effort. Thus, new technology determines the efficiency with which labor inputs (work) are extracted from hired labor, both through its impact on social conflicts (consciousness) and on the productivity of labor.

Finally, a fourth level at which social conflicts affect technological change is through their impact on factor and product prices, and hence on surplus control. In contrast to the partial equilibrium theories of induced innovations where prices are exogenous, technology itself must be seen as an instrument to counteract changes in factor prices. Labor-saving mechanization reduces employment and counteracts rising wages. Hence, its impact is not only to save on the expensive factor, but also to offset the scarcity and price increase of that factor. By changing the nature of the labor process, mechanization allows a shift in labor demand towards unskilled and unorganized segments of the labor force. Further, mechanization allows the promotion of economies of size and hence the concentration of economic power and, through this, the development of monopolistic positions, for example growers' associations. This, in turn, allows for control over the terms of trade and hence control over the surplus generated in the production process.

It is clear that the relative importance of social conflicts versus economic efficiency goals (in any case inextricably related) in determining the rate and bias of technical change very much depends on the social and structural context within which production occurs. In Midwestern agriculture, for instance, where the family farm dominates, the potential conflicts between labor and capital are internalized within the family. Social conflicts are thus not internal to the labor process, but external to the farm, and expressed, in particular, in terms of the struggle over the terms of trade between agriculture and the nonagricultural economy. The populist movement in the U.S. at the turn of the century thus defined its radical demands in terms of opposing monopoly control of the merchants and big government over price formation. While we do not investigate this subject here, it is nevertheless quite possible that the unleashing of technical change and the industrialization of agriculture that followed served to countervail these radical demands by integrating agriculture in the broader industrial process and thus creating a stabilizing identity of purpose (McConnell, 1953).

In California, by contrast, where the role of seasonal hired labor is dominant, the dynamics of technical change are strongly conditioned by social conflicts and by the continuous search for sources of cheap labor. As we will see, the dialectic between the quest for economic efficiency and the quest for control and reproduction of control over labor have been the essential determinants of the inducement and diffusion of technical change.

THE STATE IN THE EVOLUTION OF CALIFORNIA AGRICULTURE: AN OVERVIEW

California's large-scale agriculture, its extensive use of hired management, and its dependence on seasonal workers for hand-labor tasks is not a recent development. Indeed, California's agricultural structure has been "industrialized" for well over a hundred years. Two factors account for this unique structure: (a) the original distribution of the land when California became a state in 1850, which resulted in a highly concentrated pattern of ownership; and (b) the coincidental availability of large numbers of foreign workers willing to accept low wages, poor working conditions, and seasonal employment, which permitted the land to be farmed more profitably in large units than in the traditional family-sized units. Fuller (1939) has convincingly demonstrated that family workers were unwilling to exploit their own labor to compete with the foreign seasonal workers.

It should be noted that elsewhere in the U.S., land speculation also led to the accumulation of large holdings, and at the turn of the century large-scale industrialized farming appeared in several midwestern states. California's agricultural system was unique in its ability to reproduce this industrial pattern. Elsewhere, large-scale farming ceased to be competitive with family farming and large holdings generally were broken up into smaller units. California was able to maintain its industrial farm structure because it could reproduce the supply of workers willing to accept low wages and seasonal employment; other regions could not and reverted to family labor.

The origin and reproduction of California's industrial agriculture is also the result of a variety of State policies. Both federal and state government policies contributed to the initial distribution of land. In many instances, this distribution was allowed in the face of laws that were intended to insure the broad distribution of the land. California courts and the legislature failed to enforce the law.

Once the agricultural system was established, strong vested interests developed to protect their holdings. Without seasonal labor, the system could not reproduce itself and the substantial investments in land would not be maintained. Therefore, proponents of California's agricultural system became preoccupied with the maintenance of a large supply of workers who were willing to accept low incomes. During the 19th century it was possible to maintain such a labor force, relying on the Chinese who originally came to work in the gold mines and to build railroads, and on white workers who could not find adequate employment in the cities. However, with the development of the nonagricultural economy and the rise in wages, California farmers were increasingly pressed to find workers outside of the usual labor markets. Rising wages eventually eliminated large-farm experiments in the midwest; however, California landowners were able to maintain adequate labor supplies to insure the continued industrial pattern.

The success of California agriculture depended on its ability to maintain a large supply of foreign-born workers. Foreign workers were desirable because they were socialized in much less prosperous environments and, at least initially, were willing to accept the low wages and poor working conditions that domestic workers would not accept. In addition, foreign workers did not speak English, were not educated, and were unfamiliar with California's political, social and economic institutions. Because of their foreign status and racial differences, these workers were also the object of considerable racial antagonism. This antagonism worked to the advantage of agricultural employers, serving to restrict the opportunities for alternative employment in higher-wage jobs.

Until World War II, the supply of workers was handled largely through private recruiting efforts. However, agricultural interests increasingly came to depend upon various State policies to assist them in securing and, even more important, in controlling labor. State intervention became increasingly apparent in the latter functions. For example, when Japanese workers began to escape the farm labor markets by purchasing land and becoming competitors of American farmers, California passed acts that forbid aliens to own land. Later, when immigration laws threatened the supply of Mexican workers in the 1920s, California interests prevailed on the federal government to allow the continued temporary import of workers for agriculture.

Equally important, agricultural interests came to rely upon the State to aid them in preventing the successful organization of workers. Local law enforcement officials worked with agricultural interests to prevent strike efforts. Such efforts were especially evident during the depression years, when large numbers of whites were in the fields. At the level of the federal government, agricultural interests, particularly from California, were successful in exempting agricultural labor from New Deal legislation that supported nonagricultural workers in their efforts to organize against capital.

With the coming of World War II, the labor situation in California changed. No longer was it possible to count on waves of foreign immigrants to fill the jobs, and wartime prosperity allowed "Dust Bowl" immigrants to escape to the nonagricultural economy. Once again, the State was called upon. This time, a government-to-government program was designed to insure a supply of Mexican workers for California during the wartime emergency. However, after the war, this arrangement was perpetuated in the Bracero Program, which lasted until 1965. The Bracero Program was legislated to allow agriculture, both in California and in other Southwestern states, to retain its access to cheap labor while providing some protection to industrial unions from unwanted competition that might threaten hard-won benefits. With the coming of the Bracero Program, the State took on the responsibility of securing and managing the seasonal labor supply; this included setting wages and working conditions.

Recognizing that labor would become increasingly difficult to obtain and control, California interests turned their attention to the development of labor-saving innovations. Of course, even the availability of cheap labor did not preclude the possible benefits of such innovations; indeed, California farmers were among the earliest adopters of many of the innovations for planting, cultivating and harvesting grain and hay crops. However, these innovations did not eliminate the dependence on seasonal workers who were still required for cultivating and harvesting fruit and vegetable crops. These crops could not be easily mechanized. Successful mechanization required not only the development of new machines, but also substantial biological manipulation and the development of new cultivation practices. Few private firms had the incentives to underwrite the costs of such complex packages of technology, especially in view of the relatively limited markets for the technology. As a result, California interests turned to the public sector, and particularly to the University of California, for labor-saving innovations that could be called upon in the event of a failure of the efforts to preserve the labor force.

The University responded to the increased interest in mechanization by developing a series of labor-saving innovations and then selling the production rights to the private sector for nominal fees. The list includes the segmented sugar beet seed which allowed precision planting and harvesting (1941); the "in-place" sugar beet topper (1943); the sugar beet harvester (1943); the shake-catch harvester for nuts (1946); corn heads for grain combines (1954); bulk handling of fruits and vegetables (1959); the fig pickup machine (1959); the shake-catch harvester for grapes (1961); the tomato harvester (1963); the lettuce harvester (1965); and the electronic sorter for tomatoes (1976). In the case of the tomato harvester, not only was a machine developed, but so too was a new breed of tomato that could withstand machine harvest. The machine was developed in a relatively short period; the tomatoes required 20 years to perfect.

In 1965, pressures on the Federal government by organized labor and the lack of support for the continued Bracero Program from states like Texas proved too strong for California interests and the program was terminated. Loss of support for the program in states like Texas was, in part, a reflection of the successful mechanization of cotton which drastically reduced the demand for labor in those states after 1960.

At the same time, California was experiencing the first sustained efforts to organize agricultural labor-saving innovations led to the rapid mechanization of tomatoes and wine grapes, which were taking their case to consumers, successfully carrying out boycotts of grapes and lettuce. By the early 1970s, agricultural labor legislation was extended to the fields in California, further strengthening union efforts. In this new environment, the past efforts of the university to develop harvest labor-saving innovations led to the rapid mechanization of tomatoes and wine grapes which were the two most important labor-dependent crops.

The success in mechanization of these crops allowed increasing cultivation of both grapes and tomatoes in spite of rising real labor wages. In addition, labor displacement helped other crops that were not able to mechanize, since mechanization renewed the pool of available workers and helped to restore competition to the labor markets. At the same time, adjustments were made in many nonmechanized crops that permitted the use of higher-paid labor without impact on profits. For example, lemon growers, with help from the University, "rationalized" their labor markets by offering higher wages and stable employment with vacations to those highly productive workers willing to tolerate the demands of a dramatically increased pace of work. Output per worker quadrupled, allowing increased wages without penalty to employers. Similar adjustments were made in strawberries and lettuce (where it still does not pay to use the University-developed harvester).

In part because of mechanization and related developments discussed above, and also because of the rapid increase in the use of illegal workers from Mexico, whose numbers rapidly increased during the 1970s, the initial advantages of labor during the late 1960s have been largely lost. The unionization of workers has ceased to grow, as union leadership tries to find a solution that will improve the welfare of its own members while not taking a reactionary position toward illegal workers, many of whom are relatives of union members. Indeed, most labor pressures on California producers have been eliminated. However, the increasing use of illegal workers throughout the entire economy again threatens organized labor, which is fighting for a closed border. Interestingly, the reaction to this spill-over will most likely be the reinstatement of a Bracero-like program.

TECHNICAL CHANGE IN CALIFORNIA AND THE THEORY OF INDUCED INNOVATION

Using this brief historical overview, we now return to the general discussion of technical change and examine our hypothesis that social conflicts are an essential underlying determinant of the process. We discussed four levels at which social conflicts condition the choice and rate of change of technology; California affords clear examples of the interactions between social relations and technology at all four levels.

First, the relative prices of inputs used in California agriculture do not simply reflect relative scarcity of the resources. The most important example here is the relative abundance of cheap labor that has characterized most of California's development. This abundance derives, in part, from exclusive land tenure patterns that restricted ownership opportunities and created a class of owners with incentives to exploit seasonal workers. The abundance also reflects deliberate private and public efforts to augment and contain the labor force within agriculture so as to maintain a condition of excess supply and intense competition for work that keeps wages low.

The drive to mechanize labor-intensive crops came only after California landowners temporarily lost some of their power to control the immigration process, that is, when the Bracero Program was terminated. At that time, prospects for increased wages may well have caused many farmers to seek alternative methods of producing labor-intensive crops, just as neoclassical theory suggests. The point is, however, that rising wages are a reflection of a shift in social relations, not scarcity. California landowners were isolated by developments outside their sphere, such as the mechanization of cotton in the South and the rising political power of organized labor.

The second level at which social conflicts influence technical change is through the research apparatus. In this respect, the California case is particularly revealing. Publicly-supported research in California has performed two important functions: it has provided intellectual legitimacy for the exploitative social relations within agriculture, particularly during periods of crisis within the system, and it has assisted in the process of capital accumulation by anticipating crises and developing the means for accomodating unavoidable changes in a manner that preserves the dominance of landowner/grower/agribusiness interests.

A full development of these two propositions is beyond the scope of this analysis, so we present only a few examples. Regarding the legitimizing function, a crisis developed in California in the late 1930s and 40s as a consequence of the construction of federal irrigation projects that required water recipients to own no more than 160 acres and to reside on their land. Such "family farms" did not exist in California and to impose these requirements threatened basic land tenure/labor relationships. To make matters worse, a small number of agricultural economists working for the USDA and the University of California began investigating the viability of smaller farms and their impact on rural communities, finding that, contrary to prevailing belief, smaller farms could be efficient and produced significant benefits to rural communities.

This work was not acceptable to the dominant agricultural interests, who convinced the state legislature to establish an oversight committee to evaluate the research at the University. The committee sat for four years, reviewing research and recommending new directions for funding.

As a result of these recommendations, funding for the University was increased substantially, and the direction of research was shifted. Agricultural economists concentrated on "economies of size" studies which all supported the notion that 160 acre farms were not efficient units. Rural development research was eliminated (Fisk, 1978). The key provisions of the Reclamation Act were not enforced, allowing large farms to continue receiving subsidized water. In very recent times, the issue has again emerged, with agricultural economists pressing for efficiency and the "social benefits" of the water subsidy. However, growing fiscal deficits and interregional competition within the agricultural economy have shifted some of the power away from California landowners, who are now in danger of losing their subsidies.

Another example of the legitimizing function of the research apparatus concerns the efforts by university officials and professors to provide intellectual support for the continuation of the Bracero Program, and when this failed, to support the University's efforts to develop labor-saving alternatives. During the debate over the Bracero Program in the early 1960s, the University of California sponsored research on the effects of termination on the agricultural economy. These reports concluded that loss of the program would cause substantial economic damage to the California economy and, most importantly, would threaten consumers with higher food prices (Turner, 1965). The results of the research were well-publicized. Later, it was learned that a 17-page section of the report, which argued that careful rationalization of labor could provide an alternative to the Bracero Program, was censored by the University administration (Draper and Draper, 1968).

Once the Bracero Program was terminated, research focused on the social benefits of mechanization. The purpose of this research was to argue in favor of public subsidies for mechanization research. Consumer benefits from lower food prices were found to offset any losses arising from displaced workers. Interests outside the University have strongly disputed the conclusions of the many papers and reports coming from the University, but, with minor exceptions, ongoing research justifies continued or increased subsidization of mechanization projects.

In both of these examples, social scientists supported ways of problem-solving which benefited the dominant interests in the agricultural economy. But justification of the status quo is only part of the role of the research apparatus, for it must also help the agricultural system to anticipate and accomodate crisis when justification is not enough. In the conventional model of induced innovation, the line of causation is from the farmer's perception of a potential problem or profit possibility to demands on the research institution. In California, the research institutions appear to have taken the initiative in defining long-run problems and in developing methods of solving these problems in advance of their appearance. Mechanization research at the University is an excellent example of this anticipation.

At the end of World War II, facing the prospect of the termination of the special labor program with Mexico (which was eventually extended in the form of the Bracero Program), the influential Commonwealth Club held a series of seminars regarding the future of California agriculture. One of the speakers, an agricultural economist from the University of California, argued that in the future California would have to learn to do without cheap labor as a result of the limits on immigration and the growing power of organized labor in the nonfarm economy. He argued in favor of "labor rationalization" strategies whereby workers could be made much more productive with careful development of different labor skills. Such strategies had been tried in the past but were generally rejected because of the tendency of the rationalized work force to organize into collective bargaining units. The University economists nevertheless proceeded to work on rationalization strategies. Such strategies were given little attention, however, until the termination of the Bracero Program.

The other approach to the coming crisis in labor was developed by the agricultural engineering departments of the University, which identified the major crops according to their use of labor and proceeded to develop a series of projects in the early 1950s to reduce the need for labor. The tomato harvester project was one example of this effort. Long before the Bracero Program terminated in 1965, the tomato harvester was operational. Indeed, if one reviews the labor-saving innovations produced by the University, such as the sugar beet harvester, the fruit and nut pickers, as well as the tomato and lettuce harvesters, it is significant that all were ready and available long before they were actually adopted. In some cases, they still have not been adopted. It thus appears that the University researchers anticipated the loss of control over the labor force and, long before growers were anxiously demanding more mechanization research, had already made major efforts to reduce the costs of such a loss.

To argue that the University research aparatus functions within an extended time frame is not to imply that the researchers operate outside the important social relations we have already identified. On the contrary, the efforts of the agricultural engineers evidence true class-conscious behavior. They correctly understood the importance of labor and the possibility of the loss of control over labor by landowners. As one engineer openly states: "machines don't strike."

The strong interest in helping to protect landowners and growers is not difficult to explain. The University depends on a political process for its budget; without political support there is no budget. Consequently administrators are anxious to develop programs that meet with the approval of politically powerful constituents who will provide the support necessary to maintain or increase budgets. In addition, there are mechanisms outside the university that insure that certain kinds of projects will be developed. Private funds, though small relative to the funds provided by the public sector, help to set agendas by supporting the nonsalary components of projects. Internal socialization procedures reinforce the budgetary constraints. Those who are not adequately socialized are not promoted.

Before leaving this discussion of research, we think the issue of the division of agricultural research between public and private sectors deserves additional comment. The accumulation function of mechanization research clearly produces profit, and it may be asked why the University is necessary when profits should attract the private sector. The answer is that mechanization in California specialty crops requires more than engineering; it requires a long period of biological innovation which is risky and costly. By socializing these costs, the University has lowered the costs of labor-saving technology to California producers and thus helped to foster less dependence on labor than would have resulted without public support.

Recent developments in the manipulation of DNA may change this logic, for it will be increasingly possible to shorten the time required for breeding new varieties. This, combined with recent patent law changes, may well encourage the increased privatization of biological research. Such a shift in the locus of research may imply a new division of the profits within the agricultural business sector as seed companies increase their potential to extract surplus from the other sectors of the agricultural economy. Perhaps more important, privatization will allow landowners to accommodate new labor demands without need of support from public institutions that are increasingly under attack for helping to displace workers in times of rising unemployment.

However, if the accumulation function can be privatized, the private sector cannot provide the legitimizing function, which by definition can be provided only by "disinterested" university researchers. The need for such a function remains, for as long as relations between capital and labor in agriculture are substantially more exploitative than elsewhere in the economy, there will be a need to defend practices, especially in justifying future immigration policies such as are now under consideration. In other words, the legitimizing functions of the University will not be eliminated by the tendencies toward privatization of research.

The third level at which social conflicts condition technological change concerns the organization of the labor process by farm owners and managers, in order to gain additional control over labor. This set of conflicts operates at the level of the individual farm. Having set out the important control issues previously, we now simply give an example of how technology provides this control in the case of the tomato harvester.

The mechanical harvester offered tomato growers a new form of control over the labor process. Braceros had been tremendously efficient workers; young men, coming from a very low-wage environment, paid on a piece-rate basis, and subject to being sent home if they did not perform up to a standard, had considerable incentives to work hard. We have noted that when the Braceros were gone, not only were growers faced with a possible "shortage" of workers, but more importantly, the remaining workers now found themselves in a stronger position to organize and demand both higher wages and better working conditions. Growers were thus faced with loss of control and the very real prospect of having to close the large gap in working conditions that existed between agricultural and nonagricultural economies.

The adoption of the harvester not only reduced dependence on hand-labor, it helped growers to re-establish control over the work force. Growers substituted women for Mexican and Mexican-American male workers as sorters on the machines. Women were largely inexperienced and willing to accept low wages in order to earn supplemental incomes for families. The discipline of the machine replaced the discipline of the piece-rate wage system; pace of work was dictated by the conveyor belt. Growers were thus freed of their dependence on the short supply of skilled pickers. The harvester also changed the nature of the work force, "de-skilling" most of the workers, but demanding an increase in skill for those few driving the machines.

In spite of the advantages conferred by the harvester, the labor movement in agriculture grew during the 1970s, leading to relatively greater increases in farm wages than those in the nonfarm sector. The introduction of the electronic sorter on the tomato harvester came at the height of union activity. Within three years, all tomatoes were electronically sorted, reducing the labor requirement by about 15,000 workers. This innovation virtually eliminated dependence on unskilled labor.

We now turn to the fourth level of social conflicts, namely the use of technology to counteract changes in the prices of factors. In the tomato harvester case, technology reduced dependence on skilled hand labor. By reducing the number of jobs in tomato picking, workers who were displaced were available for work on those crops that could not be mechanized. In other words, the harvester not only allowed tomato growers to maintain their profits in the face of the apparent labor shortage, but also replenished the labor pool for other sectors of the agricultural economy, helping to increase competition, reduce labor organization and keep wages low. Clearly, there is a dynamic interaction between the availability of technology and the relative prices of inputs.

Another dimension of this interaction concerns the impact of technology on the relative power of tomato producers *vis a vis* the rest of the agricultural economy. Initially, producers were unwilling to adopt the harvester because it represented a large fixed investment that reduced their flexibility to choose different crop rotations. The harvester embodied substantial economies of size, since its effective use required at least 150 acres of tomatoes. The typical grower had 35 acres at the time of introduction. As a consequence of its introduction, the number of growers was substantially reduced, even though tomato acreage doubled.

Thus, the introduction of the harvester had three important and dramatic effects on the tomato economy. First, its introduction favored farmers who could afford the harvester, accept the greater risks of increasing their dependence on tomatoes, and who could acquire the necessary land for effective employment of the technology. Coincidentally, at the time the harvester was introduced, almost a million acres were being brought into cultivation by two large irrigation projects on the west side of the San Joaquin Valley. The land irrigated was well-suited to the new tomato economy since it was owned and farmed in large tracts averaging well over two thousand acres each. There was a major relocation of tomato acreage within California, from the relatively smaller vegetable farms in traditional tomato growing regions to these newly irrigated regions. Tomato production proved more profitable than most other crops, conferring large economic benefits on landowners in these irrigation projects.

Second, there was a relocation of tomato acreage within the U.S., that of California expanding due to superior technology, at the expense of family producers in the Midwest. Large canneries increased their investments in new facilities in California, furthering the incentives for greater concentration of production.

Third, as a consequence of the increased concentration of production on a few very large farms in California, the tomato growers found themselves able to form a more effective bargaining association and increased their prices. Thus, while the initial benefactors of the harvester appear to have been tomato processors, because of increased concentration of production, growers have been able to increase their return also.

In sum, these shifts within the tomato economy not only increased profit by reducing the grower's dependence on hired labor but also contributed to a major redistribution of surplus between growers, both within California and between California and the rest of the nation. The power of the growers to extract surplus through better organization in the commodity market place gave them a new advantage over processors. Such dynamic effects are not captured by a static theory of induced innovation.

SOME CONCLUDING THOUGHTS ON THE RELEVANCE OF CALIFORNIA TO LATIN AMERICAN AGRICULTURAL DEVELOPMENT

Most of the theorizing about technical change in the United States is based on examples from midwestern family farm agriculture. The U.S. family farm is unique, but its time is passing; it is different from farms in most Latin American countries in that the labor, management and ownership functions are all carried out within a single family unit. Conflicts between land, capital and labor are not apparent in such a setting, even though it is possible that such conflicts still play an indirect role in the determination of technical change in family farm settings.

The industrialized nature of California's agriculture, with its concentrated land tenure and dependence on a disorganized and highly exploited labor force (which has many of the characteristics of labor throughout Latin America) would seem to provide a far more realistic model for analyzing the forces underlying technical change than do the more well-publicized family farm models that appear in the neoclassical interpretation of technical change.

Perhaps the most important difference between the model developed here and those of Latin America concerns the role of the research institutions. In California, the research institutions function both in a short-term, crisis management mode and also in a long-term rational planning mode. The latter has led to considerable attention being paid to increasing productivity, both labor and land. The technical achievements of California are certainly related to this long-term planning ability. This capacity is probably lacking in most Latin American institutions; its presence in California suggests the true achievement of a fully developed capitalist research institution. Why it exists in California and not in less developed nations is an interesting question that deserves more attention.

TECHNICAL CHANGE IN LATIN AMERICAN AGRICULTURE: A CONCEPTUAL FRAMEWORK FOR ITS INTERPRETATION

Martín Piñeiro, Eduardo Trigo and Raúl Fiorentino

ABSTRACT

The authors present a critical analysis of the ideas dominating current discussions on technological developments in Latin America, and introduce a methodological framework of their own. Their perspective views technology as a social phenomenon, and incorporates social conflicts and the role of the state as variable in an analysis of technical change in agriculture.

INTRODUCTION

The figures for production and land productivity of major crops in Latin America show relatively low aggregate growth rates, particularly when these rates are defined in per capita terms. However, the most outstanding feature of past agricultural performance in the region is the considerable *variability* of growth rates between crops within any one country, and among countries for most major crops. Furthermore, a careful consideration of production conditions tends to suggest that this variability cannot be explained by differences in natural resource endowments or strictly technological elements, such as available technology.

This article provides a conceptual framework for the analysis of technical change which, by incorporating social conflicts and the role of the State as explicit and important variables, attempts to account for the observed variability in production and productivity performance. The basic analytical shortcomings of the most important ideas that dominate the present discussion on technological development in Latin America will be analyzed. A methodological perspective will then be presented which attempts to provide an adequate interpretative framework for understanding technological change in Latin America. This perspective implies that technology is basically a social phenomenon and as such it must be studied and interpreted as an endogenous element to the general behavior of the social system.

FOOD PRODUCTION: THE ROLE OF TECHNOLOGY

Food production in Latin America has expanded during the past 25 years at an annual rate of about 3.2 per cent, a rate just about equal to population growth, but lower than that experienced in other continents. About one third of the growth is due to horizontal expansion rather than to increments in crop yield (PREALC, 1976).

Technological change has been a basic component of this production performance, both by allowing yield increases in already cultivated areas and by making possible and profitable the incorporation of new land of less productive capacity or with higher production and distribution costs.[1] The intensity of the innovative process is therefore a basic element of production increases.

On the other hand, it is important to emphasize that technical change may have varying qualitative characteristics which will determine its biases with respect to factor use. It is quite clear that different technologies will have different degrees of capital or labor intensity, and thus different effects on the overall growth capacity of the agricultural sector given the existing resources.

The possibility of alternative growth effects, and the concomitant income distribution effects of new technology, suggest the need for developing mechanisms which assure maximum effectiveness of the process of technology generation and adoption in contributing to food production and to overall economic growth. Both the intensity of innovative activities—which depends basically on the development of appropriate institutions—and the qualitative elements of the generation process—such as the relative factor intensity of new technology—must be considered as central elements of any strategy for food production.

Induced Technological Development

The importance of technical change in the process of economic growth was recognized early in economic thinking and has received considerable attention by all schools of thought. However, and in spite of these efforts, there is still little understanding of the mechanisms that determine the intensity and nature of technical change.

In the case of the agricultural sector, the most recent thinking with a considerable influence in theory and practice is that represented by the important works of Ruttan, Hayami, Binswanger and others.[2] Their work expanded and provided an empirical basis for the induced technological development model introduced by Hicks, Fellner and others (Hicks, 1964; Fellner, 1960). The model's basic proposition is that market economies—in particular the advanced decentralized economies—have a set of institutional mechanisms, including the market, by which technological development is induced in the required direction for maximum economic growth. Hayami and Ruttan show that relative intensity of labor saving technology (farm machinery) and land saving technology (fertilizers), and consequently the factor productivity implicit in the technological paths followed by a number of countries, is highly correlated to the factor endowments of those countries. Although the data presented for a worldwide comparison seems to uphold this proposition, a more detailed observation of empirical evidence in Latin America seems to suggest that inducement mechanisms have not resulted in the described innovative process.

Initial evidence in this direction is provided by inspection of aggregate factor productivity figures. In the Latin American countries included in the Hayami-Ruttan analysis (Argentina, Brazil, Chile, Colombia, Mexico, Paraguay, Peru and Venezuela), estimates of labor productivity[3] vary—including Argentina—between 5.0 for Paraguay and 12.9 for Chile. The low variability of these estimates is in sharp contrast with those of India and New Zealand, for example, which vary between 2.1 and 141.8 respectively. Homogeneity in resource productivity is also apparent for land. Extreme figures[4] for the Latin American countries in the analysis are 0.27 for Mexico and 0.94 for Paraguay, whereas for Taiwan the figure is 10.24. The homogeneity of these data, in spite of considerable differences in factor endowments (agricultural land in Brazil and Peru), suggests that the inducement mechanisms were not able to guide technical change in the direction of saving the relatively scarce resource.

Production Performance

Disaggregate information, in turn, shows an extreme variability in agricultural production performance and land productivity changes which can by no means be explained by differences in resource endowments. Production increases of individual crops vary widely between different countries.[5] An example of this is potato production in Colombia, showing a growth rate over 3 per cent per annum, and in Peru, where production increased under similar ecological conditions (Andean hillsides) at an annual rate of less than 1 per cent. Additional remarkable examples (among many) of uneven comparisons are: (a) wheat and corn in Mexico and Brazil (about 4%) and in

Colombia (-2.7% for wheat and 0.6% for corn); and (b) beans in Argentina (over 7%) and in Brazil (about 2%).

However, the point to be emphasized is the considerable variability in land productivity (yields) growth rates between different crops in one country, and between different countries for each particular crop. Colombia, for example, had high yield increases for rice (3.2%) and cotton (3.5%) and very low ones for corn (0.2%). Brazil had rapid increases in corn (4.7%) and low increases in rice (0.2%), beans (-0.8%) and cotton (0.9%).

On a product basis, potato yields increased rapidly in Mexico (5.1%) and Argentina (2.8%), and hardly at all in Peru (0.9%). Corn yields increased in Brazil (4.7%) and Argentina (2.2%) and very little in the Andean region. Rice yields increased in Colombia (3.2%) and México (2.4%) but not in Brazil (0.2%), while bean yields increased in Mexico (4.3%) and not in Argentina (1.3%) or Brazil (0.8%).

Part of this variability can be explained by the relative intensity of factor use.[6] For example, rice in Colombia is produced with highly capital-intensive technology, while corn is not. In Mexico wheat production has expanded on the basis of capital-intensive technology, while corn is still a labor intensive crop. However, this relationship does not always hold. An example is beans, which are produced with capital intensive technologies in Argentina and labor-intensive technology in Brazil, and which in both cases show small yield increases. On the other hand, the figures presented indicate that factor intensity of the technology used bears no clear relation to relative factor endowment in the countries analyzed.

The considerable variability shown by the above figures strongly suggests that inducement mechanisms have been relatively weak in guiding technical change in the direction of an economically efficient use of available resources. It is clear that, with infrequent exceptions, the nature of the technology that has prevailed in each situation has not been related to the relative scarcity of land or other factors of production.

The above arguments do not necessarily imply that inducement mechanisms do not exist. They certainly do, but in Latin American societies they have operated in a different context and, as is discussed below, have different effects. On the other hand, the above argument also suggests that, in the past, Latin American societies have been unable or unwilling to implement effective strategies for food production. The understanding of this failure and its consequences for food production is a subject which requires the development of a more comprehensive analytical framework that will allow for the explicit consideration of other causal relationships. Such a framework is discussed below.

TECHNOLOGY AS A SOURCE OF SOCIAL CONFLICT

The Role of the State

Income distributional effects of technological change are heavily dependent on the factor intensity of the adopted innovations, on the price elasticity of product demand, on the relative factor endowment and on the access to new resources, including information, of the production units. Technical change primarily affects the distribution of income between producers and consumers, but it also affects the welfare of laborers and other segments of society (Piñeiro, Martinez and Armelin, 1975; Schmidt and Seckler, 1970; Scobie and Posada, 1976; Hewitt de Alcantara, 1976; Cleaver, 1972). Consequently, different social groups, and particularly those directly related to agricultural production, will have different attitudes towards technology depending on their expectations of the effects of the technology and their capacity to appropriate the potential economic benefits derived from its utilization (de Janvry, 1977).

On the other hand, the rather high levels of investment required, the complexity of biological research, the atomistic structure of agricultural producing firms, and the low possibilities for keeping private any benefits resulting from the innovative process act to reduce, in general, the interest of and possibilities for private efforts in agricultural research activities. As a consequence, a considerable proportion of research efforts are public. This gives the State an important role in the determination of the intensity and direction of technological change. State actions materialize through two sets of instruments: (a) through government policies (mostly economic) which determine the socioeconomic context of agricultural production and limit the private profitability of technological adoption; and (b) through research resources allocation and institutional control, which determine the direction and intensity of innovative activities.

The complex role played by the State in the process of technology generation and adoption makes an understanding of the way by which State decisions take place, and are affected by groups with vested interests in agricultural production, a necessary and prior step for the explanation of technical change (Oliveira, 1975:120–123). State decisions should be regarded in this context as the result of the interaction of different social groups, with different interests and political power. State actions with respect to the technological process are determined by the relative power and the specific interests of those groups participating in the political process (de Janvry, 1977, Ref. 10; Gutman, 1978). Under certain (and frequent) conditions of social relations and concentration of political power in the hands of special economic interests, inducement mechanisms (in the Hayami-Ruttan sense) will be strongly affected in specific directions. Inducement mechanisms will contribute to technology generation which is consistent with the factor endowments of the most powerful rural groups, which may differ from those of the majority of farmers or of the economy as a whole. In this way technical change will be dominated by innovations which are "congruent" with the private needs of power groups, and research activities will be directed towards crops produced by these groups and to fit their specific resource availability.

In the absence of a social group exercising undivided hegemony, capable of imposing a specific overall economic and political strategy, the particular social relations (and consequently political conditions) prevailing in each agricultural production situation will determine the characteristics of technical change. In this way technical change will be specific, and its characteristics will vary according to variations in agrarian structure, at national, regional and product levels.

The contention of this article is that the uneven production and productivity increases of a number of different crops under a wide variety of production conditions (as shown above) can be explained on the basis of the social forces that characterize each of these production situations. These causal relationships between social relations and technical change may be fruitfully analyzed with the aid of a simple conceptual framework.

CONCEPTUAL FRAMEWORK FOR ANALYSIS OF TECHNICAL CHANGE

Basic Elements of the Model

In order to present the model's analytical categories and their interrelations, it is useful to define what may be called the socioeconomic space where technological change takes place. The socioeconomic space (a concept which refers to a specific production location) is defined by a set of structural elements and a set of social relations, both of which define the characteristics, intensity and direction of innovation. The structural elements are: (a) the predominant farm type, which defines the basic production relations, the characteristics of the demand for technology at farm level and the sociopolitical importance of that specific farming group; and (b) the institutional characteristics of technology generation, in particular the size and organization of private and public activities in technology production. The social interrelations determine the attitude (interests) of the different social groups related to production, circulation and consumption of the product, their interplay within the State mechanisms, and the nature of the resulting government policies towards the agricultural sector.[7]

With regard to social groups and their interplay at the level of the State, it is necessary to include not only those groups related to the production process (agricultural groups) but also those related through product and input markets, and those who relate to agriculture only as consumers. All of these groups will have specific interests in relation to agricultural production and will be affected, in very different forms, by the intensity and characteristics of technical change.

Four main social groups affect the innovative process: (1) the domestic urban industrial sector composed of two highly differentiated classes—the laborers and the entrepreneurs—which, in spite of their class differences and conflicts, tend to favor low price policies for agricultural commodities;[8] (2) the agricultural production sector composed of widely diverse social groups such as peasants and owners of commercial farms and *haciendas,* who have different interests in technical change and who express, both at the micro level and at the State level, different demands;[9] (3) the agricultural marketing-processing sector; and (4) the producers of agricultural inputs and technology. Through their interaction with the State, these social groups affect the process of policy generation and, ultimately, technological change.

Policies which affect the innovative process are broadly divided into economic policies and scientific policies for the agricultural sector.

Supply and demand for innovations. On the other hand, the process of technical change may be seen as being composed of two interrelated processes of supply and demand for innovations. The demand for innovations depends heavily on: (a) agricultural policies—such as price, credit and taxation policies—which affect the profitability of innovations; (b) the structural characteristics— production relations and factor endowments—and relative importance of different farm types, which determine the nature of demanded innovations; and (c) exogenous economic conditions that determine conditions for accumulation through agricultural production. The supply of innovations, in turn, is a direct result of: (a) the structure and conduct of the public institutional model for technology generation; and (b) the actions of private organizations. These two items are strongly dependent on the scientific agricultural policy. A central point is that demand and supply are interdependent through the role played by the State in the determination of model components which affect both sides.

Levels of articulation. The basic working hypothesis suggested here is: (1) an active process of technical change will require adequate articulations among the different components which define a particular socioeconomic space, and (2) this process of technical change has frequently been ineffective in Latin America, due to inadequate articulations between those components. There are at least four "levels" of articulation which are of particular importance.

First is the articulation of social groups in general, and dominant groups in particular, to the overall process of technology generation. This articulation explains the extent to which technology is a recognized social issue in the sense that research is a socially accepted and valued activity and consequently receives resources and an adequate institutional setting in relation to the overall State organization.

Second is the articulation of government policies at large, and economic policies in particular, to production conditions in such a way as to promote and permit adoption of innovations. An effective innovative process requires, in addition to new technologies, a whole array of complementary services (such as marketing, transportation, etc.) and the definition of an appropriate economic context (prices, credit) to induce farmers' adoption of innovations.

The third level of articulation is that of technology generation to actual demand for technology. The basic characteristics of available technology must be congruent with what is required at the farm level in terms of resource endowments and production relations. This articulation's efficacy will depend on the ability of the institutional mechanism to "read" technological demands of different types of farms, and then respond to those demands.

Finally, there is the articulation among the different components of the institutional model of technology generation, which determines the appropriateness of these components to specific functions (such as basic research and technology creation) and to the overall system, and the efficiency of the mechanisms for coordination and diffusion of information.

Interpretation of Technological Change

An analysis of technical change in U.S. agriculture, along the lines presented by Owen (1966), suggests the ways by which articulation of the technological process took place in that country and in other developed economies. The tremendous development of the industrial sector during the first half of this century gave way to a clear supremacy of industrial and consumer interests in the definition of public policy. Under these conditions, where a good portion of the surpluses generated from technological change in the agricultural sector were appropriated by consumers, the interest of the dominant industrial groups (and consequently the State) in technological progress was assured. A natural outcome of this was both an economic policy which induced an increase in agricultural production, and the creation of efficient institutions responsible for the generation and diffusion of new technology.

In addition, the agricultural sector of the U.S. can be characterized by commercial entreprises with professionalized management, efficient factor markets, access to information and adequate social organization (Owen's Western Paradigm).[10] Under these conditions, research resource allocation was geared by factor scarcities, and relative factor and product prices, and research institutions were induced to work in the desired direction with respect to production needs. Producers, pressed by the need to maintain or improve the profitability of their business, quickly adopted the set of techniques generated by public and private institutions, thus ensuring an extraordinary increase in agricultural production.

Disarticulation in Latin American Agriculture

The material development of the Latin American agricultural sector has generated, as has been pointed out by de Janvry (1975) and others, the emergence of a series of structural conditions that inhibit the replication of an articulated development process similar to that identified as the Western Paradigm.

The conceptual framework discussed above is useful to emphasize three interrelated conditions that strongly influenced the process of technical change in Latin America. First, in most Latin American countries, in spite of the important industrial development of recent years, rural groups preserve considerable political power which implies the absence of a clearly hegemonic class or social group (O'Donnell, 1977; Portantiero, 1977). Therefore, the basic element required for a development model based on the transference of agricultural surplus through a "treadmill" strategy, such as the one proposed by Owen in reference to the development of the U.S., is missing.

Second, the lack of adequate recognition of technological change as an instrument for economic development, and the incapacity of the State to mediate in the social conflicts that technology may generate, has resulted in low levels of investment in research, inadequate institutional development and insufficient coordination of research activities with other State actions (fourth level of articulation).

Third, the relative weakness of the State and the influence of institutional models and ideologies of the developed world has inhibited the development and growth of organizational patterns adapted to local conditions and with large participation of interested social groups. In this context, three reasons explain why the inducement mechanism failed to provide an adequate guide to the generation of technology, resulting in a disarticulation between demand for and supply of technology (third level).

- A great proportion of the available new technology is generated in the developed world and adapted to its economic conditions. Thus, inducement mechanisms were unable to influence the characteristics of this part of the supply of new technology.
- The considerable diversity of farm types within any one country, where large producers with relatively extensive systems of production stand side by side with very intensive small farms *(minifundios)*, introduces an element of heterogeneity into the required technology and required mechanisms of technology diffusion. Diversity makes the operation of inducement mechanisms considerably more difficult and the generation of new knowledge more costly. In general, research institutions have been inadequately prepared to be responsive to all research needs and as a consequence they have concentrated their efforts to satisfy the demands of those groups which had stronger social integration and more political power.
- Inducement mechanisms require that factor prices reflect their relative scarcity at a global level. In Latin American agriculture, factors of production—especially labor—are tied to inflexible farm structures; thus, social factor scarcity differs from private factor scarcities associated with the different farm structures. Consequently, there is a whole map of factor prices which hampers the inducement process (Mellor, 1977). In addition, economic policy tends in many cases to deliberately distort relative prices (for example, subsidized credit).

The levels of disarticulation presented in this model do not imply that technical progress has been absent in Latin America. It has already been shown that certain crops in some countries have achieved rapid increases in production. The hypothesis advanced here is that those cases have always implied a set of conditions such that: (a) a substantial part of the benefits derived from production increases and technical change were appropriated by a specific social group; and (b) this social group had sufficient political power to impose economic policies that permitted high profits through production increases.

In this respect, it is possible to indicate a few cases where technological change has been promoted and instrumented by one of the four major social groups identified. For example, the broiler chicken industry is one case of technical change instrumented by the processing sector; sugar in Peru by the large commercial farm sector (Flores Sáenz, 1977); livestock improvement in Argentina by livestock breeders (producers of technology); and wheat and soybeans in Brazil by the urban industrial sector for balance of payments purposes.

CONCLUSION

This article presents the highlights of a conceptual framework developed for the interpretation of the uneven and erratic performance of agricultural production and technical change in Latin America. The basic elements in the model can be summarized as follows.

The high risk, heavy investment characteristics of agricultural research, as well as the low possibilities of keeping private any resulting research benefits, determine that the State play a central role in the innovative process. This is carried out not only through the generation of new technology by public institutions, but also through the creation of an appropriate economic context for technological adoption at the farm level.

In many of the industrial societies, and in the U.S. in particular, the existence of politically stable coalitions that were interested in technical change made possible the consistent resolution of potential conflicts generated by such change. This explains the relative homogeneity of State policies across products and regions, and consequently the homogeneity of production performance.

In Latin America, the diversity of agricultural production, the association of special interest groups with specific agricultural activities, and the relative weakness of the State in the application of economic and scientific policies have caused the policies instrumented in each particular case to be the result of special circumstances. This depended on the relative political power and special interests of social groups related to each case.

We believe that this set of conditions explains the failure of inducement mechanisms to guide technical change, uniformly across products, in the direction of increased production and appropriate factor biases, and explains the incapacity of the State to instrument coherent strategies for food production and technical change.

NOTES

[1] Horizontal expansion on land of relatively high productive capacity was the most important source of agricultural growth in Latin America until the beginning of the 20th century. As the development process progressed and most countries completed their occupation of these lands, it became less attractive and possible to rely on the incorporation of new land for production increases. A more intensive use of marginal land requires, in most cases, not only important (public) investment in the form of economic infrastructure (roads, irrigation, etc.) but also new technology adapted to the particular ecological and productive conditions. For this reason, it seems difficult to imagine that significant increases in production can be obtained in the absence of new knowledge, even in situations where the overall strategy is based on the utilization of presently unused agricultural land. It is interesting to note that, probably as a consequence of the influence of developed countries in research activities, relatively little work is done in this direction compared with research which is oriented to increasing the productivity of agricultural land already in production.

[2] The principal exposition of the Theory of Induced Innovation is Hayami and Ruttan's book, *Agricultural Development: an International Perspective (1971)*. Its publication has been instrumental in stimulating the interest of researchers, including ourselves, in the analysis of technical change as a product of social relations.

[3] Expressed as productivity of male workers accounted in wheat units. These are an index of value of production converted to wheat units using relative prices of the different commodities. See Hayami and Ruttan (1971), Ref. 3. Appendix A.

[4] Land productivity is also expressed in wheat units.

[5] Growth rates and other figures cited through the rest of the paper are calculated from Food and Agriculture Organization (FAO) data and cover the period 1950–1975. For a detailed analysis of these figures see Trigo, Fiorentino and Piñeiro (1978).

[6] At least in some cases, it would seem that major technological breakthroughs are only adapted to restricted conditions. For example, new rice varieties developed in recent years are adapted to irrigated conditions but not to dry land farming typical of Brazil. However, this explanation cannot be generalized without raising perplexing questions. For example, why have there not been major research breakthroughs applicable to the original ecological environments of some at those crops which have great importance in the people's diets (for example cassava in Brazil, potatoes in Peru and corn in the Andean region), while other crops of secondary importance (such as wheat in Mexico and soybeans in Brazil) *have* had them?

[7] We want to stress the dialectic interrelations among productive forces, social relations, the nature of the State and back (through State policies) to productive forces.

[8] Examples of this can be found in the coincidence of stated positions in agricultural policy between labor unions and representatives of industry in Argentina. See, for example, O'Donnell (1977).

[9] For a detailed classification of farm types, their structural characteristics and their demand for technology, see, for example, Piñeiro and Trigo (1977).

[10] Owen's description of the Western Paradigm conditions is similar to the description of a capitalist economy in neoclassical theory. Without discussing the exaggeration of this characterization, we point out that it should be taken restrictively with respect to the U.S. agricultural sector during the period 1940–1960.

PART II

A LATIN AMERICAN PERSPECTIVE: SOME EMPIRICAL EVIDENCE

INTRODUCTION: A Few Words on Methodology

The first part of this book has presented three theoretical perspectives regarding broad interpretation of agricultural technical change. One of these perspectives (see Chapter 3) has been used as a general conceptual framework in developing seven case studies in six countries of Latin America. This part presents two of these case studies (Chapters 4 and 5), and a comparative analysis of the studies (Chapters 6 and 7).

The seven case studies were developed with the idea of providing detailed empirical evidence describing specific innovative processes that could illustrate the explanatory power of the proposed interpretative framework. The case studies represent specific production situations characterized by the existence of a dominant agricultural product and a relatively homogeneous farm structure, which define identifiable social and technical production relations. The case studies were selected in a way that would permit a certain degree of representativeness in relation to the agricultural production conditions in Latin America, and that would allow us to characterize the explanatory variables included in the theoretical model.

The two main dimensions used for the selection of case studies followed a specific line of reasoning. The existence of a dominant product assured reasonably homogeneous relationships between the agricultural producing sector and the rest of the economy, and consequently the possibility of readily identifying social groups related to the production and circulation process of that product. This in turn allowed for a more precise description and analysis of the institutional process of policy making, and for identifying the major sources of social conflict and the ways in which they had been mediated by the public sector.

A second dimension is related to the agrarian structure determined by the type of productive unit that characterizes each case study. The importance of identifying the dominant farm structure is that it in turn defines the production relations and, indirectly, the way in which the economic surplus is divided within the productive sector.

Case studies were selected with an eye toward an appropriate representation of main farm structures in Latin America, as illustrated by different production situations. In this way, the seven case studies illustrate the hacienda (milk production in Ecuador), agrarian capitalism and family farms (beef in Uruguay, corn in Argentina and rice in Colombia), and peasant economies operating independently (potatoes in Peru) or operating within and in relation to large enterprises (Northeast Brazil). Finally, a relatively new agrarian structure, characterized by an organization resembling industrial production with a high degree of vertical integration and economic concentration, was represented by sugar production in Colombia[1].

The seven case studies provide the empirical basis for the development of the comparative papers presented in the last three chapters of this second part. However, two studies, sugar production in Colombia and corn production in Argentina, are presented in full to give a feeling for the structure and content of the seven case studies.

The two selected case studies represent clear and extreme situations of innovative processes. Sugar cane is a situation with high economic concentration and vertical integration, leading to the formation of a cartel which is able to assume full control of the innovative process. Thus, through an agrarian initiative, a whole process of social articulation leading to technical change is set in motion. Economic policy and innovations adopted are tailored to suit the needs of the dominant social sector and to ensure that they receive most of the benefits of technical change.

The study on corn production in Argentina illustrates a different situation. Here the innovative process is set in motion by policies initiated with the support of an industrial sector interested in increasing agricultural production and exports as a means of solving a balance of payment crisis. Innovations adopted are less controlled by the agrarian sectors and the benefits of technical change are distributed widely among different social groups.

These two cases also illustrate the importance of certain social and economic dimensions in defining different processes of social articulation. It is the nature of these social processes that determine the rate and biases of technical change.

Chapters 6 and 7 present a comparative analysis of the seven case studies. Chapter 6 concentrates on an analytical description of the development of research institutions, the social forces that led to their creation, and reflects on their early success and the problems that emerged subsequently. A number of policy recommendations regarding research organization and technologi-

cal policy emerge, and are later discussed in detail in the third part of this volume.
volume.

Chapter 7 attempts to reconstruct the different processes of social articulation illustrated in the case studies, in order to find a general logic or dialectic relationship between production relations, social conflict, public policy and technical change. The chapter gives the background for a number of propositions related to the interpretation of technical change that are advanced in Chapter 8.

NOTES

[1] A study on tomato production in California, USA was used as a comparative case in a developed country, and as supporting evidence in the paper by de Janvry and LeVeen presented in the first part of this volume (Chapter 2).

SOCIAL RELATIONS OF PRODUCTION, CONFLICT AND TECHNICAL CHANGE: THE CASE OF SUGAR PRODUCTION IN COLOMBIA

Martín E. Piñeiro, Raúl Fiorentino, Eduardo J. Trigo, Alvaro Balcázar and Astrid Martínez

ABSTRACT

The introduction of new technology in the sugar cane production process in the Cauca Valley, Colombia, coincided with the appearance of social conflicts among the members of that agricultural sector. The authors view the modernization process as a result of pan-societal trends, and discuss the reciprocal relationship between changing technology and the evolving social order. A brief history of sugar cane production in Colombia is presented, focusing on the changing modes of production within the sector and how relations with other sectors of society varied thorugh time.

INTRODUCTION

This article presents the most important findings from a study of the modernization process in sugar cane production in the Cauca Valley of Colombia (Piñeiro and others, 1979). The study views the overall process as a result of broader social phenomena, and makes an attempt to relate the events that ocurred in the technological arena with the genesis and definition of social conflicts that developed within the sugar cane productive process.

The article has three sections. The first is a very brief review of the sugar sector's historical evolution, including a discussion of the socioeconomic structures within which sugar cane production has taken place since 1950. The second section describes and analyzes the technological behavior of the sugar sector during 1960–1978. Finally, the third section summarizes the technological process studied, and relates these technical characteristics to the predominant production relations in the sugar sector, as well as describing the relations between this sector and other social sectors.

SUGAR PRODUCTION IN THE CAUCA VALLEY: HISTORICAL DEVELOPMENT AND STRUCTURE CIRCA 1960

The Beginnings of Sugar Production

Although the first sugar mill in the Cauca Valley appeared at the end of the last century, sugar production was not significant until the 1940s, when the industry was consolidated with the construction of a number of mills. The sugar industry was slow to develop because of competition with the production of *"panela,"*[1] which was displaced only as an indirect consequence of the country's urbanization process.[2]

The consolidation of the industry clearly introduced capitalistic production relationships. Furthermore, evidence indicates that sugar activity began as a capitalist venture and that, as a consequence, the use of wage labor and the presence of suitable conditions for capital accumulation prevailed from the start. This study thus begins when the dominant sector, the sugar mill owners, had already become large capitalist enterpreneurs.

The consolidation of the industry, through the establishment of a significant number of sugar mills, the occupation of the land in the Valley, and the attainment of a certain productive scale took place during the 1940s and 1950s. During this period, production was still geared exclusively to domestic demand, which maintained sustained but limited growth, achieved mainly by expanding the areas under cultivation.

At this time, the sugar mills acted individually, competing with each other to capture the limited domestic market and for access to the best farmlands (FEDESARROLLO, 1976:178). There are clear indications of severe isolation among the various mills, and little contact between the mills and the official institutions set up to improve productive techniques. A sign of this was the complete lack of communication among the sugar mills regarding their technological discoveries and the care they took to prevent information leaks about their individual production processes.[3] The ICA–Palmira station maintained a certain level of activity in the 1930s, particularly in plant breeding, and in the 1940s, the mills reported some research findings and the adoption of exogenous innovations. However, cooperative programs between the experimental station and the mills did not begin until the end of the 1950s.

The experimental station operated with extremely limited resources during this period. It is not surprising, then, that before 1960, very little technological innovation was incorporated into sugar production; that introduced was limited to a few techniques with minor productive impact. At the same time, the absence of any strong relations among the mills, the limited nature of the market, and the relative inmaturity of the industry prevented the emergence of any corporate organization to oversee the relationship between the sector and the State.

Not until the end of the 1950s did the industry begin to develop into an oligopoly. The process continued over the last twenty years, leading to the sector's present structure. In 1959, ASOCAÑA (Colombian Association of Sugar Cane Producers) was created as a producers' association. After 1960, the external market began to open, and ASOCAÑA became a fundamental element in the industry's subsequent development.

Structural Characteristics

Growth and organization of production. The sugar industry in Colombia entered the 1960s as a slow growth industry controlled by the small number of families or economic groups that owned the sector's twenty mills. It was in a state of relative stagnation, with eighty per cent of the cane being produced directly by the mills on their own land, but underwent an abrupt turnaround after 1960. In the first place, production began to grow sharply, climbing from 328,000 tons in 1960 to 963,000 tons in 1978 (see Table 1). At the same time, marketing targets shifted radically as export percentages began to climb. The figures in Table 1 show that raw exported sugar grew from nearly nothing in 1960 to a peak of 238,000 tons in 1968, a year when exports made up over thirty per cent of total production. In the following years, the percentage began to slide, eventually settling at around twenty per cent.

A second type of change that took place after 1960 affected the organization of production. Table 2 shows land area used for cane crops and the supply sources for production in eight mills. It reflects a sharp fall in the percentage produced on wholly owned land, with a corresponding rise in the importance of independent suppliers, who were producing nearly fifty per cent of the sugar cane by 1977.

These changes in the organization of production have not greatly affected the structure of the industry. Table 3 groups the sugar mills according to size and shows their participation in sugar production. As can be seen, the rate of concentration, calculated as the percentage of production controlled by the four largest mills, held steady throughout the period at levels of fifty per cent

TABLE 1 Colombia: Area Cultivated and Harvested, Agricultural and Industrial Yields and Production of Sugar, White Sugar and Molasses, 1960–1977 (Total Country)

YEAR	SUGAR CANE									SUGAR PRODUCTION / THOUSAND TONS			
	AREA/HA[1]		YIELDS							RAW		WHITE[5]	MOLASSES[6]
	Cultivated	Harvested	Acres Harvested[2]		Sugar Yield[3] %					Total[4]	Exported		
			a	b	c	d							
1960	61,600	39,630	88.0	129.8	10.1	9.4				328.3	0.13	308.2	104.6
1961	62,519	40,847	84.6	—	—	—				362.6	48.70	294.7	103.7
1962	63,787	42,020	92.8	116.0	10.4	10.3				401.9	65.6	315.8	117.0
1963	63,636	36,410	95.0	—	—	—				368.1	40.8	307.3	103.8
1964	64,201	39,361	100.0	115.6	11.0	10.8				427.6	25.7	377.4	118.1
1965	70,363	42,230	103.1	—	—	—				485.2	94.6	366.8	130.6
1966	78,707	49,588	100.1	110.8	9.6	10.8				537.4	113.9	397.6	148.9
1967	82,335	52,518	101.0	—	—	—				596.6	200.0	372.4	160.5
1968	86,050	56,109	105.8	114.9	10.6	11.2				663.3	238.7	398.7	178.1
1969	91,745	60,093	104.3	—	—	—				708.7	171.3	504.6	188.0
1970	91,982	58,741	102.4	108.2	10.4	11.2				676.2	129.4	513.4	180.5
1971	97,960	63,675	112.4	—	—	—				744.0	165.2	543.5	214.7
1972	102,250	67,737	110.0	113.7	10.4	11.0				823.7	202.8	583.0	223.5
1973	113,767	68,172	108.0	—	—	—				809.9	142.5	626.6	220.9
1974	116,216	69,637	115.0	126.5	10.0	11.2				894.8	128.6	719.5	240.2
1975	118,450	74,054	120.0	—	—	10.9				969.7	197.8	724.8	266.6
1976	122,728	84,820	104.0	—	—	10.6				934.6	100.2	784.4	264.6
1977	127,889	76,465	98.8	—	—	10.7				810.6	—	—	226.6
1978	—	—	—	—	—	—				963.0	—	—	—

SOURCES: FEDESARROLLO.

METHOD:
[1] *Area cultivated and harvested:* Area planted includes canals, roads, camps, highways, etc., equivalent to 8 per cent of the total area (ASOCAÑA).
[2] *Cane yield by acre, 1960–1974:*
 a) ASOCAÑA.
 b) According to survey by FEDESARROLLO. Information covers 4 large mills, one medium sized mill, one small-to-medium mill and one small mill (FEDESARROLLO, 1976:302).
[3] *Sugar yield, 1960–1974:*
 c) Figures for 17 mills average (FEDESARROLLO, 1976:308).
 d) Estimated by the authors, dividing column 7 by the product of column 2 and column 3.
[4] *Total sugar:*
 1960–1974 (FEDESARROLLO, 1976: 415).
 1975–1977, Banco de la República.
[5] *White sugar:* White sugar production was calculated by applying a 93.9 per cent recovery factor to the difference between total sugar and raw sugar exports.
[6] *Molasses:* Authors' estimation.

TABLE 2 Colombia: Area Harvested and Production of Sugar Cane in Eight Surveyed Mills According to Source of Supply

SOURCE OF SUPPLY / YEAR	WHOLLY OWNED LAND				RENTED LAND				ADMINISTERED LAND				SUPPLIERS			
	Area/ha		Production/ton		Area/ha		Production/ton		Area/ha		Production/ton		Area/ha		Production/ton	
	No.	%	No.	%	No.	%	No.	%	No.	%	No.	%	No.	%	No.	%
1960	11,308	76.2	1,535,647	79.8	925	6.2	122,609	6.4	—	—	—	—	2,609	17.6	267,129	13.8
1962	16,613	78.1	1,963,748	79.8	1,273	6.0	148,051	6.0	—	—	—	—	3,378	15.9	349,846	14.2
1964	15,821	72.9	1,847,826	73.3	2,667	12.3	293,315	11.6	—	—	—	—	3,203	14.8	380,082	15.1
1966	14,867	51.2	1,602,220	49.8	5,449	18.8	635,213	19.7	—	—	—	—	8,735	30.1	983,667	30.5
1968	16,648	47.6	2,021,014	49.5	9,392	26.8	1,087,493	26.6	—	—	—	—	8,942	25.6	973,111	23.8
1970	17,699	46.0	1,935,475	46.2	10,548	27.4	1,188,188	28.4	596	1.6	70,419	1.7	9,600	25.0	992,107	23.7
1971	16,634	39.0	1,973,847	41.8	8,529	20.0	946,497	20.0	2,352	5.5	252,887	5.4	15,161	35.5	1,550,290	32.8
1972	18,870	44.5	2,253,098	46.5	10,471	24.7	1,095,623	22.6	1,251	3.0	129,187	2.7	11,805	27.8	1,569,572	28.2
1974	15,611	42.3	2,144,314	44.0	5,818	15.8	568,228	11.7	3,265	8.8	331,618	6.8	12,262	33.2	1,679,738	34.5
1976	20,246	38.7	2,144,474	38.9	2,938	5.6	313,317	5.7	4,546	8.7	438,718	8.0	22,946	43.9	2,453,399	44.5
1977	17,701	37.8	1,834,696	36.1	3,043	6.5	315,096	6.2	4,160	8.9	450,248	8.9	22,093	47.0	2,481,225	48.8

SOURCES: ASOCAÑA AND FEDESARROLLO.

Harvested area and production of ground cane according to source of supply for the eight surveyed mills (Providencia, Manuelita, Castilla, Riopaila, Mayagüez, San Carlos, Pichichi y Balsilla), 1960–1977 (incomplete).

TABLE 3 Colombia: Percent Participation in Sugar Production of Mills, by Size: 1961–1974

YEAR	GROUP I Large	GROUP II Medium	GROUP III Small to Medium	GROUP IV Small	GROUP V	TOTAL
1961	54.20	16.74	9.90	19.6[a]	–	100.00
1962	56.94	16.41	9.42	17.23[a]	–	100.00
1963	58.08	21.07	10.25	10.69	–	100.00
1964	60.22	17.85	9.36	12.13	0.44	100.00
1965	55.87	19.34	8.39	16.40[a]	–	100.00
1966	53.06	22.30	8.82	15.53[a]	0.29	100.00
1967	50.46	25.20	8.72	15.36[a]	0.26	100.00
1968	51.14	25.67	10.18	12.85[a]	0.16	100.00
1969	52.79	25.09	10.67	11.23[b]	0.22	100.00
1970	52.23	25.70	10.54	10.91	0.62	100.00
1971	50.03	25.99	11.66	11.69	0.63	100.00
1972	50.13	27.30	11.16	10.87	0.54	100.00
1973	52.19	25.38	11.16	11.02	0.25	100.00
1974	49.94	26.02	13.10	10.61	0.24	100.00

SOURCE: FEDESARROLLO (1976:235).

NOTES:
[a] Includes San Fernando and María Luisa Mills.
[b] Includes María Luisa Mill.

GROUP I: (Large)	Castilla, Manuelita, Riopaila, Providencia
GROUP II: (Medium)	Cauca, Mayaguez, Pichichí
GROUP III: (Small to Medium)	La Cabaña, Tumaco, Bengala, Meléndez, El Papayal, El Porvenir, Carmelita, La Industria
GROUP IV: (Small)	Balsilla, Oriente, El Naranjo
GROUP V:	Sicarare, Santa Cruz

Classification taken from FEDESARROLLO (1976:171,236).

or more. The sector can thus be described as a high concentration industry. A situation which was exacerbated by the extreme vertical integration between processing and distribution. Eighty per cent of total sugar production was distributed through channels that were either entirely or partially controlled by the mills. The institutional structure encouraged this situation by dictating mechanisms for control and stabilization based on the production percentage controlled by each of the participants in the sugar process.[4]

Despite this economic concentration and highly vertical integration, a situation of conglomeration did not develop. While in some cases the economic groups which control the process have made investments in other sectors, they nevertheless engage in sugar production as their principal activity. Because of their concentration and specialization in sugar production, these groups swiftly attained a position of political dominance in the region. This region, in turn, is a focal point of national political power. It is also important to recall the traditional consumption of *"panela"* in Colombia, which places sugar in a position of secondary importance as a wage-good.[5]

Prevailing production relationships. Production relationships in both stages of the process (farming and industry) are clearly capitalistic, although certain elements exist which complicate the flow of services and money within the sector. The exchanges that take place within the productive process are described in Fig. 1, which clearly illustrates the dominance of the mills in

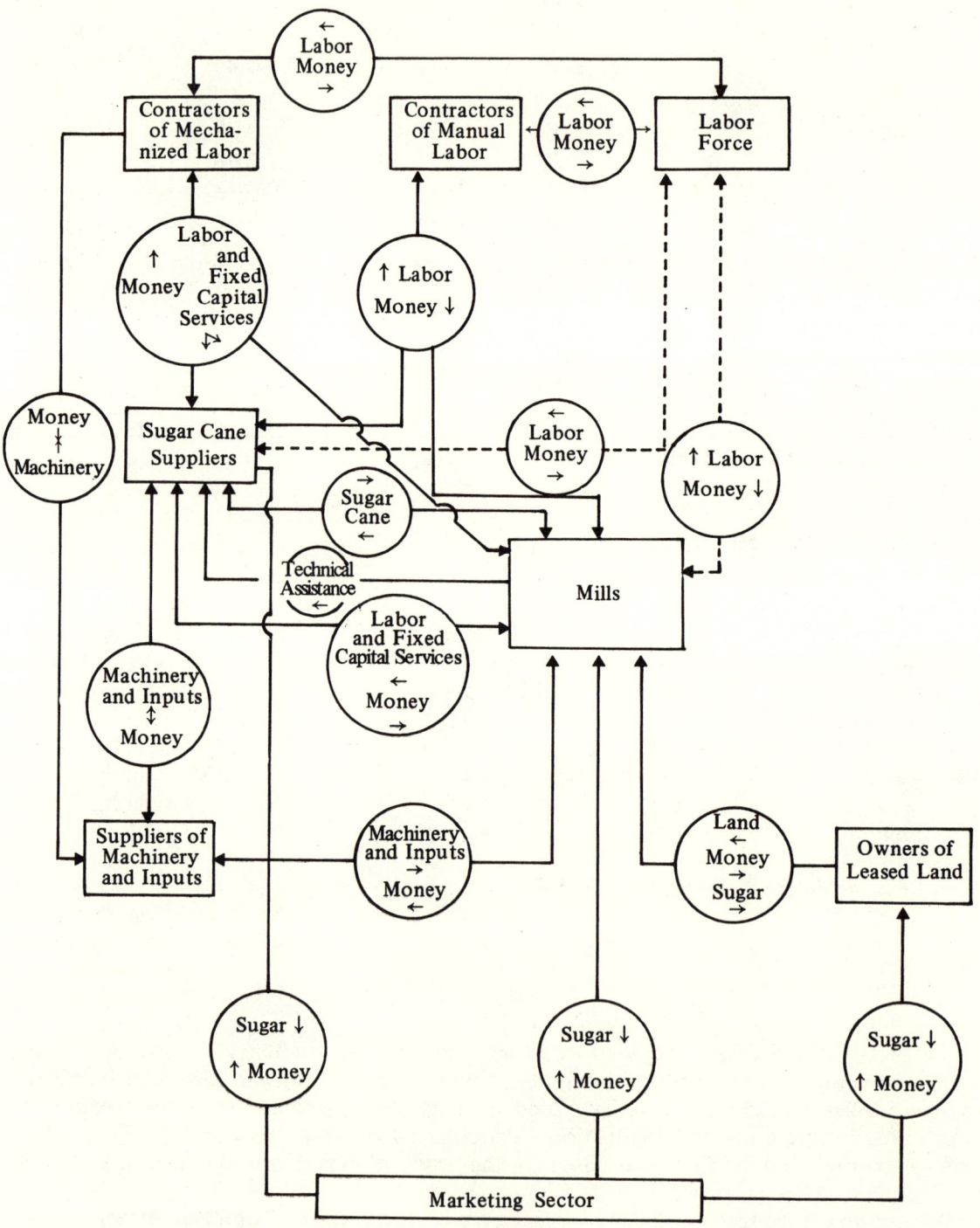

Fig. 1. Colombia: flow of goods and services in sugar cane production in the Cauca Valley

production and distribution, given the elevated rates of concentration and vertical integration. This situation is even more pronounced when relations between the mills and the independent suppliers are analyzed. The latter receive capital and technical assistance services from the mills, which in turn control many of their production decisions.

TECHNOLOGICAL BEHAVIOR FROM 1960 TO 1978

The Sixties

The decade began with a sharp rise in the demand for sugar. On one hand, the domestic demand for sugar grew markedly (by approximately four per cent) because of the considerable increases in population and urbanization, and the substitution of sugar for panela. Simultaneously, the possibility of exporting sugar to the protected market of the United States arose, as a consequence of its economic blockade on Cuba. The quotas awarded to Colombia by the United States increased significantly after 1962, and became the main force behind the industry's development. Within a few years, exportable production had exceeded the ability of the United States market to absorb sugar. Nevertheless, it is important to note that these increases, and the fact that the needed export marketing systems had been created, made it possible for the mills to enter the international free market. Although prices were lower in the arena, it allowed the mills to maintain a high rate of production growth.

The benefits to be derived from the higher prices offered by the United States' quota system led the mills to combine their efforts, splitting the export quotas among themselves and excluding new potential competitors from this highly profitable market. Responding to this situation in April 1962, ASOCAÑA created the sugar pact as an institutional mechanism for allocating quotas among the participating mills.

The first step was to create a supply cartel for foreign markets, and a powerful block quickly emerged. The mills created jointly owned enterprises for exporting molasses (COLMIELES), and distributing and exporting sugar. With the establishment of the cartel came the development of strong producer organizations to negotiate with the State. Colombian production was able to respond in this way to the rising demand because of excellent ecological conditions that produced comparatively high yields, even with rudimentary production techniques. Production increases required available land and labor, both of which were abundant in the Valley during the 1960s.

Because the productive techniques already in use generated significant benefits, there was no strong incentive for modernizing. The mills invested in expansion by incorporating new lands into production, using familiar technological patterns. For this reason, during the first half of the 1960s, agricultural production was slow to modernize.

As production became further consolidated and significant portions of the market were captured by the large mills, the erratic nature of the world sugar market became clear, and the mills recognized the need for common action to mitigate the market fluctuations. This new sectoral awareness led directly to the creation of ASOCAÑA, established to defend the interests of the mills and to coordinate these interests with those of other economic sectors in the country. The trend continued throughout the decade. Consequently, as the 1960s opened, the mills joined forces to institutionalize actions in areas in which common efforts had not been made previously. This was particularly important for export promoting policies and the negotiation of domestic prices.

At the same time, the economic and class structure of the sector, which had been taking shape during the first part of the century, provided the mills with freedom of action. The mills clearly comprised the dominant social group in the sugar sector and, in general, in the Cauca Valley. During this period, no counter-proposals were made by the independent sugar cane suppliers. The suppliers were still a minority group, and PROCAÑA, the Association of Cane Producers, created to defend their rights, was not to appear until the following decade.

Similarly, the sugar workers had not yet become organized enough to oppose the controlling interests. Although labor had made sporadic attempts to organize over several decades, they were poorly coordinated and focused on periodic demands for wage increases. Even when the unions were at their strongest, the repressive nature of the system made it practically impossible to introduce any proposals for worker participation.

Another sector interested in expanding and modernizing sugar production activities was composed of agroindustrial companies that sold machinery and inputs. This group took no active interest in the institutional aspects of sugar production. Although they sought to promote their products, as in any rapidly growing high capital market, they did not assume a clear role in sugar policy formulation.

The weakness of these sectors as compared to the mills facilitated the effective implementation of a number of proposals put forth by the mills. Most were channeled through ASOCAÑA, which since its founding had successfully united its constituents. In order to understand how the mills responded to the process of technical change, it is necessary first to understand how they normally operated. Specific technological policy efforts should be compared and contrasted with the magnitude of ASOCAÑA's endeavors in key areas of economic policy, such as marketing and prices or credit.

Early efforts sought to increase profitability and promote sugar exports. During the early 1960s, the record shows no State action directed specifically to sugar production. It was only in 1967 that three important measures were taken to stimulate exports: (a) a progressive exchange rate adjustment mechanism; (b) the tax guarantee certificate (as a means of reincorporating non-traditional Colombian exports); and (c) the establishment of an Export Promotion Fund to handle a major line of credit for this purpose.

During the same period the mills began to recognize the new opportunities offered by foreign trade. The 1962 Multilateral Pact, which involved only the mills, was set up to regulate the supply of export sugar through quotas assigned to each mill on the basis of their previous output. It is important to note that because the provisions of Multilateral Pact regulated the export quotas of all the mills, they also dictated sugar supplies for the domestic market. It must also be kept in mind that any institutional organization based on a quota market implicitly establishes barriers that prevent new companies from entering the activity. Thus, the Multilateral Pact was established to ensure a stable market regulated by the mills.

A second area that should be considered in this analysis is domestic prices. Until 1960, the mills had complete freedom in setting prices. Beginning that year, however, the government established fixed prices for mill sales and, a few years later, for retailers. From then until the present, mill prices have been established with the active participation of the mills, based on production costs. It is also important to note that domestic prices tended to offset the cyclical oscillations of export prices, a phenomenon which was very evident during the second half of the decade[6] (see Fig. 2).

The appearance of the wage policy is also an important component in evaluating the institutional actions in the sector. In the first place, the general sugar labor force was not consolidated into a single union that could have negotiated wages and working conditions for the sector as a whole. Although collective agreements for negotiating wages had existed since the sixties, they were made independently for each mill. Since the mills were widely dispersed throughout the Valley, communication among workers was difficult, a situation which favored the negotiation of agreements with highly disparate wage levels from one mill to another. The effect of this dispersion was so great that differences of up to 40 per cent existed within a given worker category.[7] The lowest wages were paid at the smallest mills, where the small work force made it difficult to impose demands, and where the owners had an unstable potential for accumulation because of higher costs.

Commercial relations between the mills and the suppliers were also stepped up during the 1960s. Fragmentary information from PROCAÑA leaders suggests that these relationships were very unequal and favored the mills in commercial matters, thus forcing the suppliers into a position of confrontation. It is useful to note that when the suppliers created PROCAÑA to avail themselves of an exclusive representative base, they expressly stated their desire not to be absorbed by ASOCAÑA, which had been willing to accept them. Clearly, this "expressed desire" stemmed from concrete objective conditions, such as the suppliers' control of a large amount of sugar land in the Valley.

The concerted institutional action taken by the mills through ASOCAÑA was successful in influencing domestic prices, controlling wages and regulating export activity. Finally, with other groups effectively barred from active participation, the mills began to play a role in the modern-

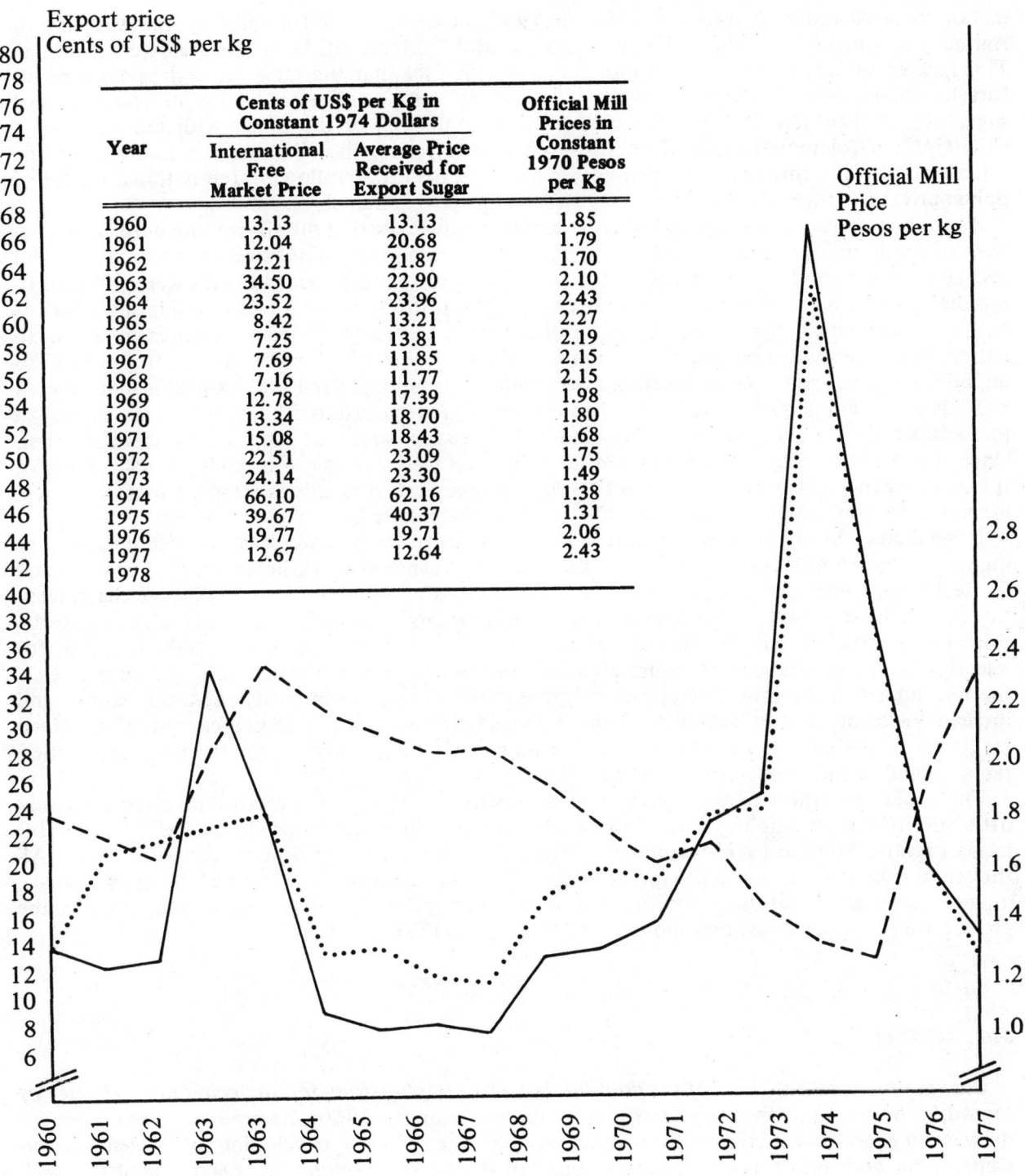

Year	Cents of US$ per Kg in Constant 1974 Dollars		Official Mill Prices in Constant 1970 Pesos per Kg
	International Free Market Price	Average Price Received for Export Sugar	
1960	13.13	13.13	1.85
1961	12.04	20.68	1.79
1962	12.21	21.87	1.70
1963	34.50	22.90	2.10
1964	23.52	23.96	2.43
1965	8.42	13.21	2.27
1966	7.25	13.81	2.19
1967	7.69	11.85	2.15
1968	7.16	11.77	2.15
1969	12.78	17.39	1.98
1970	13.34	18.70	1.80
1971	15.08	18.43	1.68
1972	22.51	23.09	1.75
1973	24.14	23.30	1.49
1974	66.10	62.16	1.38
1975	39.67	40.37	1.31
1976	19.77	19.71	2.06
1977	12.67	12.64	2.43
1978			

Key
A ___ International free market price.
B .... Average price received for export sugar.
C _ _ _ Official mill prices in constant 1970 pesos per kg.

Fig. 2. Colombia: Ratio between export sugar prices and official mill price.

ization of agricultural production. Prior to 1960, official efforts (initially by the Ministry of Agriculture and later by the Colombian Agricultural Institute—ICA) in this area were minimal. This lack of official support was compounded by the fact that the mills themselves made no effort to obtain support, in sharp contrast with ASOCAÑA's concerted action in other areas of sugar policy. This attitude began to change towards the end of the 1970s with the creation of CENICAÑA (Colombian Sugar Cane Research Center), once the ICA program had ended. The initial impulse for this center had appeared as early as 1963, following the national model of public-private institutions such as that formed for the coffee industry.

A sugar research center was not created earlier because, during the sixties, the need for incentives to modernize production and to introduce labor-saving techniques was not very great, land and labor being amply available. More importantly, during this period the mills were concentrating their efforts on controlling tremendous expansion in the sector and on consolidating ASOCAÑA. Any economic activity that expands by 10 per cent annually obviously requires close coordination, even when using known production techniques. Equally important, the control of marketing policies, prices and incomes in an industry that has been organized into a cartel is more urgent and has a conceptually higher priority than efforts at modernization. These policies are actually a prerequisite for modernization efforts, and it is not surprising that they were established first. Once the "economic" policies were solidified, the mills were assured of receiving a major share of the profits generated by the incorporation of technological innovations. This led to new interest in modernization, which characterized the latest developmental stages of the industry.

Nevertheless, by the end of the sixties, the sugar sector had already incorporated a significant number of new production techniques (Table 4). It became evident, however, that the demand for technology expressed by the mills was not particularly selective as to qualitative characteristics (factor biases), perhaps because all production factors were abundantly available. Thus, the principal techniques incorporated do not have a clear factor-use bias. The only criterion for selecting the incorporated techniques appears to have been relative efficiency.[8] In spite of this, the net impact of the innovative process during the sixties was to intensify the use of capital and increase yields per unit of land (Figs. 3 and 4). This heightened use of capital cannot be explained by trends in the relative factor prices, which remained nearly constant for the three production factors: land, capital and labor (Table 5).

The rapid expansion of production which occurred at this time therefore involved relatively little adoption of technology and consequently minimal increases in productive efficiency. This had a negative effect on mill income as compared to the income of other production agents. Although the income of the mills increased by 60 per cent between 1960 and 1970 as a consequence of their agricultural activities, the incomes of other social groups participating in the productive process increased by more than 250 per cent (Table 6).[9]

The Seventies

Changes in social relations of production and their implications for the economic behavior of the mills. As the sugar industry grew and matured during the 1960s, its productive structure underwent changes that increased the concentration of the industry, developing an efficient institutional framework for operating the cartel. Signs of this process include the weakening of the smaller mills that were not closely tied to the dominant economic groups, the control of small mills by large ones, the increasing vertical integration of the principal economic groups, and the development and reinforcement of mill-based enterprises for the distribution, exportation and processing of sugar by-products.

At the same time, the rapid expansion of production that took place during the sixties, and the qualitative characteristics of this expansion, produced a series of changes in the economic conditions under which the industry was developing. The principal changes were:
  (a) Increased occupation of available land for sugar cane production, primarily by the independent suppliers. As a consequence of this process, land became a scarce resource for some mills in the 1970s.
  (b) The growing unionization of the labor force and their resulting ability to negotiate with the mills on remuneration and working conditions.

TABLE 4 Colombia: Chronological Order of Technological Innovations Adopted by Mills

INNOVATION	1960	1961	1962	1963	1964	1965	1966	1967	1968	1969	1970	1971	1972	1973	1974	1975	1976	1977	1978
1. Varieties[1]	ABJ		CE	DF			E					B		C		H	DK		
2. Fertilization with "N"	AB										G		G	H					
3. Construction of deep wells						C	D						G					I	
4. Deep Subsoil	J					AB		K	D[2] E									DH	
5. Fertilization with P and K	A							E			B	G	F						
6. Chemical weed control			AB			A		CK	F		G				H			F	
7. Mechanized harvesting				E		AD			B		C					H		FI	
8. Microleveling					B	A	D	CEGJ											
9. Uprooting					B	A	D				E								
10. Redesigning lots						C	E				BJ			G	D	C HK			
11. Fertilizer dosage adjustment											ACJ FGK			H		A			
12. Seed treatment									K		ABC	BDJ		EJK	E	FH		I	
13. Deep drainage									K		C	C						FI	
14. Biological control											AB		A	J	K	E		DI	
15. Spray irrigation											B		G	G		G		H	I
16. Cane burning											D			GA					
17. Use of nurseries (seedbeds)													K	E			F		
18. Chisel plough													G						
19. Australian cutting															C				
20. Harvest program																E	E	ABF	
21. Gravity irrigation, small prefabricated dams[3]																	AE		
22. Combined weed control[4]																	BA	E	

SOURCE: Data developed from information supplied by eight mills and three independent suppliers covered by the survey.
[1] All mills before 1960.
[2] Prior to 1960, Manuelita dug its first deep well in 1953.
[3] Substituting bamboo and mud for canvas.
[4] Mechanical and manual post-emergent control, and chemical pre-emergent controls.

KEY: The mills are identified according to the following system:
A,B: large mills
C,D: medium large mills
E,F: medium small mills
G,H: small mills
I,J,K: independent cane producers.

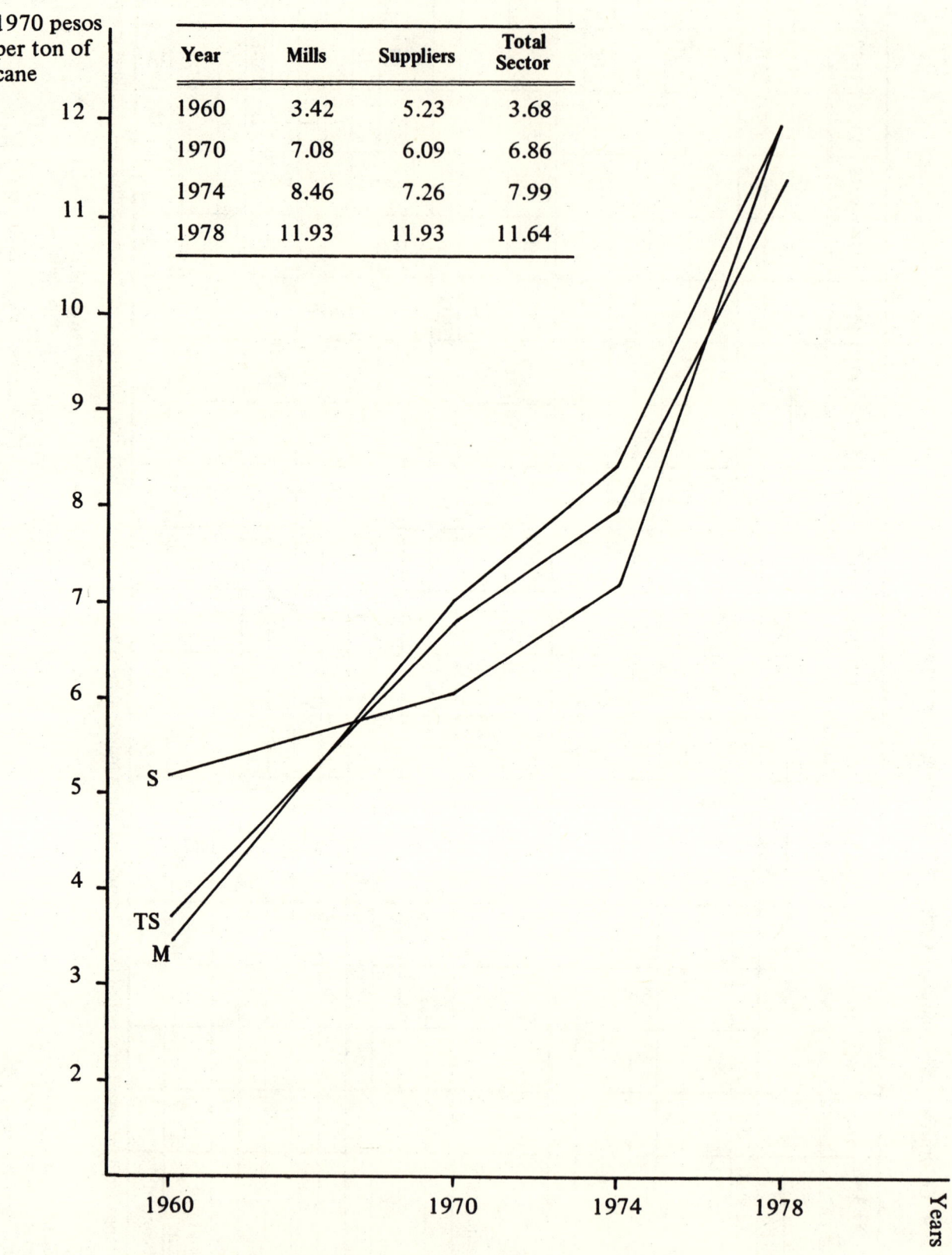

Fig. 3. Colombia: Use of fixed capital per ton of cane produced, excluding harvest. Average for the mills, the suppliers and for the total sugar-growing sector.

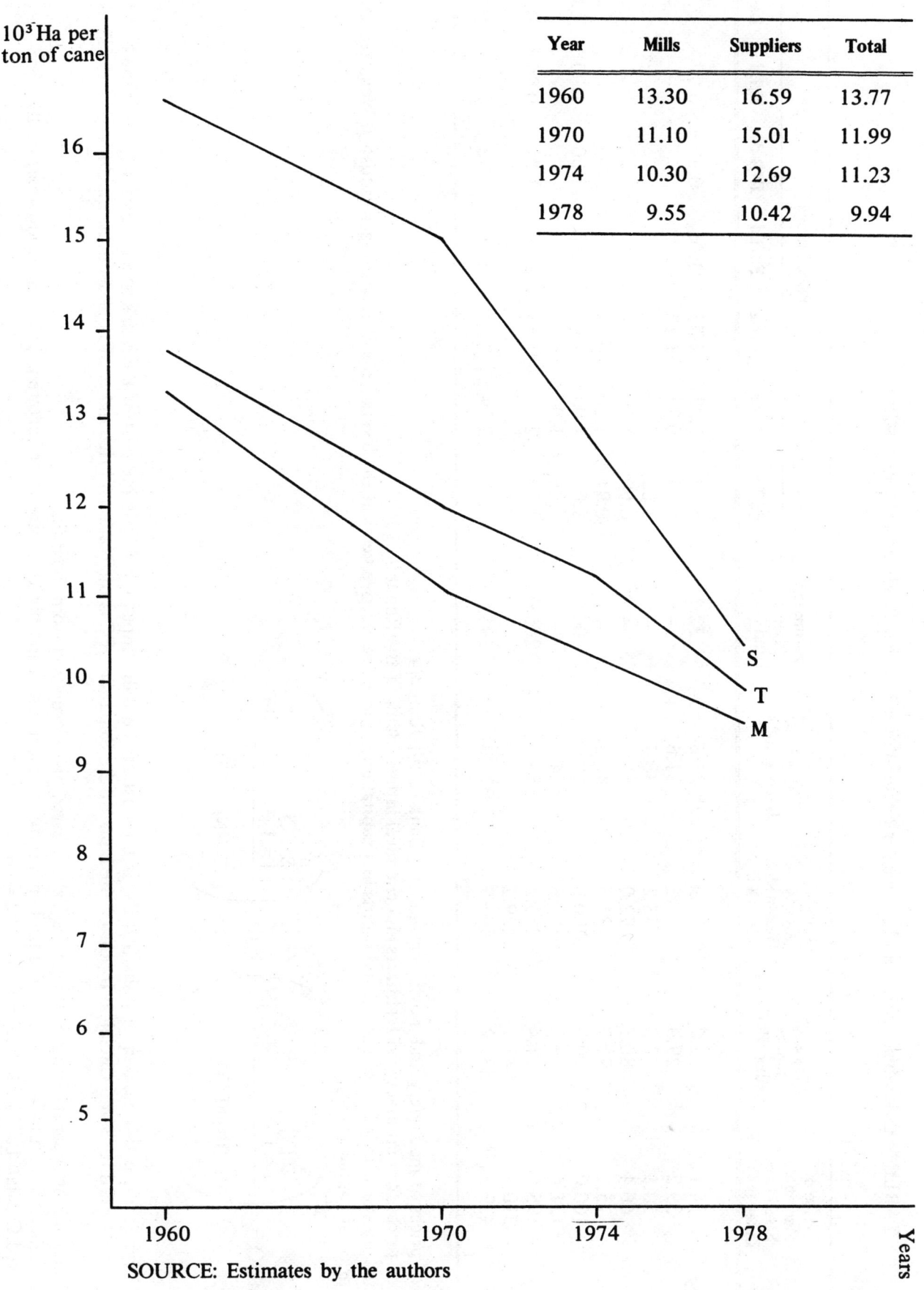

Fig. 4. Colombia: Land use per ton of cane produced. Average for mills, suppliers and for the total sugar-producing sector.

TABLE 5 Colombia: Indices and Ratios for Factor Prices in Sugar Production 1960–1978

Year	Price of Machinery Fixed Cap. Index (FC)[1]	Price of Inputs Index (IN)[2]	Capital Goods Weighted Index (K)[3]	Wage Index (La)[4]	Estimated Price of Land (Ld)[5]	Price ratios[6]						
						FC/La	IN/La	K/La	FC/Ld	IN/Ld	K/Ld	La/Ld
1960	34.1	28.7	31.9	23.6	32.2	1.44	1.22	1.35	1.05	0.89	0.99	0.73
1962	39.2	29.4	35.3	31.6	—	1.24	0.93	1.12	—	—	—	—
1964	50.5	62.3	55.2	48.7	—	1.03	1.28	1.13	—	—	—	—
1966	66.4	80.3	72.0	67.2	—	0.99	1.19	1.07	—	—	—	—
1968	75.1	90.7	81.4	82.3	—	0.91	1.10	0.99	—	—	—	—
1970	100.0	100.0	100.0	100.0	100.0	1.00	1.00	1.00	1.00	1.00	1.00	1.00
1972	131.6	129.3	130.6	124.4	—	1.06	1.04	—	—	—	—	—
1974	195.6	331.4	255.2	156.8	265.0	1.24	2.11	1.63	0.74	1.25	0.96	0.59
1976	310.4	581.1	429.2	—	—	—	—	—	—	—	—	—
1978	439.2	687.9	547.9	270.6	396.0	1.62	2.54	2.12	1.10	1.73	1.38	0.68

NOTES:

[1] Index of prices of machinery and moving equipment; Bank of the Republic.
[2] Index of prices for manufactured fertilizers and chemical inputs; Bank of the Republic.
[3] Weighted by the share of fixed capital and inputs in a capital unit, for the respective years at the level of the Valley's Sugar Industry: Weighted index for the price of capital goods,

$$(IPK) = \left(\frac{\sum_{j=1}^{n} IN_j}{\sum_{j=1}^{n} (IN_j + FC_j)} \right) \cdot iIN + \left(\frac{\sum_{j=1}^{n} FC_j}{\sum_{j=1}^{n} (IN_j + FC_j)} \right) \cdot iFC$$

Where: j varies up to the number of mills (n); IN_j is the cost of the mill's input (j); FC_j is the cost of the mill's fixed capital (j); iIN is the National Price Index of Fertilizers and Chemical Inputs of the Bank of the Republic; iFC is the National Price Index of non-electrical machinery of the Bank of the Republic.
[4] Wage Index of the sugar industry of the Cauca Valley; based on information from the mills.
[5] Corresponds to the estimated profit of the land, as a result of sugar prices and the kg of sugar per hectare paid by the landowners' mills.
[6] FC = Fixed Capital; La = Labor; IN = Inputs; Ld = land; K = Capital.

TABLE 6 Colombia: Distribution of the Income Generated by the Sugar Producing Sector Among the Social Groups that Participate in the Pre-Harvest Activities (Constant Prices Expressed in Thousands of 1970 Pesos)

Social groups making payments \ Social groups receiving income	Years	Distribution & Mills in Industrial & Harvest Stages[a]	Mills Pre-Harvest Stage[b]	Independent Suppliers	Land Owners	Workers	Suppliers of Capital Goods	Suppliers of Inputs	Total Payments
Consumers and Government (CAT)[f]	1960	731,520							731,520
	1970	1,551,030							1,551,030
	1974	3,335,720							3,335,720
	1978[e]	3,017,900							3,017,900
Sugar Distribution Sector and Mills in Industrial and Harvest Stage	1960		280,194	62,760		93,719[c]	53,572		490,297
	1970		563,078	209,320		257,791	103,636		1,133,826
	1974		968,553	507,899		188,449	128,138		1,993,040
	1978		619,946	513,128		150,141	130,333		1,403,550
Mills in Pre-Harvest Stage	1960				5,593	31,047	9,083	12,376	58,099
	1970				57,185	72,991	33,557	44,126	207,861
	1974				70,746	57,232	49,660	104,427	282,066
	1978				50,118	52,038	57,207	89,323	248,687
Independent Suppliers	1960					6,496	2,189	2,825	11,511
	1970					24,350	12,087	15,346	51,783
	1974					28,565	24,594	50,667	103,827
	1978					31,110	48,347	78,252	157,710
Total Gross Income	1960	731,520	280,194	62,760	5,593	131,262	64,844	15,202	559,907
	1970	1,551,030	563,078	209,320	57,185	355,133	149,280	59,473	1,393,470
	1974	3,335,720	968,553	507,899	70,746	274,247	202,393	155,094	2,378,933
	1978	3,017,900	619,946	513,128	50,118	233,289	255,887	167,576	1,809,947
Total Intersectoral Payments	1960	490,297	58,099	11,511					559,907
	1970	1,133,826	207,861	51,783					1,393,470
	1974	1,993,040	282,066	103,827			d	d	2,378,933
	1978	1,403,550	248,687	157,710					1,809,947
Net Income	1960		222,094	51,249	5,593	131,262			410,200
	1970	d	355,217	157,536	57,185	355,133			925,072
	1974		686,485	404,072	70,746	274,247	d	d	1,435,551
	1978		371,259	355,418	50,118	233,289			1,010,086

NOTES:
[a] As a consequence of the high degree of vertical integration, these groups can be considered the mill's industrial and commercial interests.
[b] Represents the pre-harvest agricultural activity of the mills.
[c] These payments correspond only to the agricultural pre-harvest stage.
[d] Data that cannot be estimated with the available information.
[e] The figure corresponding to 1978 is the value of the sale of the 1977 harvest.
[f] Includes molasses.

(c) The steadily falling supply of labor. Various factors were responsible, especially the boom in cane production during the preceeding three decades that, even when considered in isolation from other social components, created structural difficulties in obtaining the specialized labor (such as cane cutters) needed for increased cane production.

These strictly economic changes implied qualitative alterations in the social relationships within the industry, reflected mainly in changes in the relative importance and potential negotiating power of the different social sectors. As an example, by the end of the sixties all the social groups competing with the mills had improved their share of the income (Table 6).

The mills responded to these changes by taking actions to maintain their ability to negotiate the distribution of income and the direction that the sector's development was to take. During the 1970s, primary efforts were directed at financing and developing the structure of the cartel. This was reflected in the growing importance of the institutions representing the industry, as they expanded their activities to areas beyond simple export regulation, their primary activity during the previous decade. They played a particularly key role in the process of defining State policy, a phenomenon increasingly important in the later years of the decade. This participation of the mills was manifested in actions designed to consolidate the joint control by ASOCAÑA and the State of the diverse aspects of sugar activity, such as production standards, circulation and consumption. As a result, the Office of Production Control was created in 1972 and the National Sugar Commission in 1978.

The Office of Production Control was responsible for inspecting the activity of all participants in the sugar process and had the authority to impose sanctions on those not complying with established performance standards. The National Sugar Commission attempted to generate a uniform sugar policy that favored the mills. The principal objective of this Commission was to guarantee administered prices, which would finance and protect the ongoing process of strengthening the cartel, as well as the economic concentration and institutional power of the mills.

As noted previously, a number of social groups with the potential for challenging the mills developed greater negotiating power during the 1960s. Special mention should be made of the labor sector, which sought to unionize workers through awareness programs, and the independent producers and land owners (rapidly disappearing from the scene), who attempted to promote more favorable distribution of the economic surplus.

The mills responded to the labor sector by taking actions to regulate the labor market in two important ways. The first was through coercion, with direct assistance from the State apparatus in controlling the workers' movements. The second was by reforming the mills' internal administrative structure, reflected in the increased activities of "contractors,"[10] and the type of contract labor with which they are associated. These contractors hire laborers for temporary work at wages that are frequently lower than those paid by the mills, without any of the social services which the mills were normally required to provide. The contractors thus assisted in reducing labor costs at the mills.[11] Combined with the general decline in urban and rural wages during the 1970s in Colombia, the result was a precipitous drop in wages for the work force engaged in sugar production in the Cauca Valley.

Although independent suppliers were increasing their participation in the distribution of the income from the sugar sector, they had no direct confrontations with the mills that were comparable to those between mills and the wage-earning sector, due mainly to three factors. In the first place, the mills and the sugar cane growers represented the same social class; consequently, their conflicts were not deeply rooted and were more amenable to solution through negotiation. In the second place, a considerable number of suppliers were either mill owners or close relatives of the owners. Thirdly, and this is particularly important for the technological process, the mills and the suppliers pursued similar objectives for modernizing the sugar industry. Although these facts tended to ameliorate conflicts between the mills and the independent suppliers, the suppliers still developed other corporate activities to improve their negotiating capacity.

The creation of PROCAÑA was clearly the most significant effort made for ensuring the fair distribution of any surplus beyond the payment of labor, inputs and replacing capital goods. The confrontation between mills and suppliers over how to distribute this surplus did not, in general, affect the social forces that influence how the surplus is produced, nor the technological characteristics associated with it. In fact, the two groups reacted in similar ways to the technological mechanisms in use.

The suppliers acted on several fronts to protect their incomes. In the first place, they sustained sugar cane prices by publishing bulletins with "revealing" information, instructing PROCAÑA members, and in general projecting a united attitude. In addition, they acted at an institutional level to acquire "their" share of the advantages obtained by the mills. Typical of this action was their struggle to be assigned export quotas, previously the exclusive domain of the mills.

Technical change and the social relations of production. The 1970s were considerably richer in technological activity than the previous decade. Demand and price conditions continued to be highly favorable, especially from 1974 to 1977, although a decline was noted during the last year under study.[12] As noted, the sharp changes in the availability of production factors served as a direct and powerful incentive for modernizing production.

The first effect of the mills' new priorities at the institutional level was the development of local facilities for generating sugar cane production technology. The Colombian Agricultural Institute (ICA) dismantled its sugar cane research program in Palmira in 1973. This action was in accordance with the "decentralized" agricultural research model gradually adopted in Colombia during the seventies (ICA, 1974). The model stems from an awareness that the mills could, with certain flexibility, meet their own technological needs and thus control the quality of the process. In addition, the technological question in the sugar sector was not of a highly political nature, given the fact that proposals for modernization were not likely to mitigate any pressing social problems (as would be the case if the *campesino* systems were modernized), and sugar production had increased sufficiently to meet both export and domestic needs. Because of this and a passive attitude on the part of the public sector, the mills consolidated their own activity in the technological sphere.[13] Their efforts took place along three succeeding lines of action: (a) the importation of foreign techniques and innovations; (b) the development of the necessary institutional organization within the mills themselves for performing adaptive research; and (c) the creation of CENICAÑA. This gradual institutionalization of sectoral activity for technology generation was a result of an increased awareness that imported foreign technology had to be adapted to the characteristics and requirements of the mills' ecological and economic conditions.

The most outstanding feature of this process was that it increasingly became a joint endeavor of the mills. Reversing their attitudes of the sixties, the mills allowed their technical staff to exchange information and even to work together on information-gathering field trips. These new developments show how, as the industry matured into a cartel, it became both unnecessary and undesirable to maintain competitive mechanisms, including technological aspects. In this way, innovation became a joint activity and its increased importance as an instrument for productive growth meant that the technology variable became a barrier to entrance into the industry. The creation of CENICAÑA was one of the final steps by the cartel to gain control over this phase of the productive process, an inevitable last measure after the more fundamental aspects of organizing economic activity had been resolved.

The structure of technological change in the industry during the 1970s clearly shows that the mill owners were the social group that encouraged, strengthened and directed the process. They were able to impose their specific demands for the kind of technology to be introduced from outside, and how it would be adapted to local conditions in the industry as a whole. It is important to recall, however, that the technological requirements of the independent suppliers were similar to those of the mills. Thus, toward the end of the sixties, institutional coordination grew strong and the mills were able to provide efficient channels for transferring technology to the suppliers. This institutional coordination was founded on contracts between the mills and the suppliers, by which the mills committed themselves to perform specific, often important agricultural tasks on the suppliers' lands, frequently using the mills' machinery. The suppliers, in turn, agreed to pay for this service in sugar or in cash and to comply with recommendations made by the mills on technology use.

This operating alliance was a key element in the rapid technological innovation process that took place during the seventies as a rational response to the new economic circumstances surrounding the industry. It is not surprising that the number of innovations incorporated by the mills and the suppliers increased dramatically during this period. These innovations included: leveling and redesigning lots, improving drainage, deep sub-soil, chisel plowing, seed control, new strains, mixed fertilization, spraying and biological pest control. Technological innovation increased sugar cane yields per hectare, but it also fundamentally altered the relative importance of the different production factors in the preharvest activities, modifying the use-ratio of the various factors.

In general, the innovations were capital-intensive, reducing labor and land use despite the fact that the price of capital increased in relation to both factors (Table 5). This capital-intensive bias was evident with regard to preharvest activities for both mills and independent suppliers (Figs. 5 and 6), although it was less important for harvest activities. As a result of the capital-intensive bias and its effects on labor during this period, the larger mills were faced with a strong unionization campaign. At the same time, the independent suppliers began an active process of technological incorporation, though which they intensified both their use of capital (although to a lesser degree than the mills) and their use of labor per land unit. The suppliers, given their small size, did not have to deal with the unionization of farm laborers that was affecting the mills; they also had less access to the supply of new capital.

The principal qualitative elements of the technological process in the seventies can be stated as: (a) the imbalance between changes in the relative factor prices at the market level and the biases of the incorporated technology; and (b) the gradually widening gap between the technology adopted by the mills and that used by the suppliers, especially as capital became the dominant factor in the innovative process.

In sum, the nature of the technological process pursued by the mills permitted them to reverse the income distribution trends of the 1960s. As indicated in Table 6, the per cent participation in total income by the wage-earning sector and by the land-owners declined between 1970 and 1978, despite the fact that the latter social group controlled the scarcer production factors.

SUMMARY AND CONCLUSIONS

This empirical analysis of the sugar sector in Colombia clearly shows that the production modernization process was subordinated to the social relations that prevailed inside the sector. For this reason, technology could not assume great importance until the dominant social sectors had developed appropriate conditions in the productive structure and had taken control of the mechanisms for generating and appropriating production surpluses. This ensured the full coordination of the technological process.[14] Once technology had begun to serve as an important element in the productive performance of the mills, the modernization process was carefully manipulated in order to ensure that the mills would continue to dominate the sector.

Within this general context, four characteristics of the innovation process in the sector are worth noting: (a) the subordination of technology to the development of a strong cartel by the dominant sector; (b) the considerable capacity of the dominant social sector for negotiating with the State to obtain economic conditions favorable to the sector; (c) the gradual assumption of control by private interests over the generation of technology; and (d) the biases of the adopted technology in terms of the use of production factors. These characteristics are discussed below.

Regarding point (a), the subordination of technology to cartel development, it is important to recall that during the first decade of the period under study, the mills directed their principal efforts towards consolidating their individual competitive status in the industry. This competition involved obtaining market segments and controlling the land. During this time, technology was a secondary variable, viewed as one of the many factors in the competition between mills jealously guarding the technological information at their disposal.

However, once the cartel was formed, it became strong enough to place control over the technological process in the hands of the mills, which then appropriated the resulting economic surplus. Technology increased in its relative importance and became an instrument used by the cartel for competing with external economic interests, thus becoming a barrier to entrance into the industry. This explains the rapid liberalization of the policy on exchanging information which culminated with the creation of CENICAÑA, a joint effort by the cartel to stimulate the development of technology.

Point (b) is the ability of the sugar sector to negotiate with the State for establishing a favorable economic context for the accumulation of capital and for solidifying the dominant position of the mill owner over other social sectors involved in the production and distribution of sugar. In the initial process of production growth, the mills concentrated on guaranteeing favorable prices and protecting themselves from wide price swings on the world market. This objective was

Year	% Fixed Capital Land (FC. Ld.)	% Inputs Land (I. Ld.)	% Labor Land (La. Ld.)	% Fixed Capital Labor (FC. La.)	% Inputs Labor (I. La.)
1960	2.57	3.51	1.44	1.79	2.44
1970	6.38	7.06	1.57	4.07	4.50
1974	8.21	8.55	1.33	6.17	6.43
1978	11.94	9.33	1.13	10.55	8.75

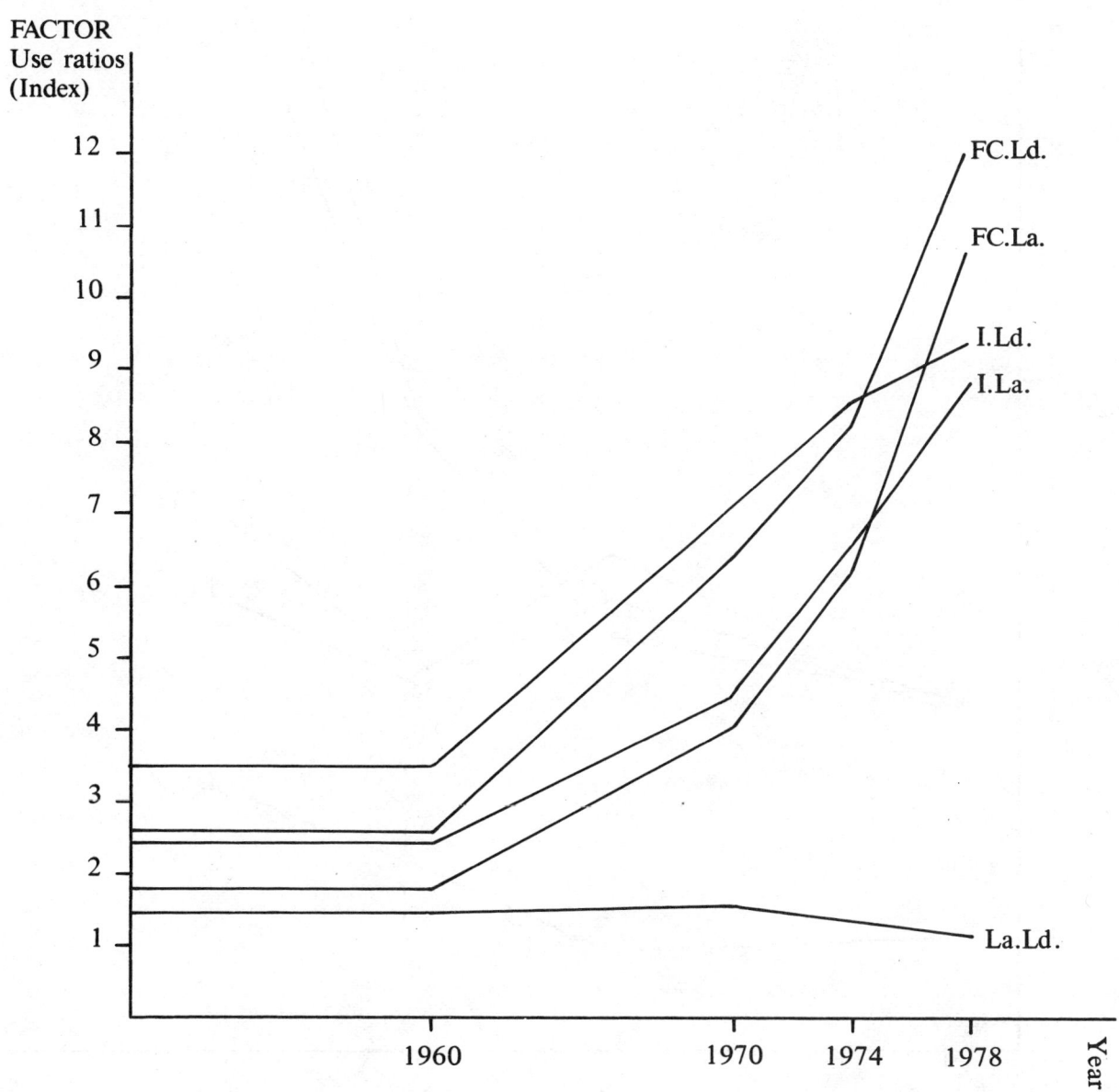

SOURCE: Estimates by the authors.

Fig. 5. Colombia: Factor use ratios in the production of one ton of cane, excluding harvest, for the eight mills analyzed.

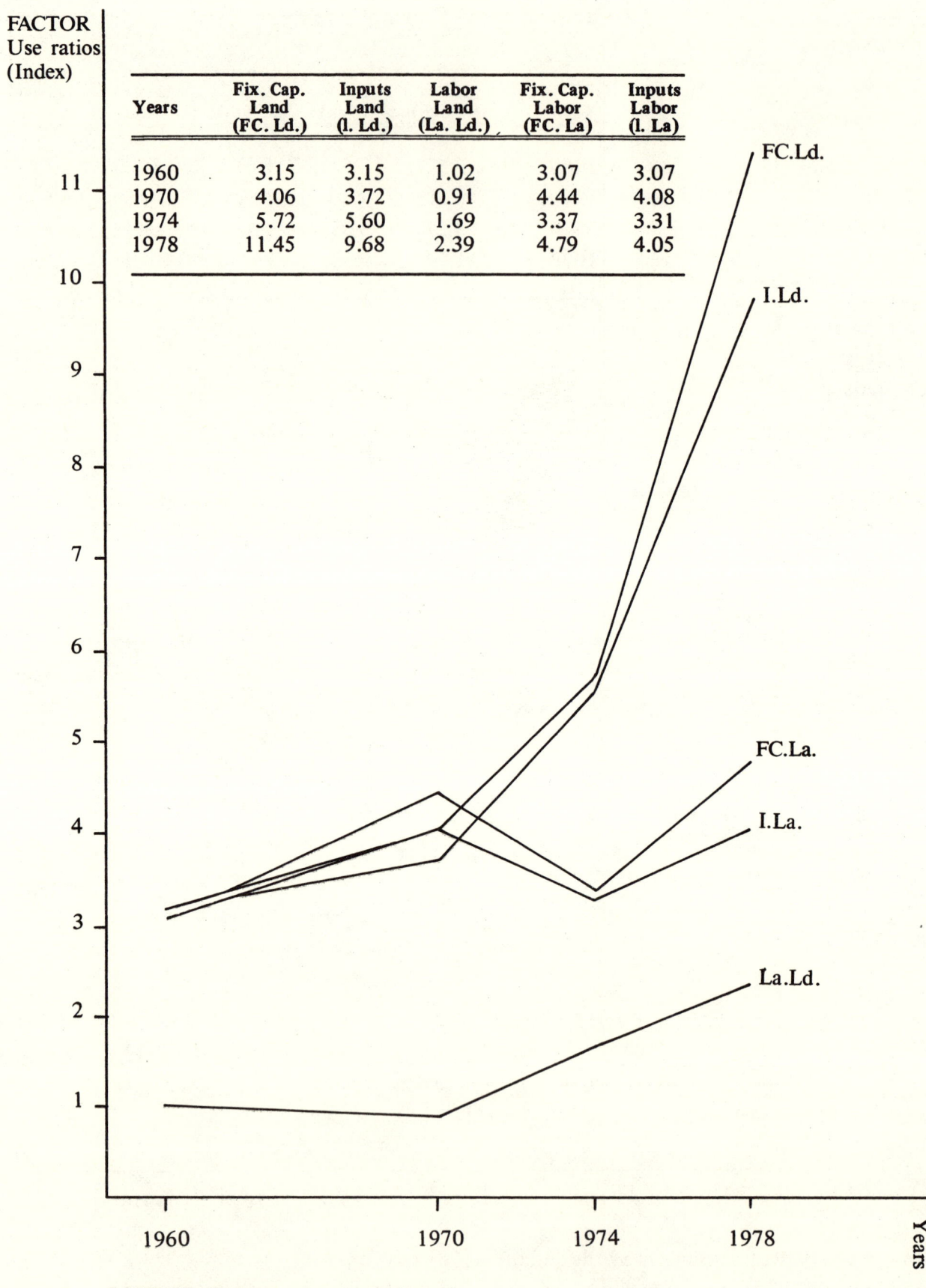

SOURCE: Estimates by the authors.

Fig. 6. Colombia: Factor use ratios for average independent suppliers.

achieved, as shown in Fig. 2, through a complex domestic price-fixing mechanism to offset the international prices received by the mills; it guaranteed the growth potential of the sector. The price policy did not seek to remove surplus from sector control, in spite of the fact that sugar could be considered a wage-good[15] (although not a strong wage-good in the case of Colombia).

It is important to emphasize that the policies adopted by the State reflect two types of conflicts. First, friction exists between the sugar producing sector and the urban consumers, and second, a conflict inside the sugar sector divides the mills from the rest of the social groups involved in production (wage-earners, landowners and independent suppliers). In the case of the intersectoral conflicts, current policies provide for a negotiated balance based on three main conditions for the production of sugar and its introduction into the rest of the economic system: (1) sugar is neither a point of capital accumulation in the overall economy, nor a wage-good of major importance; (2) sugar is a widespread product of central importance to the economy of the region involved, which has of considerable political power; and (3) production is concentrated and enjoys a high level of corporate organization. Under these conditions, and given certain general characteristics of society and State in Colombia, the sugar sector has sought and obtained considerable penetration into the public institutions involved in the policy decisions affecting it. This situation leads to a new form of fragmentation or balkanization of the State bureaucratic apparatus, what Cardoso has called the bureaucratic rings of the State (Cardoso and Faletto, 1969). As for the sector's internal conflicts, the policy approach to finding solutions implies a certain degree of sectoral autonomy for the resolution of conflicts within the industry. These conflicts are resolved by and for the dominant sector by strengthening the existing conditions of domination.

Private control over research (point c) appears to be reinforced by the State's minimal role in generating technology, and the mills' assumption of greater control over the generation process. This is particularly true in the case of the larger mills, which lead most of the activities of the cartel.

This trend is the result of two converging forces. On the one hand, the public institution responsible for agricultural research (ICA) gradually lost its operational capability, especially in a number of crops which, like sugar, are primarily in the hands of highly commercial corporate enterprises. This trend stems from a redefinition of the role of public institutions and the private sector. On the other hand, as technology assumed growing importance in sustaining the cartel structure, the mills became more interested in developing institutional mechanisms for controlling the process of generating, importing and adapting technology. CENICAÑA was an appropriate measure for gaining control of the process, and at the same time maintaining a certain level of social participation in the research costs.[16] Yet the sugar sector did not work to establish State research institutions, or to influence the research decisions of existing organizations.

Point (d), biases in the use of production factors, has to do with the capital intensive nature of the technology adopted by the sector. This was true even during the seventies, when the cost of capital grew relatively higher. It should be recalled that, if the five possible ratios are examined (fixed capital/land, inputs/land, labor/land, fixed capital/labor, inputs/labor) for each of the three periods (see Figs. 4 and 5), only four of the fifteen possible cases show use ratios changing in inverse proportion to the price. In the case of independent suppliers, the number of exceptions is considerably greater as a result of the heavier use of labor during the 1970–1978 period, when wages fell off sharply. In other words, suppliers increased their use of labor as relative labor prices fell.

The diverging tendencies in factor use by mills and by suppliers lead to two interrelated observations. First, the cost of factors for each mill may not have been faithfully reflected in prices, due to the rigidity and imperfections of the market price structure. At the same time, the real costs of the labor factor include not only wages paid, but also the costs of mobilizing and feeding work crews, particularly in the case of harvesters. These costs may be subject to variations different from those of wages. In addition, labor unionization produced strikes and demands which had to be internalized by the mills as real costs. The second observation is that mill use of capital as a mechanism for replacing other factors had at least a partial effect on the price of land and labor. This skews any interpretation based on the causal relations between the two variables. Due to the nature of the technological process after 1970, the mills were able to maintain, and even to reduce, the percentage share of total sugar industry income[17] received by wage-earners and the landowners. This was in spite of the fact that workers and owners controlled what seemed to be the scarcest factors.

From this viewpoint, the technological route actually adopted by the mills appears rational for their particular economic interests. It is consistent with the specific types of conflict that emerged with the distribution of surplus among other social groups participating in the productive process, and with the conditions needed for prolonging the existing forms of domination.

The role of the mills in the adoption of technology, and the consequences for technology distribution, should be interpreted in conjunction with the serious flaws in the factor markets. Taken together, both phenomena illustrate some of the difficulties in interpreting final results in light of the theory of induced innovation. More particularly, they cast doubt on the advisability of expecting that such mechanisms are appropriate institutional means of guaranteeing an efficient technological approach to the macroeconomic use of productive factors.

This study clearly shows that, while the dominant sectors adopt technologies consistent with their production conditions, they do not necessarily wield their institutional and political power so as to strengthen the public institutions in charge of technology generation. Nor do they channel or "induce" the research process toward creating technologies that reflect the relative availability (or prices) of factors. Quite the contrary, the study shows that under conditions of cartel establishment and high levels of bloc power, precisely the opposite effect may occur.

NOTES

[1] *"Panela"* is a semi-refined derivative of sugar cane which is obtained through very rudimentary technological processes, usually small presses installed on farms or in small sugar mills. The bulk of panela production is in the hands of small-scale hillside farmers. Historically, panela has met much of the demand for sweeteners in Colombia. However, its unsuitability for industrial purposes, candies, soft drinks, pastries, etc., and changes in demand patterns as a consequence of rural-urban migration, have reduced its importance, although it continues to be an important product among low-income consumers.

[2] At the same time, the consumption of *panela* reduces the demand for refined sugar as a wage-good, which explains the relatively passive attitude of industrial sectors toward price-fixing policies.

[3] Personal interviews with Mariño and Poveda, 1978.

[4] Of these, the *"Pacto Multilateral de Productores de Azúcar"* is the most interesting.

[5] As will be shown below, these two facts join forces with the special characteristics of the overall Colombian social structure, in which bourgeoisie sectors compete in the political realm, but cooperate in controlling and exercising power. This explains why sugar cane mills were able to enter into negotiations with other social sectors to reach a relatively favorable economic policy for sugar's interests.

[6] Based on the available information, it is difficult to evaluate whether or not this policy was actually promoted by the mills (rather than the State), as appears to be the case. It is nevertheless clear that the mills did intervene actively in policy implementation, especially after the early years of organizing and learning how to initiate effective corporate action in the area of economic policies.

[7] This became clear in the surveys taken in eight selected mills.

[8] For a theoretical discussion on the efficiency of the techniques, see Brown (1970).

[9] Changes in the distribution of income among different social sectors are the result of a number of variations in the relative use of the production factors supplied by each sector and the price of these factors. Table 6 is calculated at constant prices, and general trends do not change. In addition, the prices of land and labor were partially determined by technological developments within the sector.

[10] Contractors are individuals who hire workers to perform agricultural tasks by contract with the mills.

[11] It is interesting to note that sugar production in Northeastern Brazil was strongly affected by the contractor *(empreitero)* during the sixties. But in this case, we are dealing with an institutional innovation associated with replacing precarious forms of production. See Fiorentino (1977).

[12] However, 1980 was once again a year of high prices.

[13] The "weak" nature of sugar as a wage-good in Colombia should not be overlooked, and the considerable political power, at the national level, of the dominant sectors of the Cauca Valley should also be recalled.

[14] See Piñeiro and Trigo (1977) for a description of possible breakdowns in coordination and the conditions needed to ensure that the process will be coordinated so that technical change will occur.

[15] The weak nature of sugar as a wage-good in Colombia should be emphasized.

[16] This is because CENICAÑA's financing is determined by a rate applied to marketed sugar, and thus the real costs are shared by producers and consumers.

[17] In the case of labor, it should be recalled that wage levels are determined to a large extent by the general conditions of the Colombia economy, characterized by high unemployment and low wages.

AGRICULTURE IN THE ARGENTINE PAMPAS: TECHNOLOGY ADOPTION IN CORN CULTIVATION FROM 1950 TO 1978

Jorge Federico Sábato

ABSTRACT

This paper summarizes overall agricultural development in the Argentine pampas, advancing several broad hypotheses which attempt to explain producer behavior. The relations between producers and the public sector, as well as the interaction of demand and supply of technology, are also examined, using corn production as an example.

INTRODUCTION

The Argentine pampas are an immense plain covering 55 million hectares of fertile land. They have a pleasant temperate climate, and rainfall is favorable for over 25 million hectares. In short, this is one of the world's most richly endowed regions for agricultural production. The use of this vast grassland, which was still practically deserted only a century ago, lay the foundations of the wealth of the Argentine nation. The ups and downs of the grain crops and livestock herds in the pampas were closely associated with the fortunes of the entire country.

For a hundred years, production development in the pampas has been uneven. From the end of the last century to the beginning of World War II, growth was strong, with characteristics and levels of productivity similar to those of the vast plains of North America. However, around 1940, the similarities abruptly ended. Massive adoption of new technology in the United States of America and Canada triggered a second phase of strong production growth, while in the Argentina pampas technical progress came to a standstill during two decades, production stagnated and dropped. Only as recently as 1960 did a slight recovery begin to take place, followed by striking production growth based on the use of new techniques. However, three characteristics continue to prevail:

(a) The techno-productivity gap between the pampas and the North American grasslands, which began in 1940, has not increased, but, in the past 15 years, neither has it narrowed. narrowed.

(b) The new technologies incorporated in the past 25 years were not strongly oriented toward the use of land, which was the scarcest factor.

(c) As a result of these two facts, the pampas today are producing considerably less than their potential would indicate.

This situation gave rise to three key questions that prompted the present research:

(1) Why did technical progress stand still for nearly 20 years, and what made it start again?

(2) When the adoption of new techniques resumed, why was no emphasis placed on making better use of land, the one productive factor that was becoming relatively scarce?

(3) Why, in short, did producers apparently have neither the intention nor the economic incentives to make full use of existing productive capacity in accordance with available resources and techniques?

These questions made the broad viewpoint adopted by the PROTAAL Project appropriate for analyzing the evolution of agriculture in the pampas, due to a combination of factors. In the first place, the development of production in the pampas was very uneven, and technological considerations played an important role in this unusual behavior. At the same time, production units in in the pampas could legitimately be viewed as capitalist business enterprises, whose behavior in relation to technical change could be modeled theoretically. Finally, however, these theoretical models, which assign rural business a decisive role in technical progress, give no satisfactory explanation of what happened in the case of the Argentine pampas. It is quite clear that certain political conditions had a heavy influence. Although these factors are not always examined in detail, the PROTAAL Project has paid them particular attention. They include: government action, the interests of input suppliers, and the specific circumstances of production under which entrepreneurial reasoning by the producers does not necessarily lead to technical change thus contradicting a basic consumption of the classic models.

The magnitude of the subject and the complexity of the approach demanded certain methodological precautions. The most important, based on the general criteria adopted by PROTAAL, was to use the analysis of one product as a guide to investigating the problem. In this case, corn was chosen, and the method proved useful and satisfactory. However, it is important to recall that the characteristics of this particular crop were not under study; rather, corn was used as a means of understanding the development and overall features of agriculture in the pampas in general. Indeed, the research made it possible to interpret the two phenomena in a new way which, in several areas, was different from earlier views. This new interpretation is comprised basically of two different but complementary groups of ideas. One has to do with the operational model of production units in the pampas and the behavior of rural producers. The other concerns the historical patterns of agriculture in the pampas.

The reasoning process led to several natural divisions of this article. The first gives a summary of the overall development of agriculture in the pampas, methods used for studying the behavior of producers, and key findings of the study in several departments in the pampas. The second section examines the behavior of producers and puts forward broad hypotheses helpful for understanding it. The third section expands the field of analysis and examines the relations between producers and the public sector and between the demand and the supply of technology, information helpful for rounding out the discussion of overall agricultural growth in the pampas. Finally, conclusions are presented.

EVOLUTION OF AGRICULTURE IN THE PAMPAS

The Comprehensive Profile: Facts and Attitudes

Agricultural production in the pampas grew steadily after the turn of the century. However, with the outbreak of World War II, it came to a standstill, and until the beginning of the 60s, it stagnated and even regressed. After that time, major technological changes began to occur, and a new phase of growth began, lasting until the present.

The period when pampas agriculture ceased to progress has been studied extensively, various technical and political theses being put forward to explain what happened. However, very few analyses have been done of the renewed growth, the nature and effects of which are unclear. It is possible, therefore, that an examination of these issues will cast light on the earlier phase and contribute to a better understanding of it. In order to provide background material, a brief recapitulation of facts about the stagnation, its effects, and interpretations will be given.

Agriculture in the pampas stands still. After 1925, despite the crisis of 1930, production in the pampas experienced slow, steady growth which topped out during the war. For ten years it slipped, finally beginning to recover in the mid-fifties. Nevertheless, by 1960/1964, it had not yet resumed levels achieved in 1940/1944 (see Table 1).

TABLE 1 Indices of Physical Volume for Production in the Pampas (Five-Year Averages, 1935–1939 = 0) (Reca, 1967).

Five-year period	Total	Total Livestock	Total cattle	Total Farming	Wheat, corn and flax
1925–1929	93.3	87.9	96.5	93.9	97.9
1930–1934	95.0	88.6	89.8	97.5	97.1
1935–1939	100.0	100.0	100.0	100.0	100.0
1940–1944	108.4	128.6	115.2	102.1	96.2
1945–1949	97.9	135.8	129.0	75.3	61.5
1950–1954	89.3	133.6	135.8	65.8	52.2
1955–1959	102.2	144.3	148.0	80.4	64.7
1960–1964	105.4	146.7	153.5	83.6	67.1

If production in the pampas is broken down into components, it becomes clear that the stagnation was the result of a *farm recession,* particularly affecting corn, wheat and flax. Statistics show that, of all these products, corn was hit the hardest. Livestock, especially cattle, experienced sustained growth throughout the period.[1] It should be recalled, however, that in the early twenties, a deep crisis gripped the livestock industry, and production fell drastically from earlier levels.

The economic and social effects on Argentina of stagnation in pampas agriculture. Sluggish production in the pampas had severe effects on the economy and society of Argentina, in both general and specific terms. It was decisive in the strangulation of the foreign sector of the economy (see Table 2). Agricultural production in the pampas provided over 85 per cent of the

TABLE 2 Value of Exports (FOB), and Agricultural Exports as a Percent of the Total (Five-Year Averages in Millions of 1950 Dollars) (ECLA, 1959)

Five-year period	Total exports	Percent of agricultural exports	Percent of livestock exports	Percent of farm exports
1925–1929	1,582.7	95.0	38.8	56.2
1930–1934	1,481.0	95.6	35.9	59.7
1935–1939	1,479.4	94.7	38.2	56.5
1940–1944	1,192.5	86.5	54.8	31.7
1945–1949	1,180.1	89.8	55.7	34.1
1950–1954	937.1	93.0	46.8	46.2
1955–1957	1,047.7	92.7	46.9	45.8

country's total exports. When domestic consumption rose and production stood still, total export capacity fell in absolute terms at a moment when industrial development was being actively encouraged. After 1950, with the end of the Korean war, the quantum fall in exports was heightened by a slide in international beef and grain prices (see Table 3).

TABLE 3 Argentine Foreign Sector, by Broad Categories 1950–1977 (Millions of Constant Dollars)*

Periods	Exports	Imports	Trade balance	Net capital flow	Result
1950–1962	13,278.3	15,381.4	−2,103.1	+1,602.1	− 501.0
1963–1972	15,756.0	13,685.1	+2,079.9	−1,954.1	116.8
1973–1977	19,684.1	16,940.2	+2,743.9	− 103.1	+2,640.8
Annual averages for each period					
1950–1962	1,021.4	1,183.2	− 161.8	+ 123.2	− 38.5
1963–1972	1,575.6	1,368.5	+ 207.1	− 195.4	+ 11.7
1973–1977	3,936.8	3,388.0	+ 548.8	− 20.6	+ 528.2

*Compiled by the authors on the basis of INEC (1950–1978), BCRA (various years), OECEI (1973), FBB (1976) and Mallon and Sourrouille (1976).

Between 1950 and 1963, the foreign commercial balance was negative for nine years, slightly positive for three years and favorable for only one year in the period (1953). In constant dollars, the cumulative negative balance topped 2.1 billion dollars, although it was partially offset by a positive net capital income of 1.6 million dollars. The after effects of this situation held back the economy for ten more years. From 1963 to 1973, export growth and import restrictions built up a cumulative positive foreign trade balance of over 2.0 million constant dollars, but this balance was practically wiped out by the net capital outflow of 1.95 billion dollars.

In short, for a quarter of a century, the foreign sector of the Argentine economy was in a constant state of crisis. On several occasions, it came very close to exhausting its meager foreign currency reserves. This state of affairs placed a powerful brake on possibilities for overall growth in the economy, and it produced a continuous environment of social and political crisis.

Stagnation and growth in pampas production: immobility and technical change. By the middle of the 1950s, it became clear that the recession in pampas agriculture was one of the chief causes of national unrest. The stimulation of agriculture became a central objective of all successive administrations, and the catchword for production increases was technical change.

Half a century earlier, Argentinas's great prosperity and rapid progress had been based on burgeoning pampas production through the occupation of new lands. The expansion of the agricultural frontier reached its limit by 1920, and by the mid-50s little more could be gained by this route. In contrast, there were tremendous possibilities for increasing the productivity of land already in use, both by ecological methods and by technology. Through the 30s productivity in pampas agriculture had been equivalent to that of the great grain and beef-producing zones of the United States, Australia and Canada, where extensive production systems were commonly used. However, after 1940, these countries began to boost their productivity by incorporating new techniques, while pampas agriculture continued using the same methods as before. Productivity barely held steady and even decreased. It was clear that the only way to end the stagnation and raise production was to incorporate new techniques, many of which were already known and proven.

Explanations for the stagnation and political reactions to the problem. Because the inertia in the pampas had a dramatic effect on the entire country, and because the phenomenon, when observed from a broader standpoint, was so remarkable, explanatory studies proliferated. At the same time, the effects on the country fueled strong political disputes on the issue. The academic analyses and political positions had certain similarities, and although a list of them all would be beyond the scope of this article, it is worthwhile to give a schematic summary of two opposing views that dominated discussion of the situation.

The first view based its arguments on the fact that foreign exchange obtained for agricultural exports had gone chiefly to the government, which had used it to finance the protected development of inefficient industries and to begin a demagogic distribution of income based primarily on employment generation and on non-productive expenditures by the public sector. The resulting fall in real net income for the agricultural sector in the pampas had destroyed all motivation for production and for investment, and the result was stagnation. At the same time, as resources extracted from the agricultural sector went into inefficient or non-productive activities, it became impossible to establish a form of economic development that would offer a viable alternative for the rest of the economy. The fundamental cause of the stagnation, according to this view, lay with the action of certain government administrations, particularly the Peronists, that operated on erroneous doctrines and employed demagogic methods.

The other view was more heterogeneous in terms of both its arguments and its proponents. In a certain sense, the reasoning focused on the deadlocked land ownership structure in the pampas. Unlike the situation in similar zones of the USA and Canada, which also grew through the occupation of fertile, unpopulated territory, few land titles in the Argentina pampa were actually granted to settlers, while many went to a small urban group. The resulting large holdings were rarely subdivided, and great expanses remained as large *estancias,* while many much smaller family operations were set up as *chacras* (similar to the farms in North America), often based on land rented from the large farms. This was the predominant pattern. Such a system of land tenure produced a number of converging effects. It led to a heavy concentration of income which, on the one hand, impeded the growth of rural and urban domestic markets, and on the other converted a considerable amount of the surplus into luxurious and non-productive expenditures. One side effect of this concentration of income was that the tenant farmers were unable to increase their productive capital, and this placed a rein on any gradual, sustained increase of agricultural productivity. Accordingly, the use of land took clear precedence over other factors in pampas operations. Land values rose steadily, becoming ever more inaccessible to the *chacra* farmers. This effectively circumscribed production growth, and agriculture continued to operate within very narrow profit margins. A few successive droughts, a plague of insects, a labor or machinery shortage, or any combination of these setbacks, was enough to send agricultural production into a state of crisis and recession, as occurred at the end of World War II. In short, the concentration of property in the hands of large pampas landowners was not only the cause of the productive stagnation in the region, but was also, more importantly, the source of crucial, relentless problems in the economy and society of Argentina.

Both views recognized that stagnation in pampas agriculture was closely linked to the absence of technological progress, which stemmed from limited capital accumulation in agriculture. They also agreed that political events played a central role, both for explaining the stagnation and for overcoming it. However, they differed on the causes of the situation, and consequently offered different solutions. For some, the crisis was due to the action of populist governments and industrialists who had ruined the rural areas and, as a result, the country itself. For others, it was the disproportionate, pernicious influence that large pampas landowners had over the government. This had undermined the possibility for progress in Argentina and had held back the development of agriculture in the pampas.

In summary, the stagnation in the pampas produced general conflict throughout the economy and the society, revealing major ruptures in the very heart of the region. The political atmosphere was shaken by society-wide confrontations and others in the agricultural sector. Warring interests hampered the action of the government by subjecting it to opposing, urgent demands on key questions, such as produce prices, and proposing totally different means of solving the problems.

Both views might appear valid for the period 1940–1960, but no later. In the succeeding 20 years, production began grow once again, in spite of repeated populist governments and the absence of reforms in land tenure. This situation cannot be explained by either of the two views. It brings to light a whole new dimension of the long-debated causes of the crisis, as well as the proposals for overcoming it.

Growth in pampas agriculture. Table 4 shows annual grain production for all those years since 1930 in which the total surpassed 20 million tons. Note that harvests topped this limit twice in the first half of the 30s and twice in the first half of the 40s. Twenty more years would pass until 1964/1965, when this line was again crossed. The situation stands in striking contrast to what has happened since the end of the 1960s; every harvest, with the exception of 1971/1972 (a severe drought year) has surpassed this figure by ever wider margins. The low harvest in 1974/1975 was probably exaggerated due to a statistical anomaly. At the same time, it should be noted that the amount of land under cultivation remained approximately unchanged. This means the yield was increasing, as shown in Table 5, and it implies technical change. It can also be seen that, while wheat yields did not increase significantly, corn and grain sorghum rose sharply.

TABLE 4 Grain Harvests of Over 20 Million Tons per Year (Wheat, Corn, Grain Sorghum, Sunflower, Soy, Flax, Oats, Barley and Rye)*

Crop year	Area (million ha)	Production (millions of MT)	Yield (MT/ha)	Yield index (1930/1931 = 100)
1930–1931	19.8	20.2	1.02	100
1934–1935	20.9	22.0	1.05	103
1940–1941	20.5	22.2	1.08	106
1943–1944	19.8	20.7	1.05	103
1964–1965	17.9	21.1	1.18	115
1969–1970	20.6	23.4	1.14	111
1970–1971	19.0	21.9	1.15	113
1972–1973	19.9	25.4	1.28	125
1973–1974	17.5	25.9	1.48	145
1974–1975	17.4	20.8	1.19	117
1975–1976	18.3	23.0	1.26	124
1976–1977	20.3	29.7	1.46	143
1977–1978	19.1	28.1	1.47	144
1978–1979	19.4	29.8	1.54	151

*Compiled by L. Reca on the basis of information obtained from the Secretariat of State for Agriculture and Animal Husbandry and the Grain Exchange.

TABLE 5 Annual and Five-Year Yields of Wheat, Corn, Grain Sorghum, and Soy, in Kilograms per Hectare Harvested (1950–1977)*

Five-year periods	Wheat kg/ha	Index base 1960–1964 = 100	Corn kg/ha	Index base 1960–1964 = 100	Grain sorghum kg/ha	Index base 1960–1964 = 100	Soy kg/ha
1950–1954	1,178	79.8	1,540	87.6	(...)	(...)	(...)
1955–1959	1,309	88.6	1,772	100.8	1,767	106.5	(...)
1960–1964	1,477	100.0	1,758	100.0	1,659	100.0	(...)
1965–1969	1,223	82.8	2,163	123.0	2,006	120.9	(...)
1970–1974	1,451	98.2	2,475	140.8	2,222	133.9	1,461
1975–1977	1,564	105.9	3,014	171.4	2,909	175.3	1,969

*Compiled by the author on the basis of statistics from the Secretariat of State for Agriculture and Animal Husbandry and the Grain Exchange.

The figures in Table 5 show how land productivity grew. Increases in labor productivity were even greater, so that in overall terms, employment in the sector dropped significantly, while production climbed. For certain items, such as corn, labor needs for the most common techniques were cut dramatically. In 1950, four hours and 19 minutes of labor were required to produce 100 kilograms of corn, while by 1965, the time had been slashed to 24 minutes (Coscia and Torchelli, 1968). At present, it is around ten minutes (Pizarro and Carcciamani, 1979).

While farm production in the pampas began to recover and move into a new period of growth, cattle production continued increasing steadily, as it had since the mid-30s. This was a substantial change from the pattern of complementary swings that had typified the two activities until 1960. The growth of farm production from 1920 to 1935 was coeval with the slide in cattle production during the same period; later, the booming cattle industry (1935–1959) was offset by slumping farm production. Starting in 1960, the two lines grew simultaneously for the first time. Even though this pointed to changes in the operation of pampas productive systems, other important phenomena remained unchanged. For example, the livestock cycles continued to occur, with periods of heavy slaughter and liquidation of stock at low prices, followed by phases of plant recovery, low slaughter rates and high prices.

Agricultural recovery in the pampas had a heavy impact on the foreign sector (see Table 3). After 1963, exports began to rise, but for ten years, net capital outflow practically cancelled the effects of increased exportation through production growth. After 1973, international prices rose and exports and imports expanded quickly. For the first time since the end of the War, the economy and government of Argentina found relief from the heavy restrictions that had been imposed by the foreign sector.

Why existing interpretations of the new growth phase are inadequate. The new growth of production in the pampas that began in 1960 and increased throughout the decade had characteristics that could not be explained by former theory on the agricultural sector; it also stimulated policy changes. Figure 1 shows the changes over time of land surface under cultivation and of production for wheat, grain sorghum, corn and soy from 1950 to 1977. Table 5 gives the five-year average prices indices for beef, wheat and corn, from 1950–1954 to 1975–1977. Figure 10 shows how, until 1968, production growth of these grains closely followed increases of land under cultivation. Growth also coincided with rising farm prices from 1950–1954 to 1960–1964, as seen in Table 6. Both situations were used to justify the two views of production problems in the pampas, summarized above.

TABLE 6 Five-Year Price Averages (1960–100)*

Five-year Period	Beef index	Wheat index	Corn index
1950–1954	67.8	89.1	84.9
1954–1959	73.1	86.9	95.8
1969–1964	92.9	118.6	110.4
1965–1969	95.8	107.0	105.8
1970–1974	131.5	98.2	95.6
1975–1977	80.2	67.7	73.1

*Compiled from price lists from the Grain Exchange and adjusted for inflation, according to the general level of domestic prices.

Those who held that the countryside had been "punished" by populist, industrialist administrations, and that this had immobilized pampas farming, argued that agricultural enterprises had predictably responded to rising prices by increasing production. In their view, this was an important defense for the accuracy of their thesis. However, those who criticized land tenure structures

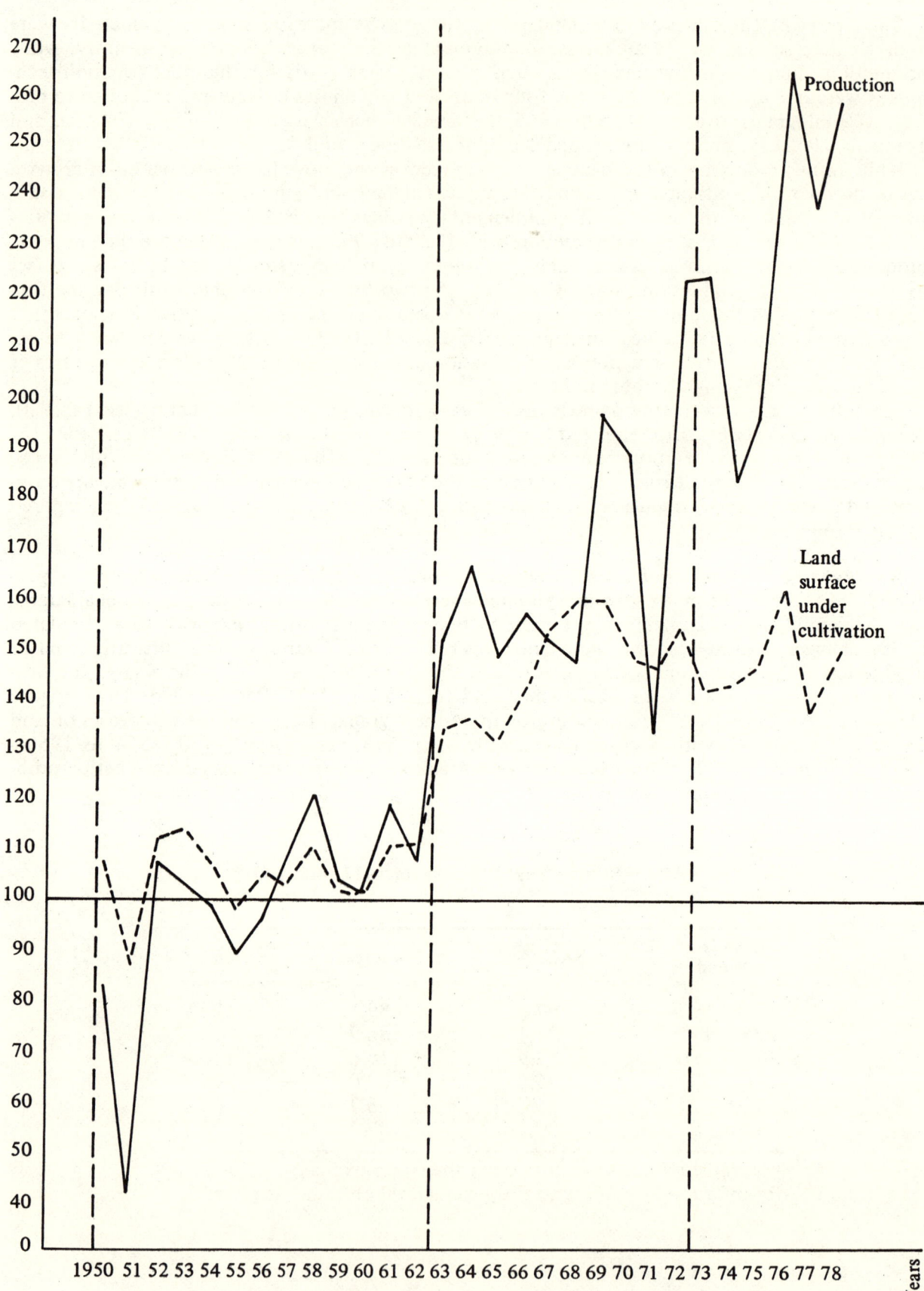

Fig. 1. Wheat, corn, grain sorghum and soy: Aggregate production and land surface under cultivation (indexed with 1960 = 100); 1950–1978.

in the pampas also felt that their arguments were justified. They recognized that production had, in fact, increased, and they agreed with their opponents that the production rise indicated a change in farming techniques. However, they stressed that this change had taken the form of mechanization, which biased factor use. It replaced labor without increasing land productivity, and land therefore continued to be the privileged factor because of the potential it offered the country.

Thus, the trends in pampas agriculture through 1968 provided arguments to support the two opposing theories. However, later developments proved contrary to both positions. Prices fell, land under cultivation held steady, and production rose sharply. Only in recent years have the fundamentals of the two positions been reexamined in light of new facts. Meanwhile, the government administrations, with their different, often conflicting policies, continued to act using analytical criteria increasingly inadequate to real circumstances. This gave rise to two crucial questions: How could agricultural enterprise in the pampas be modeled? How did the process of technical change fit into this model?

Analysis of Eight Departments in the Pampas from 1950 to 1978

Purposes and methods. The behavior of producers can be explained with the use of a microeconomic analysis that shows why and how they make their production decisions. The study must cover both routine decisions (what to produce, how to use factors) and decisions on whether or not to introduce change (adoption of technological innovations, demand for technology).

In the case of the pampas, it is immediately clear that production and market conditions are subject to violent fluctuations. The study must therefore ask how producers deal with these swings. Consequently, an empirical analysis was undertaken for the purpose of determining what criteria the producers had used for allocating and reallocating their productive resources over time and in the face of changing conditions in their environment.

It is impossible to obtain detailed data on factor use, compiled over a long period of time through representative samples of enterprises, and it was necessary to work with aggregate information. Data by department (the smallest territorial unit available in statistical information) were used, but they were limited to changes in land use. Equivalent information for labor and capital was not available.

Within these parameters, corn was selected as the subject of analysis, as it offered the most advantages for study. In general, it suffered more severe setbacks than any other agricultural product from 1940 to 1960, followed by spectacular growth rates due to the incorporation of technological innovations (adoption of hybrids). Thus, it was an ideal indicator and reflected, with particular clarity, the vicissitudes of pampas agriculture. It also lent itself to an empirical analysis organized by department, as the pampas contain a clearly defined zone that gives higher corn yields, and production is concentrated in that area (the "corn belt"). In addition, for historical reasons, certain departments in the belt are dominated by particular types of operations (either *estancias* or *chacras*). Both facts facilitate a comparative analysis among different departments of the corn belt and other departments, as well as between the departments of the corn belt dominated by *estancias*, and those dominated by *chacras*. This provided a double control system for examining production developments and producer behavior (through annual variations in land use).

Of the eight departments, three were selected from outside the corn belt, and five from inside. Of these five, two were dominated by *estancias* and three by *chacras*. The corn-producing departments range from 200,000 to 350,000 hectares in size, while the number of operations varies from 1,500 to almost 4,000. Altogether, the eight departments make up five per cent of the pampas and contain nine per cent of the operations in the region. At the same time, they produce one fourth of the country's total corn harvest, one third of the soy, and ten per cent of the wheat, and they maintain over five per cent of the livestock in Argentina. The analysis covered cattle herds and five major crops. Statistical series were compiled from 1950 to 1978 (for cattle, from 1956 to 1977) on the land area used annually for each activity, production and yields obtained, and annual value for each department.

Physical volume of production. The statistics on physical volume give a preliminary idea of how production of the selected items developed. Table 7 shows that, in terms of physical volume, farm production grew more than beef production, as did the agricultural sector in the pampas as a whole. This expansion of farm products occurred in three stages. Until the first five years of the 1960s, rapid growth indicated recovery from post-war production falls. From 1961–1965 to 1971–1975, having resumed earlier production levels, growth was slow. It again picked up from 1976 to 1979. The picture becomes more complex when Table 7 is broken down by departments in the corn belt and departments outside the corn belt (see Table 8).

TABLE 7 Indices of Total Physical Volume of Production for the Eight Selected Items (Index of Five-Year Averages, 1960–1961 to 1964–1965 = 100)

Five-year periods	All products	Beef	Farm commodities	% Variation between five-year periods
1950/1951–1954/1955	–	–	59	+ 27.1
1955/1956–1959/1960	80	88	75	+ 33.3
1960/1961–1964/1965	100	100	100	+ 10.0
1965/1966–1969/1970	110	109	110	+ 11.8
1970/1971–1974/1975	120	115	123	+ 35.0
1975/1976–1978/1979	151*	123*	166	
Annual growth rate	3.31%	1.83%	4.31%	

*A two-year average (1976 and 1977) for farm commodities and beef.

TABLE 8 The Data in Table 7 Broken Down by Departments Within and Outside the Corn Belt

Five-Year Period	Departments Inside the Corn Belt (Belgrano, Caseros, Constitución, Pergamino and Rojas)				Departments Outside the Corn Belt (Junín, Lincoln and Rio IV)			
	All products	Beef	Farm commodities		All products	Beef	Farm commodities	
			Index	% Variation between five-year periods			Index	% Variation between five-year periods
1950/1951–1954/1955	–	–	71	+ 26.8	–	–	40	+ 32.5
1955/1956–1959/1960	88	81	90	+ 11.1	71	92	53	+ 88.7
1960/1961–1964/1965	100	100	100	+ 21.0	100	100	100	– 6.0
1965/1966–1969/1970	117	102	121	+ 9.1	103	112	94	+ 14.8
1970/1971–1974/1975	126	104	132	+ 51.5	114	120	108	+ 6.5
1974/1975–1978/1979	179*	106*	200		123*	132*	115	
Annual growth rate	3.71%	1.46%	4.32%		2.86%	1.97%	4.40%	

*Average for two years only (1976 and 1977) for all farm commodities and beef.

The comparison is helpful in tracing the distinguishing features of the different zones. In departments outside the corn belt, climatic conditions appear to exert a decisive influence on agriculture. During the five years from 1961 to 1965, the weather was excellent and production was very high. During the following half-decade, weather was poor and production dropped. Thus, the impact of incorporating new technology (which is felt over the long term) was largely overshadowed by the weather. This increases risk, as will be seen below in greater detail. In the corn belt, however, the climate is more stable and production responds more closely to the introduction of innovations. This variable can be seen broadly in the quantum statistical jumps in production: from 1956 to 1960, due to mechanization of labor; from 1966 to 1970, with the massive introduction of corn hybrids; from 1976 to 1979, due to widespread adoption of soy crops.

The relationship between the effect of weather and the effect of technical change can also be seen in departments outside the corn belt. The strong growth of beef, less sensitive to climatic variations than farm production, can be associated with the production rises of forage crops, through the introduction and expansion of grain sorghum in the zone.

A more detailed look at the trends of farm production, both inside and outside the corn belt, depends on how the different products are combined. A strictly product-by-product analysis is thus inadequate. This becomes very clear if we compare the physical volume of production in the *chacra*-dominated departments with that of the *estancia*-dominated departments inside the corn belt. Increases in the former were greater than in the latter, especially during the 1970s, as greater emphasis was placed on farm growth, and more land titles were granted for farming than for livestock. In the departments dominated by *estancias,* on the other hand, the 1970s saw land transferred away from crops and converted to livestock.

Productive phases and the allocation of land. In trying to understand producer behavior, it is helpful to compare differences in the use of the land factor inside the corn belt from 1950 to 1978. For this purpose, only one department has been examined in each subzone, in order to avoid additional problems of aggregation. The departments chosen were Constitución in Santa Fe (dominated by *chacras*) and Pergamino in Buenos Aires (comparatively more *estancias*). These two departments offered certain additional advantages. First, they are similar in size (around 300,000 productive hectares); and second, they are contiguous, which reduces possible environmental differences between them.

Table 9 gives estimates of the area used for farming and for livestock in the two departments. As can be seen, the area used for livestock declined throughout the period in Constitución (although at differing rates), while in Pergamino, after an initial reduction, it held relatively steady

TABLE 9 Constitución and Pergamino: Land Area Used for Farming and for Livestock (Thousands of Hectares, Five-Year Averages)

Five-Year Period	Constitución			Pergamino		
	Farming	% of total	Livestock	Farming	% of total	Livestock
1950/1951 – 1954/1955	156	54	134	143	50	142
1955/1956 – 1959/1960	174	60	116	160	56	125
1960/1961 – 1964/1965	182	63	108	164	58	121
1965/1966 – 1969/1970	184	63	106	161	57	124
1970/1971 – 1974/1975	203	70	87	146	51	139
1975/1976 – 1976/1977	216	74	74	137	48	148

during the 60s, growing by the end of the decade, and continued to advance in the 70s. In evaluating these two different behavior patterns, it is worthwhile to recall what happened with production values in the two departments, estimated in Table 10.

TABLE 10 Constitución and Pergamino: Trends in Total Production Value by Hectare (Millions of 1960 Pesos, Five-Year Averages)

Constitución

	All products	Index	Beef	Index	Index per ha	Farm commodities	Index	Index per ha	8/7
	(1)	(2)	(3)	(4)	(5)	(6)	(7)	(8)	(9)
1950/1951–1954/1955	–	–	–	–	–	658	58	67	1.16
1955/1956–1959/1960	1,094	81	134	64	63	960	84	87	1.04
1960/1961–1964/1965	1,350	100	210	100	100	1,140	100	100	1.00
1965/1966–1969/1970	1,475	109	209	99	102	1,266	111	109	0.98
1970/1971–1974/1975	1,736	129	268	127	148	1,468	129	116	0.90
1975/1976–1976/1977	2,762	205	227	108	132	2,535	222	174	0.78

Pergamino

Five-Year Period	(1)	(2)	(3)	(4)	(5)	(6)	(7)	(8)	(9)
1950/1951–1954/1955	–	–	–	–	–	529	52	59	1.13
1955/1956–1959/1960	1,032	76	212	64	64	820	80	81	1.01
1960/1961–1964/1965	1,354	100	330	100	100	1,024	100	100	1.00
1965/1966–1969/1970	1,573	116	320	97	97	1,253	122	124	1.02
1970/1971–1974/1975	1,559	115	413	125	115	1,146	112	126	1.13
1975/1976–1976/1977	1,528	113	288	87	77	1,240	121	145	1.20

As can be seen, the total value of production was quite similar for the two departments until the end of the 1960s, although the composition was different (more livestock in Pergamino). At that time, Pergamino began to perform better, due exclusively to rising values of agricultural production, both in overall terms and in terms of per-hectare productivity. The value of agricultural production per hectare remained higher in Pergamino during the first half of the 70s. However, during the same decade, a strong divergence began to occur, as land was shifted from one use to another.

Column nine in Table 10 shows the total farm production index, divided by the per-hectare farm production index. If the five-year period from 1961 to 1965 is taken as a base figure, this quotient gives an idea of the extent to which farming, compared to the base period, was subsidized by other activities, especially livestock, through land transfers, or by contrast, how much land shifted from farming to other uses. In the second case, the quotient is greater than one, and in the former, less than one.

Column nine thus shows that the trends were parallel until 1961–1965 (farming received land from livestock), shifted slightly in 1966–1970, and diverged strongly after that. Land subsidies became clearly significant and moved in opposite directions in Constitución and Pergamino. The result coincides with observations in the selected departments inside the corn belt and in the two subzones.

These data suggest that during the 1970s, in the predominantly *chacra* farming departments of Santa Fe, increased livestock productivity was used to free more land for farming, while a ground level of livestock activity was maintained as a back-up. At the same time, in the more heavily livestock-raising areas of Buenos Aires, rising farm productivity seemed to be used for protecting livestock by devoting more land to it and simultaneously expanding the ground level of farming.

It can be assumed that increased livestock productivity in the *chacra* departments of Santa Fe could have given more freedom (more available land) to producers for adopting new types and combinations of farm operations. Thus, during the 70s, they would have been out in front in the search for new possibilities, while the *estancia* areas of Buenos Aires maintained a more conservative attitude: by using farm productivity increases to sustain livestock production levels, they had comparatively less freedom to test new approaches. However, to the extent that efforts in the departments of Santa Fe proved successful, the innovations would be adopted. All these phenomena are apparent when figures are examined on land allocations to the different products (Table 11).

Figure 2 shows annual curves for the data in Table 11, graphically illustrating the trends and phases that took place in the two departments. Superimposed curves for land under cultivation, production and yields of the three products show three different phases from 1950 to 1978.

TABLE 11 Constitución and Pergamino: Area Planted to Corn, Wheat and Soy (Thousands of Hectares, Five-Year Averages)

FIVE-YEAR PERIOD	CORN		WHEAT		SOY		TOTAL*	
	Constitución	Pergamino	Constitución	Pergamino	Constitución	Pergamino	Constitución	Pergamino
1950/1951 – 1954/1955	90.8	77.0	56.5	54.5	–	–	155.7	138.5
1955/1956 – 1959/1960	86.8	81.7	58.6	52.8	–	–	174.5	156.1
1960/1961 – 1964/1965	84.1	81.3	71.4	59.8	–	–	181.4	158.9
1965/1966 – 1969/1970	89.4	88.9	59.4	51.5	–	–	183.7	156.4
1970/1971 – 1974/1975	112.8	96.4	46.2	24.8	15.9	3.9	203.2	142.1
1975/1976 – 1978/1979	65.1	82.5	58.8	24.3	103.5	22.3	235.2	137.1

*Totals also include area planted to sunflowers and grain sorghum.

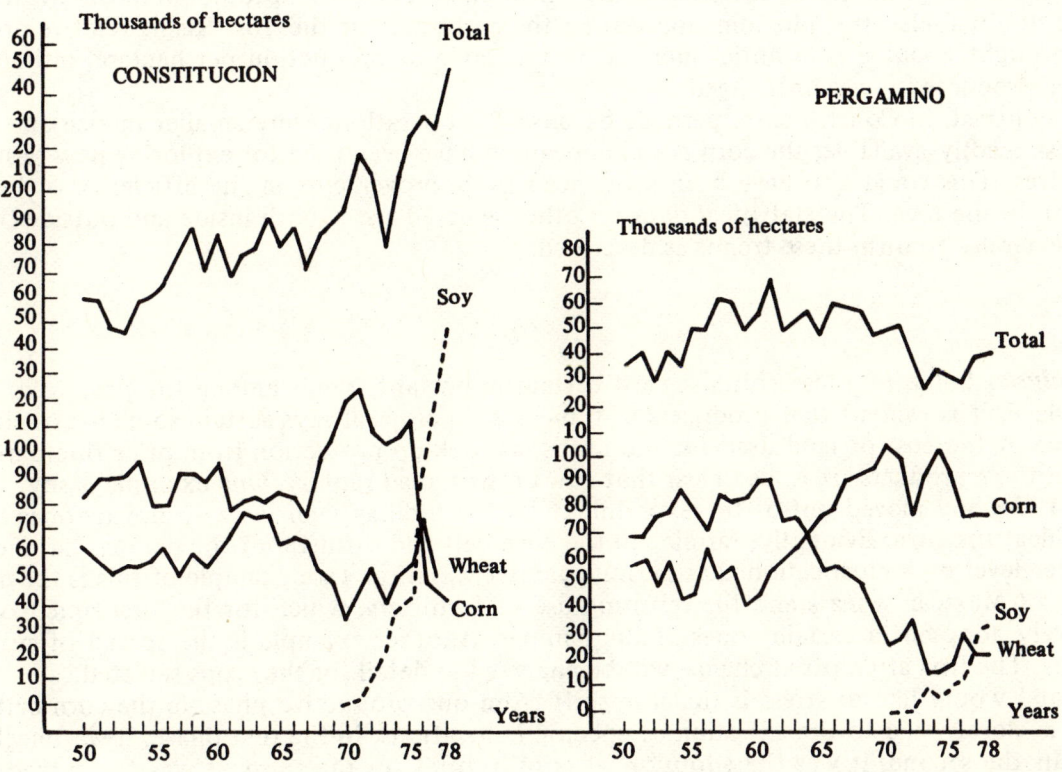

Fig. 2. Constitución and Pergamino: Land surface planted to major commodities (1950/1951–1978/1979). Thousands of hectares.

The first phase lasted from the mid-50s until the mid-60s. It was characterized by recovery from the deep crisis that had shaken the wheat and corn crops from 1951 to 1953. Production increases were apparently due to the combination of good climatic conditions (1950/1951 to 1954/1955 had been years of severe drought) and a process of mechanization, which was already having a significant impact (in early 1950, the labor shortage had become critical). The two factors resulted in better harvests in the fields, as reflected in per-hectare increases for corn and wheat in both departments. This progress in productive capacity also benefited from improved domestic prices, which brought a sharp increase in the total area under cultivation and in production for both departments. An important feature of this first phase was that the combination of all these factors appears to have had a relatively balanced impact on wheat and corn, even though, under traditional methods, corn required more labor. However, the simple fact that mechanization was not yet widespread enabled wheat crops to benefit more. Consequently, the option of allocating land to one crop or the other depended more on relative price relations and climatic conditions. During the second half of the 1950s, the scale seemed to tilt toward wheat in Constitución and corn in Pergamino, while during the first five years of the 60s the turn toward wheat was more pronounced and common in both departments.

During the second half of the 60s, a new phase began. The broad, rapid spread of corn hybrids, together with a strong drive toward mechanization which had been in progress since the beginning of the decade, suddenly upset the balance. Per-hectare corn yields leaped, and the consequences were inevitable. The land used for corn increased rapidly, and overall production soared. It is of note that, of the two departments, *estancia*–dominated Pergamino took the initiative, beginning in 1964 to expand area planted to corn. In Constitución, this was not evident until 1967.

The third stage began during the first half of the 1970s. To a certain extent, it overlapped with the completion of stage two and is at times difficult to distinguish. The new phase was typified by the widespread adoption of soy crops, which brought in their wake the recovery of wheat and the displacement of corn. This was when the behavior of producers in Pergamino and in Constitución split sharply. Data on the area planted and in production in Pergamino show that the corn production rises coincided with the transfer of land to livestock, thus reducing the total area under cultivation, primarily at the cost of the wheat crop. The prolonged boom in the livestock cycle no doubt fueled this phenomenon during the early part of the 70s. Rising relative prices for beef brought about a substantial increase in the value of production per hectare, although per hectare productivity was unchanged.

By contrast, in Constitución, perhaps because the operations were smaller in size and capital was less readily available, the corn boom appears to have been used for exploring new farming alternatives. This could also have been influenced by improvements in the efficiency of beef production in the area. The statistical data on other selected areas, both inside and outside the corn belt, generally confirm these trends as described.

Analysis. Some of these shifts suggest certain important trends among the producers. In the first place, it is evident that producers of typical grain crops always sustain some livestock activity, even at the cost of land used for the crops, as back-up protection from price fluctuations affecting their products. It is also clear that new crops spread rapidly. One example is soy. At least since 1974, soy moved out of the experimental stage and has vigorously occupied zones ecologically ideal for corn. Evidently, farming in the corn belt and throughout the pampas has moved to a higher level of sophistication that permits greater variation. One example of this is the dissemination of Mexican wheats and the resulting use of fertilizers, which for the first time have been massively adopted in certain zones of the pampas. Another example is the spread of sunflower hybrids. The first attempts at change will be analyzed in detail for the crops selected.

What I would like to stress is that the shift from one productive phase in the corn belt to another is closely tied to the adoption of specific innovations. In the first phase, it was mechanization. In the second, it was the adoption of corn hybrids. In the third, it was the spread of soy crops. Today it is probably the diversification discussed above. This is not all, however. These changes occur in a chain reaction, and for this reason, the qualitatively different levels in farming today, by comparison with 1950, can be viewed as the result of a long maturing process.

An essential component of the explanatory hypotheses to be proposed below is this linking together of innovations. Thus, it is important to clarify several ideas on the process. The adoption of corn hybrids was made possible by mechanization. In the mid-50s, seed was commercially available, but ten years were to pass before it would become widely disseminated among producers (Martínez, Fienup, and Chevallier, 1977). However, once the process of adoption had begun, the logistics of the expansion curve were similar to those experienced two decades earlier in the United States (Martínez, 1973), leading one to believe that some precondition had been absent in earlier years. This precondition was mechanization. It is clear that, without tractors or mechanical harvesters, it would have more difficult for the dissemination of hybrids to produce such an expansion. There is also a *strictly technological* reason for the linkage, often overlooked, which emerged from interviews with producers. Argentine hybrids produced much higher yields than traditionally grown varieties, but the plants were much more sensitive to problems like weeds. For this reason, they required more labor and, above all, more care than traditional varieties. Otherwise, yields would not be much increased, and seed costs would rise. The simple use of a tractor in good condition makes it possible to arrange and distribute the seeds correctly. This, in turn, facilitates mechanical weeding tasks, making them more effective, and higher yields are thus guaranteed. Without tractors and without good seeders, the hybrids would have been ineffective.

The subsequent chain reaction was from corn hybrids to soy. If the hybrids had not become widespread, the expansion of soy would not have taken place so quickly or on such a high level of magnitude. It is of interest to recall that in the 1920s, Minister of Agriculture Tomás Le Breton had supported initiatives to introduce soy crops into Argentina. Forty years later, in the mid-60s, another Minister of Agriculture, Walter Kugler, strongly pressed the issue. However, all efforts were fruitless until hybrid corn had spread, and the changes it introduced created the conditions for soy cultivation. Soy requires a certain degree of management. It places heavy demands on the producers, which could be met because the expansion of hybrid corn had taught them to work with crops requiring more intensive and tedious labor, at exactly the right moment. The necessary equipment had to be at hand, and health precautions had to be taken. Soy places demands on infrastructure endowment, which were met because the spread of hybrid corn had introduced mass production, the management of bulk harvests, the installation of chains of silos, early harvests, and the widespread use of drying (with the resulting installation of driers). This situation was entirely different from conditions prevailing through the first half of the 1960s, and it made feasible the large-scale cultivation of soy.

It appears that agriculture in the pampas today has a much richer and more sophisticated technological base than 30 years ago, and this has provided access to a diverse range of alternatives for productive progress. Nevertheless, it is evident that the production capacity potentially available in the region is not being fully tapped. The changes that did come about, while important, were too small and too slow. The producers in the pampas are not reacting against technological limitations in order to make full use of productive factors, and are not maximizing their profits.

Changes in the size of farms. The development of agriculture in the pampas has always been attended by two social and productive phenomena that kindle heated debates: the continued existence of large livestock operations on potential farmland, and the prevalence of rental farming by those who work the land. The two situations are closely linked, although the relationship between them has not always been satisfactorily understood.

It is clear, however, that the existing system of land tenure riveted the attention of both critics and defenders of events in the region, finally obscuring any examination of changes in the size of operations. This meant that too little thought was given to the shifts that actually took place in recent decades. In general, they were attributed to a crisis that had begun in the traditional system of tenure in the mid-40s led to measures and counter-measures to change the rental system, and ushered in a number of structural transformations.

The 1960 census yielded important data on the size of operations in the pampas. When these data were published and compared with those of the 1947 census, it was discovered that the strata of medium-sized and medium-sized to large operations had expanded. This was viewed as a result of the elimination of rental arrangements that had taken place in the interim. No doubt there is a relationship between the two phenomena, but the lack of data and of appropriate census classifications precludes reliable verification. The same lack of information also makes it difficult to test the influence of other factors in this situation. In any case, the trend observed in 1960 is clearly confirmed in the latest available census, that of 1969.

It is of interest to note that only one production stratum experienced simultaneous growth in all eight departments analyzed: the amount of land occupied by operations containing 501 to 1,000 hectares. These increases ranged from 6.6 points (Caseros) to 11.7 points (Rojas). Inversely, all departments with the exception of Pergamino experienced a reduction in the area used by the larger operations (over 5,000 hectares). In view of the diversity of situations and development in the different departments, these factors are of particular interest. According to the data, the zone appeared to be experiencing a general displacement toward what could be called an "intermediate sector" of operations containing from 101 to 1,000 hectares. In this sector, the "medium high" stratum (from 501 to 1,000 hectares) was gaining ground. Table 12 gives a regrouping that shows how generalized this trend was.

TABLE 12 Trends in Intermediate-Sized Farms (from 101 to 1000 Hectares) in the Selected Departments; 1947 and 1969 Census Data

Departments	1947			1969		
	101 to 500 ha (% of sector)	501 to 1000 ha (% of sector)	% of department in this sector	101 to 500 ha (% of sector)	501 to 1000 ha (% of sector)	% of department in this sector
Belgrano	93	7	55.5	83	17	60.1
Caseros	93	7	48.5	81	19	55.6
Constitución	95	5	37.5	76	24	52.6
Pergamino	79	21	42.4	61	39	49.3
Rojas	87	13	37.8	59	41	40.0
Junin	89	11	43.0	72	28	50.3
Lincoln	70	30	26.2	58	42	32.4
Río Cuarto	82	18	52.0	63	37	53.7
Pampas Region*	73	27	39.0	55	45	40.5

*According to data from Flichman (1977).

By 1969, this intermediate sector already occupied over half the land in all the selected departments of the corn belt with the exception of Rojas. The same was true in the non-corn belt departments of Junin and Río Cuarto. This fact is of prime importance, in my opinion, because it means that a very large sector, in spite of its diversity, is quite homogeneous in terms of the productive organization of the farms and the type of technology they can use. These strata habitually practice crop rotation and mixed soil use for farm and livestock production, in increasingly sophisticated combinations. Their ownership, use of machinery, and application of inputs are also more homogeneous internally than when compared with smaller or larger farm categories. It is reasonable to expect that, as this intermediate sector consolidated its demand for technical innovations, it would continue to grow more cohesive.

A contrary interpretation of the trend is that it reflects a type of development in which producers slowly accommodate themselves to a certain size of operation. From this viewpoint, producers in the preferred size range are in a position to opt for different sets of productive alternatives from year to year. Because of the nature of the technology adopted and disseminated, such an adjustment can be made smoothly. The following section is a direct outgrowth of this observation.

The question of risk. Experiences has shown that it is extremely difficult to establish lines and trends for pampas agriculture, as the standard variables (prices, production, areas under cultivation, etc.) are highly uneven. As an example, Figs. 3 and 4 show the annual index of pampas beef prices from 1923 to 1965 and from 1950 to 1977.

These data can be compared with those of similar regions in other countries, in which the more regular curves clearly show prevailing trends. By contrast, figures for the pampas need to be juggled and twisted in order to detect the lines of force hidden behind the abrupt, constant changes of direction. This irregularity is highly significant. It is illustrated in a comparative table on price variability for wheat and corn (Table 13). Note that the first period, when the coefficient of variation (CV) was high in Canada and the United States, coincided with World War II. This was also the only case in which the coefficient of variation in Argentina (corn) was less than that in the United States.

TABLE 13 Price Variability: Wheat Compared with Corn (Average Prices for Each Period, in Dollars per Quintal)*

| | WHEAT |||||| CORN ||||
| | Argentina || Canada || United States || Argentina || United States ||
	Prices	CV (%)	Prices	CV (%)	Prices	CV (%)	Prices	CV (%)	Prices	CV (%)
1935–1944	2.59	31	3.30	23	3.51	29	1.77	22	3.05	30
1945–1959	6.08	54	6.28	5	7.18	9	4.80	38	5.60	17
1960–1972	4.32	17	6.46	11	5.64	17	3.38	32	4.58	9

NOTE: CV is the coefficient of variation (standard deviation/average).

*Martínez, Fienup and Chevallier (1977), compiled from FAO Production Yearbooks price information.

This type of data led to a detailed study of the risks which price variations pose for producers. It was assumed that a comparatively high risk quota could have an important impact on producer behavior in the pampas and could explain some of the production peculiarities. At first, only price variations were considered, as they are the best indicator and major component of market risk. As the analysis progressed, it became necessary to distinguish between livestock and farm production, as production risks for farming, in terms of the variability of per-hectare yields, generally turned out to be greater than price-related market risks. Production values per hectare were also calculated. They showed that production risks and market risks did not offset one another; rather, they formed asystematic combinations. It was also found that farm production risks did not decline during the period under study (1950–1978), in spite of the incorporation of innovations which had brought about profound changes in production methods and increased per hectare production values.

Finally, data showed that farm production risks were not uniform among the departments of the corn belt. In a single five-year period, the yield variation deviated from one department to another, and in general there were no systematic trends to show that one department presented lower risks one than another. It is therefore assumed that, despite the adoption of technological innovations, yields are still highly susceptible to climatic contingencies in the zones, to changing surface areas covered by each crop, and to differences in soil quality in the departments (if more area is under cultivation, the soil shows greater qualitative differences).

It was also shown that, in general, the departments outside the corn belt were subject to higher risks of variable farm yield. The lower per hectare yields in these departments were compounded by the greater risk to which they were naturally exposed. This additional factor should be taken into consideration in evaluating the differences in the value of land, the average extension of operations, and the lower percentage of total production occupied by livestock.

The estimates showed that the volume of beef, in kilos per hectare, normally varied less than the volume of farm yields. Past experience had produced the same results, and it means that for

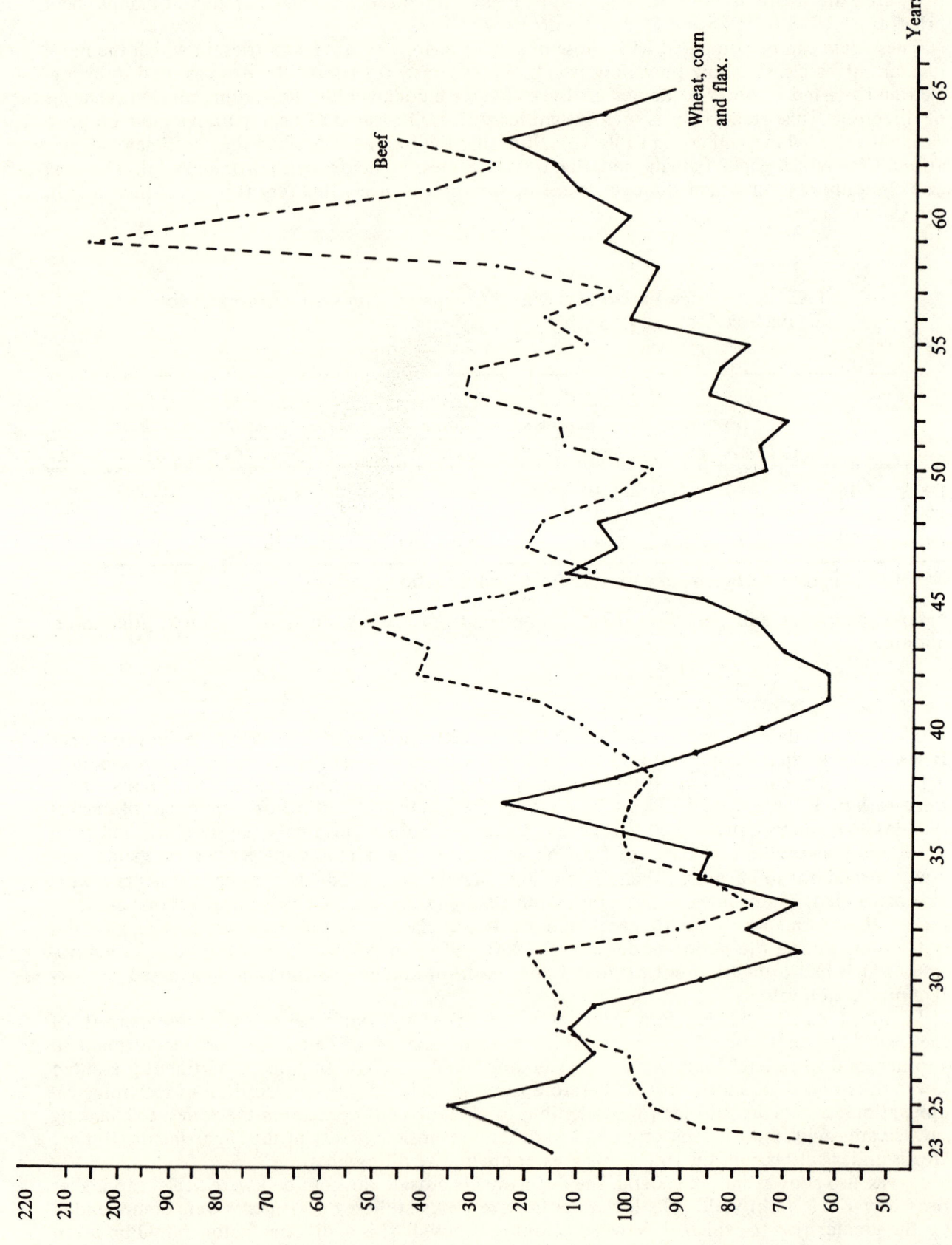

Fig. 3. Annual index of pampas farm prices (wheat, corn and flax) and beef prices, 1923–1965 (base 1935–1939 = 100).

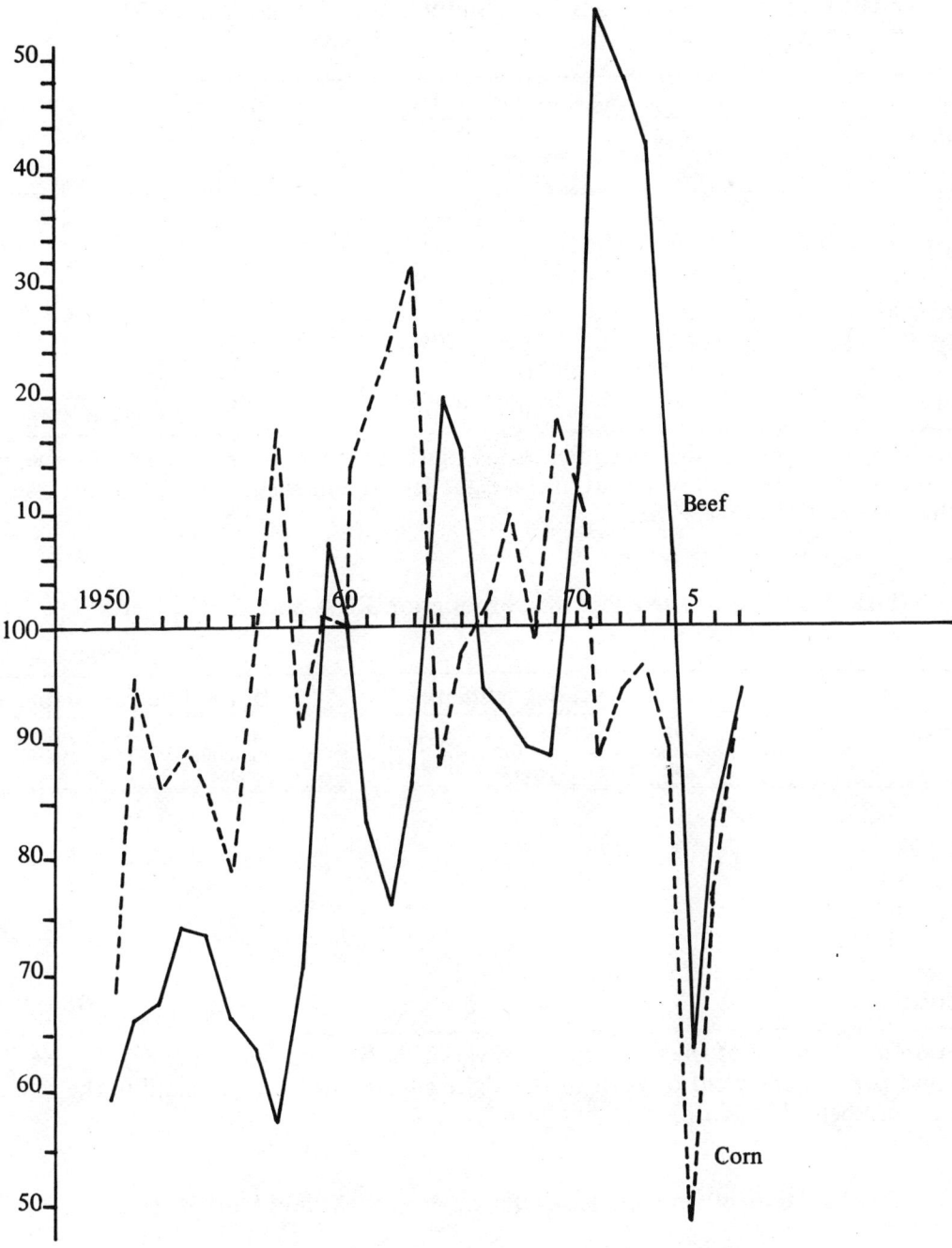

Fig. 4. Corn and beef prices (Liniers Exchange price, on the hoof).

livestock, production risks are less than market risks. This explains the producer preference for livestock in non-corn belt regions. It also explains why livestock is always kept, even in farming zones, where it provides a safety net by guaranteeing a certain level of production. Finally, these facts contribute to an understanding of why it is so important for producers to protect livestock prices, which are the most evident source of risk. All these conclusions are corroborated in Tables 14, 15 and 16. These figures classify the coefficients of variation by size into different strata.

If the tables are compared, production risks (the CV of per hectare yields) prove to be systematically greater than market risks (the CV of prices) for corn and, to a lesser extent, for wheat. However, the situation for beef is the reverse: market risks are greater than production risks. Because the two types of risk do not always balance each other for all products, income risks (the

TABLE 14 Wheat: Income Risks, Production Risks and Market Risks*

Strata	In all eight departments			In the five corn departments		
	CV of production value	CV of hectare/ yield	CV of prices	CV of production value	CV of hectare/ yield	CV of prices
Under 5%	—	1	1	—	1	1
From 5 to 10%	7	4	4	6	4	4
From 10 to 15%	5	8	1	5	6	1
From 15 to 20%	12	9	—	9	8	—
From 20 to 30%	13	16	—	6	7	—
Over 30%	11	10	—	4	4	—
Total observations	48	48	6	30	30	6

*Stratification by the range of the coefficients of variation for five-year averages of price, production value and per hectare yield of wheat in the eight selected departments and in the five departments of the corn belt (1950–1978).

TABLE 15 Corn: Income Risks, Production Risks and Market Risks*

Strata	In all eight departments			In the five corn departments		
	CV of production value	CV of hectare/ yield	CV of prices	CV of production value	CV of hectare/ yield	CV of prices
Under 5%	—	—	1	—	—	1
From 5 to 10%	3	2	2	3	1	2
From 10 to 15%	4	8	2	3	7	2
From 15 to 20%	8	9	—	8	8	—
From 20 to 30%	16	17	1	9	10	1
Over 30%	17	12	—	7	4	—
Total observations	48	48	6	30	30	6

*Stratification by the range of the coefficients of variation for five-year averages of price, production value and per hectare yield of corn in the eight selected departments and in the five departments of the corn belt (1950–1978).

TABLE 16 Beef: Income Risks, Production Risks and Market Risks*

Strata	In all eight departments			In the five corn departments		
	CV of production value	CV of hectare/ yield	CV of prices	CV of production value	CV of hectare/ yield	CV of prices
Under 5%	—	12	—	—	7	—
From 5 to 10%	6	11	1	3	7	1
From 10 to 15%	2	8	1	2	5	1
From 15 to 20%	8	1	2	6	1	2
From 20 to 30%	16	—	—	9	—	—
Over 30%	—	—	—	—	—	—
Total observations	32	32	4	20	20	4

*Stratification by the range of the coefficients of variation for five-year averages of price, production value and kilograms of beef produced per hectare in the eight selected departments and in the five departments of the corn belt (1956–1975).

CV of production value) fluctuate more wildly than the risks posed by either of the factors in per hectare income. It is also clear that income risks and production risks for corn and wheat are clearly lower in the corn belt departments than in the outside departments. This situation is reversed for beef.

Analysis. The historical data suggest that pampas producers are exposed to considerable risk, both in the market and in production. Corn production risks, generally greater than wheat production risks, are further exacerbated by the higher cost of growing corn. This is probably why corn cultivation is concentrated in the most ecologically appropriate zone, the corn belt, while wheat cultivation is much more widespread.

The relatively high risks suggest ideas and pose questions related to the objectives of this research. First and foremost, it appears reasonable that, because of the magnitude of risk, producers are more concerned with controlling the threats to their survival than with incorporating innovations that will "maximize" their income. Obviously, if the coefficient of daily risk is on the order of 20 per cent to 30 per cent of income, the adoption of new techniques that will bring about increases smaller than this would receive only secondary attention. This hypothesis is consistent with the proposal advanced by de Janvry and Martínez (1972) concerning the stages that a business operator must undergo before becoming motivated to maximize income.

This leads directly into a second question: Why was there no systematic search for technical innovations that would reduce risks? Evidently, major, widespread technical change took place among the producers in the corn belt. However, the data indicate that production risks did not decline. Therefore, because the question of risk is crucial to the survival of the businesses, the producers must have developed a strategy for controlling risk, without necessarily adopting new technologies. This strategy, and the resulting behavior, must be analyzed, so that they can guide efforts to interpret agricultural production. It is possible that this producer behavior contains an explanation of the biases and means with which producers approach the question of technical change. This will be the subject of the second section.

THE OPERATION OF AGRICULTURAL BUSINESSES AND PRODUCER BEHAVIOR

Theory and Reality: The Problem of Risk and Strategies for Controlling It

In the past 20 years, a number of theoretical models have been produced for explaining the process of technical change in the agricultural sector, based on the operation of rural enterprises. Examples include Cochrane's technological treadmill theory (1958) and Hayami and Ruttan's induced innovation model (1971). These models supplemented others that had a more general focus on rural economics and had been used to analyze the behavior of individual enterprises and of the agricultural sector as a whole.

As it happened, whenever these models were applied to the pampas, an anomaly was apparent. This brought into question the validity of the models, and an empirical problem of identifying the relevant and specific traits of the pampas region emerged. Innumerable debates and considerable disagreement ensued. For example, there was difficulty in defining the typical producer in the pampas. Some authors stated that the difference between *estancia* operators (large landowners) and *chacra* farmers (owners or renters of small or medium-sized parcels) was of key importance (ECLA, 1959; Germani, 1962; Ferrer, 1963; Giberti, 1964; de Janvry and Martínez, 1972; Flichman, 1977). Others downplayed this distinction and grouped all the producers into the single category of rural enterprises (Martínez de Hoz, 1961; Schultz, 1968).

There was a need to examine both theory and reality. As was suggested in the previous section, the major conclusion of this twofold analysis was that the question of risk was a focal point for explaining the operation of pampas enterprises and the behavior of producers. The idea emerged gradually, as it would have been difficult at the very beginning to pay much attention to a topic that was viewed tangentially or simply ignored in the standard theoretical models.

The persistence and magnitude of market risks (price variations) and production risks (yield fluctuations) were so striking in the data that it was easy to imagine how concern with risk reduction would dominate the day-to-day behavior of producers. The question of adopting technological innovations would be relegated to a secondary position, or even viewed as a function of the central problem.

Income risks in the agricultural sector can be controlled by eliminating causes or muting effects. One of the most common ways of lessening the effects is to combine different activities that are not subject to the same causes of risk. This reduces the probability of suffering losses due to production conditions and increases the chance that the profit from one will offset losses from the others. The study showed that this is the chief, if not the only, recourse used in pampas agriculture. It was shown that, with the coexistence of several annual crops, the fluctuations in overall production value were significantly less than the swings in partial values of each crop. The effect was particularly striking when beef production was added; annual fluctuations in production values overwhelmingly fell into a range of ten to 15 per cent of five-year averages, rarely passing 20 per cent. Table 17 and 18 demonstrate these effects.

TABLE 17 Five-year Fluctuations in Individual Income From Each Major Farm Crop*

Variation	Crop Production Value			Crop Prod. Value per ha		
	No. of cases	% of total	Cumulative %	No. of cases	% of total	Cumulative %
From 5 to 10%	7	4	4	13	9	9
From 10 to 15%	11	7	11	10	6	15
From 15 to 20%	15	10	21	32	21	36
From 20 to 25%	21	14	35	28	19	55
From 25 to 30%	27	18	53	21	14	69
From 30 to 50%	47	31	84	40	26	95
Over 50%	24	16	100	8	4	100
TOTALS	152	100	100	152	100	100

*Stratification of five-year coefficients of variation for total production value and per hectare production of wheat, corn, sunflowers, soy and grain sorghum, taken individually when they contribute over 5 per cent to the total value of agricultural production in the eight departments analyzed (1950–1979).

When beef production was included, another significant effect was seen. For rural enterprises, income risk is an economic problem that directly threatens financing. Even when income flux over the medium term produces a positive net balance, the enterprise needs a minimum flow of funds which is not always available. Thus, they must have financial backing. From this viewpoint, income risk becomes a financial cost that is greater when annual fluctuations increase. This is a crucial consideration, as the immediate survival of the enterprise depends on access to credit and on the medium and long-term profitability of credit costs.

This is an important point for understanding a second central function of livestock on pampas enterprises. Not only does livestock help reduce income risk from the economic standpoint, but it also provides an effective financial mechanism. Stock can be sold all year round, providing the producer with a form of ready currency. It is a type of personal bank that gives immediate access to credit for a comparatively low cost.[2]

TABLE 18 Fluctuation in Income from Farm Production and Overall Agricultural Production*

Variation	Value of aggregate production, total farm commodities			Value of aggregate production, total of all products analyzed		
	No. of cases	% of total	Cumulative %	No. of cases	% of total	Cumulative %
From 5 to 10%	5	16	16	8	25	25
From 10 to 15%	15	47	63	15	47	72
From 15 to 20%	4	13	76	6	19	91
From 20 to 25%	3	9	85	3	9	100
From 25 to 30%	3	9	94	–	–	–
From 30 to 50%	1	3	97	–	–	–
Over 50%	1	3	100	–	–	–
TOTALS	32	100	100	32	100	100

*Stratification of five-year coefficients of variation for total aggregate production value of selected farm commodities, and for the total aggregate value of production including beef in eight departments (1956–1975).

Production Combinations and Producer Behavior

The data cited above demonstrate the effectiveness of productive combinations for countering income risk. The key question is how the allocating of productive factors was affected over time, and how this conditioned the adoption of technical innovations.

The data on the eight selected departments of the pampas show that the amount of land used for each crop varied strongly from year to year. The three major crops which contributed most of the total value of farm production were selected for study, to prevent statistical distortions. Small variations occurred in only 32 per cent of the cases studied. This means that total area planted in one year differed by only five to ten per cent from the amount of land planted the year before. By contrast, in 35 per cent of the cases, the land planted to each crop was increased or decreased by over 20 per cent compared to the previous year.

To offset these powerful swings in the use of land, we would expect more stability in the overall value obtained through agricultural production. This value can be calculated for each year according to the prices received and yields obtained the year before (this is the chief source of information available to producers at the time of planting). In 66 per cent of these cases, production varied by only five to ten per cent from the overall value obtained the previous year.

These two apparently contradictory facts lead to an obvious conclusion. Every year, producers made major, swift changes in the amount of land used for each crop, but they did so by combining them in such a way that the overall income to be expected was quite stable. This interpretation is reinforced by data on the years when a decline was expected in the projected value of overall farm production. In approximately 50 per cent of the cases, livestock cycles were on the rise, and the greater value of livestock production could offset the fall in farm production value (Table 19).

Two additional issues were then explored. First, what criteria were used by the producers for deciding how much to increase or decrease the land used for each crop from one year to the next? How accurate were their decisions? The findings were inconclusive, but interesting. A clear relationship was found between recorded variations in land area planted to each crop and changes in the percentage of this crop in the total value of production obtained the previous year. In other words, if yields and prices for a given crop in a certain year increased more (or fell less) than in preceding years, it was likely that the amount of land used for the crop would be in-

TABLE 19 Annual Variations in Land Planted to Each Commodity, and Total Value of Farm Production (Projected vs. That Obtained the Preceding Year). Total for the Eight Departments from 1950 to 1978

Order of magnitude	Variation in land surface planted*			Projected value of production vs. actual results			Years of high prices of livestock
	No. of cases	% of total		No. of cases	% of total		No. of cases
Under −20%	102	15		8	4		5
From −10 to −20%	76	11		16	7		7
From −5 to −10%	64	10		18	8		9
From 0 to −5%	42	6		45	20		18
From 0 to +5%	87	13	32%	69	31	66%	
From +5 to +10%	87	13		35	16		
From +10 to +20%	82	12		20	9		
Over +20%	132	20		13	6		
TOTALS	672	100		224	100		

*Only for the three crops producing the greatest total production value.

creased proportionally the following year. However, this was not systematic, nor did it demonstrate a regular quantitative relationship. The irregularities could be attributed to the fact that, while it was reasonable for producers to increase crops that provided greater income, the variables they used (production values) were only indirect indicators of net income obtained. Data on necessary costs were not available for estimating actual income. Thus, it was possible for a crop to provide a higher percentage of total production value, but because of uneven cost rises, the contribution of the crop would not be increased, or could even fall, when considered as a percentage of net income.

The analysis uncovered another interesting fact that helped explain these anomalies: the producers obtained uneven results when they decided to increase or decrease the land used for a given crop. Increases in the area planted to a crop often corresponded to heavy falls in the final production value obtained; inversely, crops that had been cut back often yielded much more income than what could have been expected on the basis of results from the preceding year.

There is no doubt that producers have assimilated the repeated outcome of their experiences. As a result, they always maintain a certain quota of the different crop alternatives, regardless of results obtained earlier. To this should be added the need for preserving a livestock back-up, not only as an economic production alternative, but also for farm financing. All these conditions produce a complex environment for making decisions on land allocation (and on the use of other productive factors). It is very possible that excessive weight is given to estimated probabilities and to qualitative criteria, more than to quantitative measurements for deciding land use. If this is so, every year the producers would first determine how wide a range they have for variation. On this basis, they can then determine the specific amounts of land to be used for each crop. This could explain the difficulties that have been encountered in describing and giving econometric dimensions to the observed history of variations. The resulting conclusions are even more eloquent when applied to the analysis given below of the daily behavior of pampas producers.

It can also be stated that while combinations among crops and with livestock help to neutralize income swings, continuing sequences of farm products function as "pioneers" in exploring opportunities for gradual increases in income level. This is done through abrupt shifts, in a form of trial and error. If a change is successful, and income improves as a result, the producers modify their productive base and shift their center of gravity. They design a new approach for adapting to environmental conditions, and this could include the incorporation of technical innovations. Because the resulting benefits and progress are more long-lasting than those available through other, more contingent means (such as dependence on frequent, contradictory changes in relative prices), it is clear that, in the medium term, the adopted technical innovations will determine the phases of productive development.

These characteristics of producer behavior also frame a different kind of demand for technology, as the immediate, almost exclusive search for maximum benefit is no longer the overriding consideration. If the description proposed here is correct, if the major problem to be confronted is income risk, and if this risk can be lessened through productive combination, the rural pampas enterprises could be seen to act rather like the holder of a portfolio of bonds. *Their goal is to optimize the earning rate of total production and minimize risk,* rather than to maximize the rate of earnings for any particular product.

Risk and the Strategy of Combining Activities: Effects on Demand for Technology

It is clear that enterprises specialized in a certain type of production will take a different approach to incorporating technical innovations than enterprises that avoid specialization, combining several activities with different production functions in varying proportions. It is essential for the former enterprises to make efficient use of available productive factors, and efficiency is defined in terms of production increases and cost reduction. The latter enterprises, which combine activities, subordinate the pursuit of productive efficiency to one condition: they must never begin making specialized use of factors, eliminating them from alternative and variable applications to the different production lines in which the enterprise is engaged.

This is what happened in pampas agriculture, in contrast with general production situations on the North American plains, where greater specialization occurred and more in-depth technical progress took hold. The general theory is suggestive, but not conclusive, in understanding the processes underway; for this purpose, it became necessary to perform a more detailed examination of how the resulting effects came about. The research was based on three hypotheses for explaining observed facts. The first two cover the direct impact of high income risks on factors inside the enterprise. The third shows how the strategy of productive combination serves to ensure the immediate survival of pampas enterprises, as compared to other areas; it also explains how this strategy can intensify and prolong the negative events that hold back long-term growth.

Market Risks and Fixed-Capital Investments

It is a well-known fact that risk, by producing uncertainty, is reflected in higher interest rates, which represent an added cost in the capital factor. It is interesting to observe how this phenomenon unfolds in the case of pampas operations. For this purpose, the effects of wide price swings can be compared in two different situations in which other conditions are assumed to be similar.

Figure 5 gives a schematic interpretation of what would happen with production of a given commodity X. Because pampas agricultural enterprises operate in a freely competitive market in which they are "price takers," any price level P is fixed and independent of the volume of production that each enterprise can offer. Accordingly, we find two situations which obtain the same medium-term average price, $\bar{P}$, but for which the fluctuations between the limits P_{min} and P_{max} are greater in case 1 than in case 2.

It is important to observe the impact of this difference when a production function with marginal costs, as in curve C, is displaced by a function based on curve C'. In order to simplify the graph, the two curves are superimposed. This clearly shows that function C' requires more fixed capital (point A' at the origin is higher than A), and at an average price of $\bar{P}$, it yields higher earning than function C.[3]

A comparison shows that broad price fluctuations obstruct movement from production function C to production function C' more in case 1 than in case 2. In the first place, the assertion that $\bar{P}$ is the average price level over the medium term can be proven only ex-post facto, not before, which is when a decision must be made on increasing the fixed capital endowment required for C'. At that time, because habitual price fluctuations are greater in case 1, There is a

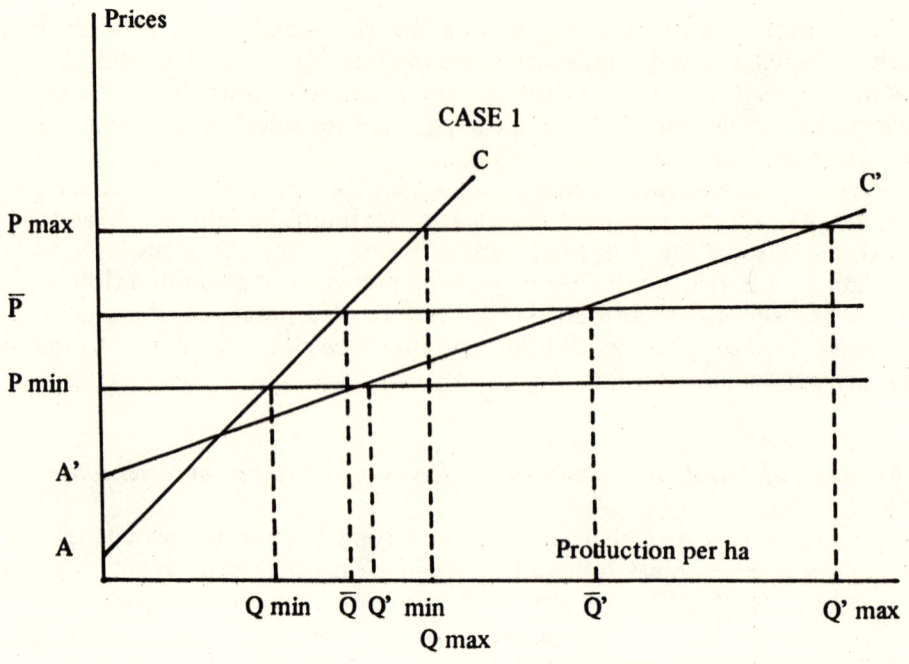
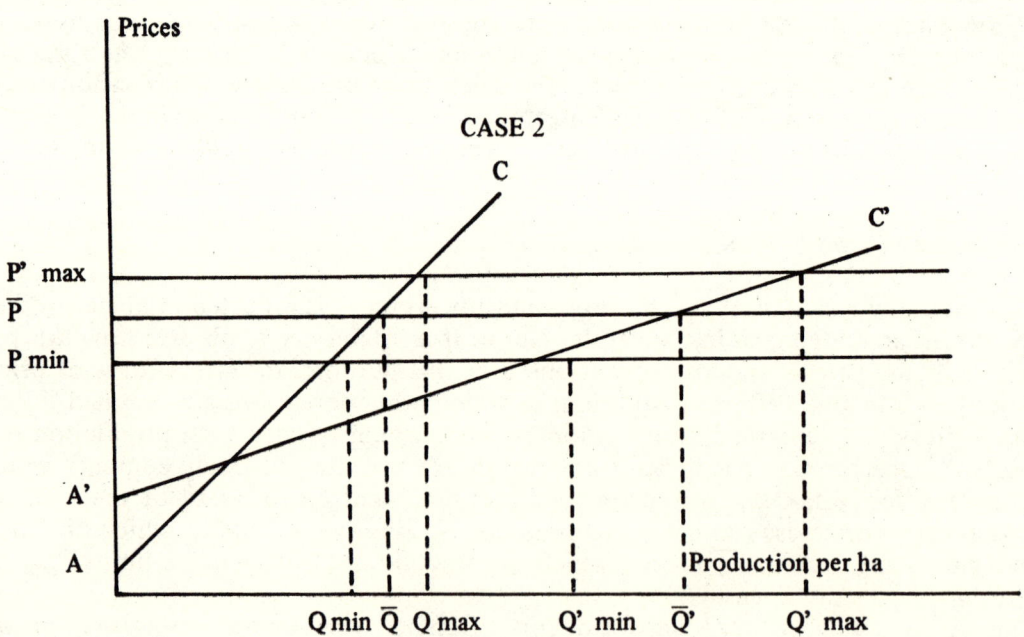

Fig. 5. Effect of broad commodity price fluctuations on the adoption of technologies requiring increased fixed capital endowments.

risk that, eventually, average price $\bar{P}$ will be less than estimated, and the investment will prove unprofitable. Thus, the decision can only be conditional. This does not occur in case 2, where the decision is clear-cut. In case 1, if prices fall to P_{min}, it is best to continue producing under function C and not move to C'. This is not true for case 2. If risk minimization tends to be preferable over profit maximization (an attitude more clearly present among producers in case 1 than in case 2), risk conditions decisions more than the perception of extra earning that may be obtained if prices should rise to P_{max}. Because of wider price swings, case 1 shows greater cost and earning fluctuations than case 2. This means that the enterprise must obtain more financial backing in order to ensure an adequate flow of funds. The additional cost discourages moving from C to C'.

In summary, broad price fluctuations (high market risks) have the following effects:
(a) They act as a brake on the adoption of technical innovations that involve higher fixed costs, particularly larger endowments of fixed capital (buildings, installations, machinery, etc.).
(b) Therefore, they set limits on the demand for this type of technical innovation, biasing it toward those which do not require more fixed capital.
(c) However, if labor is scarce and all land is occupied, as in the pampas, possibilities for achieving technological progress are much reduced, which not only holds back overall production growth, but also limits expansion of enterprises in the medium and long term.

Market Risks and Variable Production Costs

One of the fundamental characteristics of agricultural production is the long period of time between decision-making and actual production. In grain farming, for example, an average of six months must pass from planting to harvest. Livestock raising can take even longer. The breadth of price fluctuations during this waiting period also helps determine which production techniques can be used. Greater price variations induce producers to select those production functions which both reduce total expenses and provide more flexibility for postponing decisions on whether or not to make new outlays. In other words, broad price variation affects the amount and structure of variable expenditures the producer must make, because it limits the certainty of recovering costs.

However, many agricultural techniques for increasing production and for steadying or even reducing costs typically require expenditures. This is the case, for example, with the preparation of land before planting, the selection and fertilization of seeds, and preventive weed and pest control. In livestock, it involves the selection of stock, preventive health measures, feed supplements, and more careful handling.

In short, both high fixed expenses and broad price fluctuations work against the adoption of productive technologies by discouraging certain variable outlays. These effects can be analyzed with comparative graphs on standard marginal cost curves. Figure 6 shows the marginal cost curves, per unit of land surface, for two alternate production functions with the same fixed capital endowment (they both have the same point of origin on the price axis). The solid line represents expenditures the producers may choose to make or avoid from the time they feel reasonably certain about the price that will ultimately be received, while the dotted line indicates expenses that must necessarily be made before that time.

Again, it is important to note that broad price fluctuations inhibit the passage from production function C to C'. Note that the curve of function C', in addition to increasing per-hectare production at average price $\bar{P}$, provides more income and can even improve the unit cost/benefit ratio (mean production costs for the amount produced at average price $\bar{P}$ may be less with function C' than with function C). Nevertheless, producers in case 1 may prefer to maintain production function C, as the only way to reduce risks or offset the danger of price falls to P_{min}. The preference for risk minimization becomes senseless in case 2, where the advisability of moving to C' is beyond doubt.

This direct impact is compounded by an indirect, but important consequence that reinforces the restraining tendency and merits further discussion. We have seen that production functions requiring early expenditures generally stabilize production by reducing production risks and ensuring more stable yields. However, when type C production functions are used, income risks rise. In spite of the dangers involved in broad price swings, this simple fact theoretically should

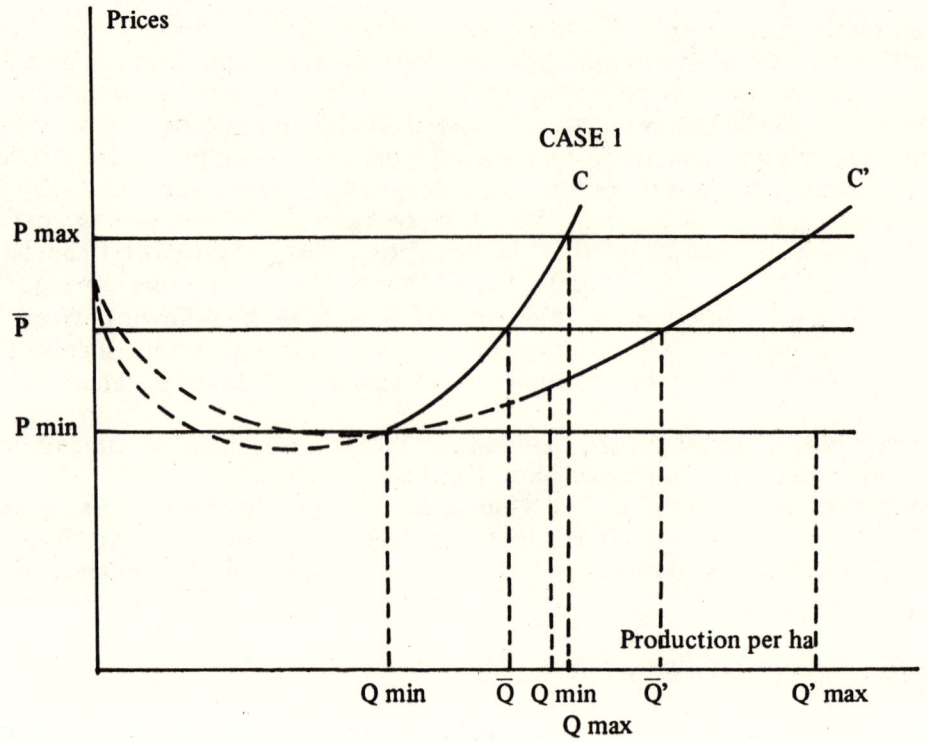

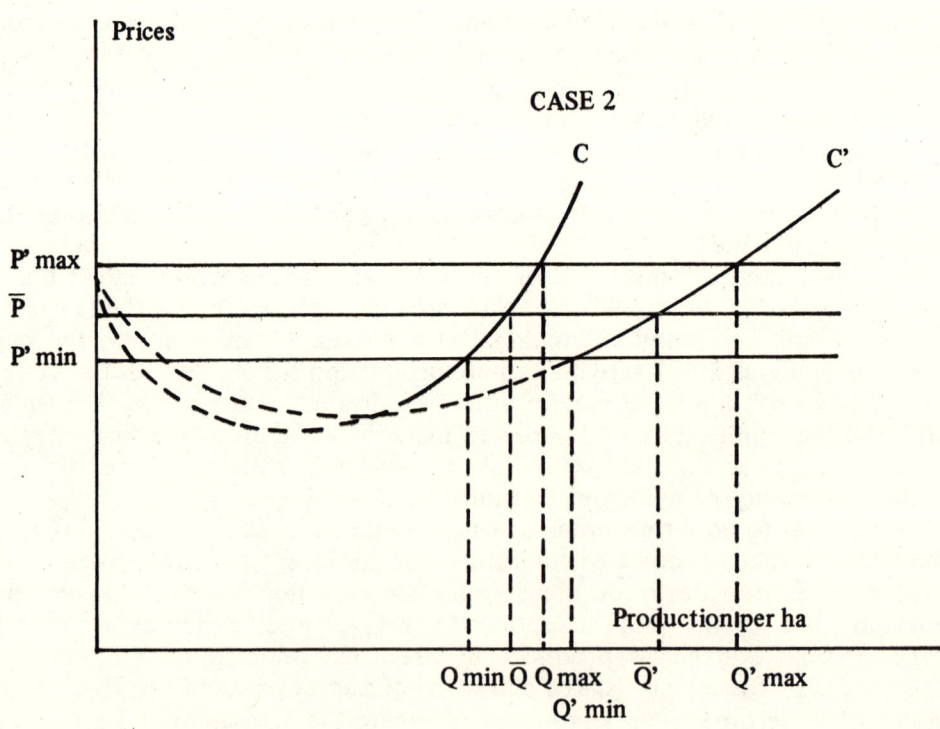

Fig. 6. Effect of broad commodity price fluctuations on the adoption of technologies with different variable cost structures.

encourage the adoption of type C' production functions, even in case 1. Paradoxically, *to the extent that broad price swings occurred,* the effect may have been the reverse, reinforcing rather than blunting the preference for type C production functions.

Introducing yield variability into the discussion confirms that curves C and C' are little more than the average marginal cost curves $\bar{C}$ and $\bar{C}'$ of a curve family limited by C_{min} and C_{max} in the first case and C'_{min} and C'_{max} in the second. These families are shown in Fig. 7, which is divided into four sections to prevent confusion of lines. The proposed graph is arbitrary, but it suggests two interesting interpretations:

- (a) In case 1, production function C, the worst possible alternative (C_{min} with P_{min}) produces far fewer *absolute* losses than the worst alternative for production function C' (C'_{min} with P_{min}), even though relative costs (as a proportion of income received) may be greater. This is decisive if, as is reasonable to expect in case 1, producers tend to give higher priority to risk minimization than to income maximization.
- (b) Production function C, under the best possible alternative for case 1 (C_{max} with P_{max}), can produce similar income and, eventually, a better cost/benefit ratio than the best alternative for function C' (C'_{max} with P_{max}), even though per hectare production is less.

These two possibilities do not contradict the fact that, *as an average,* production function C' increases the earnings and cost/benefit ratio. It should be recalled, however, that if price fluctuations are smaller, as would presumably occur in case 2, the two partial effects discussed will be weakened or even reversed.

It is important to consider the situation for case 1 from the viewpoint of the producer. If price and yield combinations are random, as could easily occur in the pampas, it is not unreasonable for producers to prefer type C production functions. This is true, in the first place, because it provides them with greater security in the worst of cases. In the second place, passing to a type C' function is less attractive if prices are high and natural conditions are favorable for high yields. Thus, the reduction of production risks with type C' does not necessarily encourage adoption of the function. The producers, in a given environment, may deem it best to play yield variations (resulting in part from technological options) against price fluctuations (generated in the marketplace, beyond producer control).

Still, the above interpretations are relative. If damage reduction is the standard used by producers for limiting themselves to type C functions (in case 1), the volume of the safety net will depend on the financial capacity of the enterprise. For this reason, if financial potential grows through time, the producer will be in a position to expand the safety net that protects from loss, adopting new production functions (preferably type C) which are more productive, even in years of poor prices and low yields, when losses may be high.

All this fits well with the situation observed in the pampas. For example, it partially explains the low level of adoption of technologies that increase variable costs and reduce production flexibility for both crops and livestock. Examples are the use of fertilizers and preventive agrochemicals, as well as many handling methods. On the other hand, it is clear that the sector is highly receptive to techniques for increasing market response capabilities in the final stages of the productive process, as occurred with the rapid spread of dry farming techniques, especially for corn. The hypothesis is also helpful in understanding why agricultural productivity increases in the last 20 years did not go hand in hand with more stable yields, as occurred in other countries.

Finally, these theories explain another, more important, phenomenon. If type C production functions persist over the long term, to the exclusion of new type C' functions, production growth is comparatively slower, and enterprise expansion is delayed because net benefits are fewer. This reinforces the hypothesis on the use of fixed capital. Together, the theories trace a peculiar profile of demand for new technologies that will restrain any increased use of the capital factor. This is the opposite of normal trends in more advanced capitalist countries. It does not derive from the abundance of land or labor; rather, it is the effect of the broader price swings so common in the pampas region. This conclusion leads to a third hypothesis.

The Strategy of Productive Combination and Technology Demand

If the broad price variations typical of different commodities entail so many dangers, if they restrict production growth, and above all, if they severely limit the expansion of agricultural en-

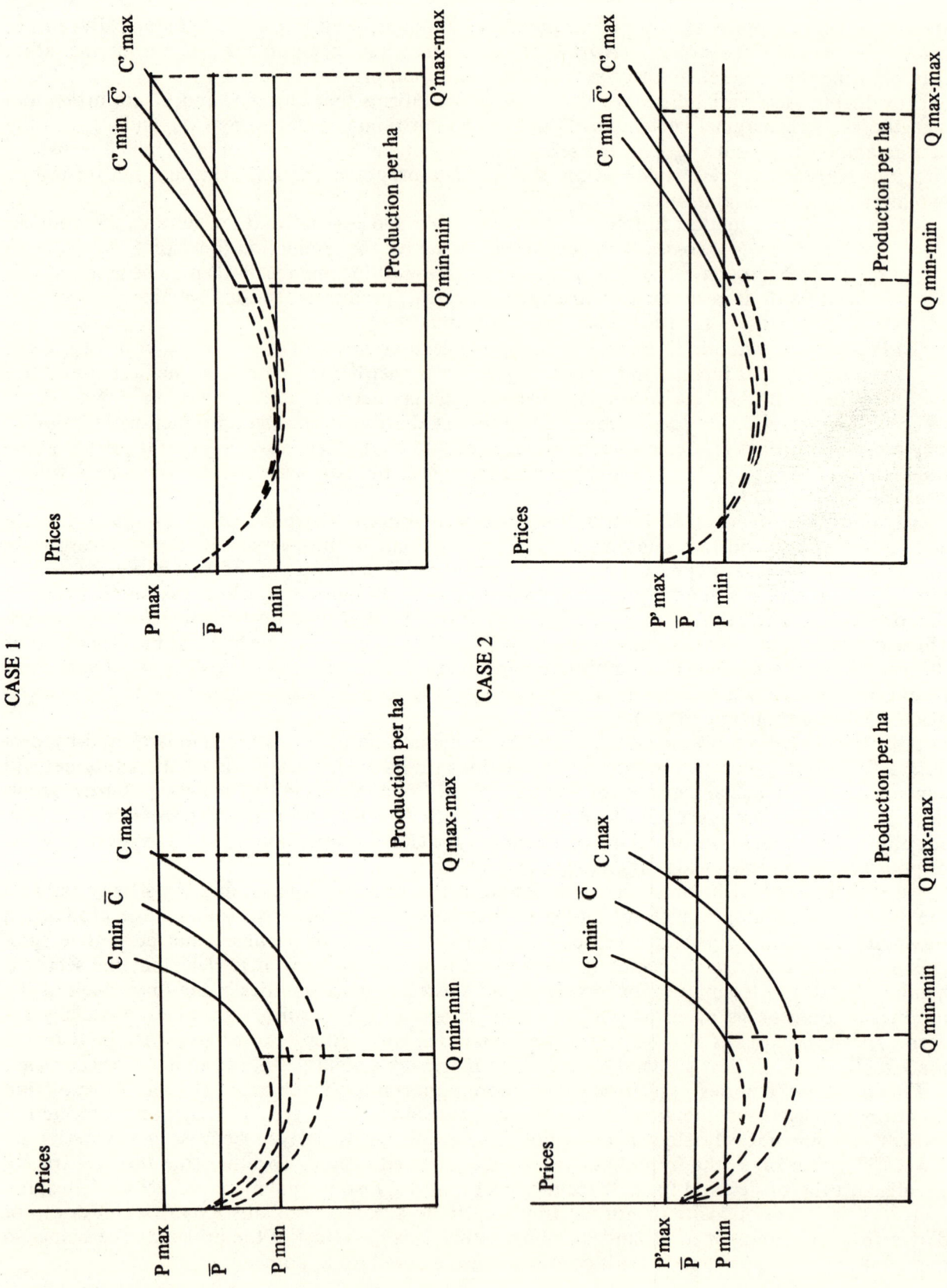

Fig. 7. Price and yield fluctuations: how they influence the adoption of technologies with different variable cost structures.

terprises in the pampas, why are producers apparently unconcerned with stabilizing prices? Federations of pampas producers rarely make energetic demands for such measures. Their attitude stands in stark contrast to events in other countries, and even in other regions of Argentina. It is even more surprising if we recall that domestic prices for grains, vegetable oils and beef fluctuated more wildly than in such countries as Canada and the United States, and sometimes were even more unpredictable than prices on international markets.

The unconcern of pampas producers seems related to the strategy of combining activities, which in the very short term sharply reduces the effects of price variations by comparison with more specialized producers in other regions. While farmers in the corn and wheat belts of the United States and Canada find it difficult or impossible to alternate grain production with beef production at a low cost, and are more exposed to income risks because of price changes, the productive model in the pampas offers a simple, low-cost alternative for offsetting market fluctuations. As a result, price variations are not so important, especially in view of the fact that stabilization mechanisms are not established quickly or easily.

This introduces the hypothesis I would like to stress: the strategy of productive combination reinforces the tendencies that stem from wide price swings and inhibits the increased use of capital in enterprises. The use of this strategy to protect against income risk establishes a new basis for the adoption of technologies: they must allow for alternative uses of resources. However, it is worthwhile to pause a moment and take a more detailed look at how this mechanism works and what its consequences are. The strategy of combining activities is more than just a means to avoid putting all eggs in one basket. It means that, for pampas producers, productive specialization is to be avoided, even when it is the only way to maximize efficiency and profits over the long term. Producers tend to reject any technique that requires productive specialization, either directly or indirectly (for example, by requiring massive capital injections that must be amortized and thus require that a certain production line be rigidly maintained). Instead, producers prefer to adopt techniques that facilitate alternative (more flexible) use of factors for different activities.

In summary, high priority is given to the production of various types of goods that use two or more different production functions; that is, they use different combinations of factors and pose different technical requirements. This creates a situation in which the access to productive factors is an evident opportunity cost for the enterprise. However, this opportunity cost is not equal for all factors if some can be adapted more easily than others to alternate uses, and are therefore more flexible. In fact, under the conditions in the pampas, land is the flexible resource *par excellence*, while capital is more rigid, particularly when used for fixed investments. This implies that, in pampas agricultural enterprises, *the opportunity cost of capital is greater than the opportunity cost of land*. This governs the relative market prices of the two factors and alters the trends for technology adoption, as can be seen if they are isolated and subjected to a Hicksian analysis. In other words, when relative market prices of land are higher than those of capital, the effect is surprisingly minor in stimulating the adoption of capital intensive technology.

The effects of the opportunity cost of capital can be analyzed, especially for fixed investments (see Fig. 5). Assuming other conditions to be equal, the capital opportunity cost acts as a capital surcharge, displacing the previous marginal cost curve C' to C", as seen in Fig. 8. This graph shows how, when curve C' is raised to C", the price does not have to drop from $\bar{P}$ to P_{min} in order for producers to reject a new production function that is more fixed-capital intensive. Much smaller price falls, such as to P'_{min}, are enough to stall change.

Thus, the phenomenon can be seen from two viewpoints. First, if the C" curve is raised above the C' curve, the internal opportunity costs for the enterprise further restrict the adoption of new production functions that require greater fixed capital endowments[4] and that would normally increase the productivity of the land factor. At the same time, when productive alternatives are available, increased internal opportunity costs on the use of capital make the biases in technology demand more sensitive to market price fluctuations for each commodity. In other words, price variations do not need to be large in order to deter greater capital use, as we saw in the first two hypotheses.

To summarize, the third hypothesis states that rural enterprises in the pampas prefer to preserve their access to productive alternatives in order to cut income risks. Therefore, they introduce internal opportunity costs on the use of productive factors and bias the demand for technology toward those systems which allow for more flexible use of factors. Because the capital

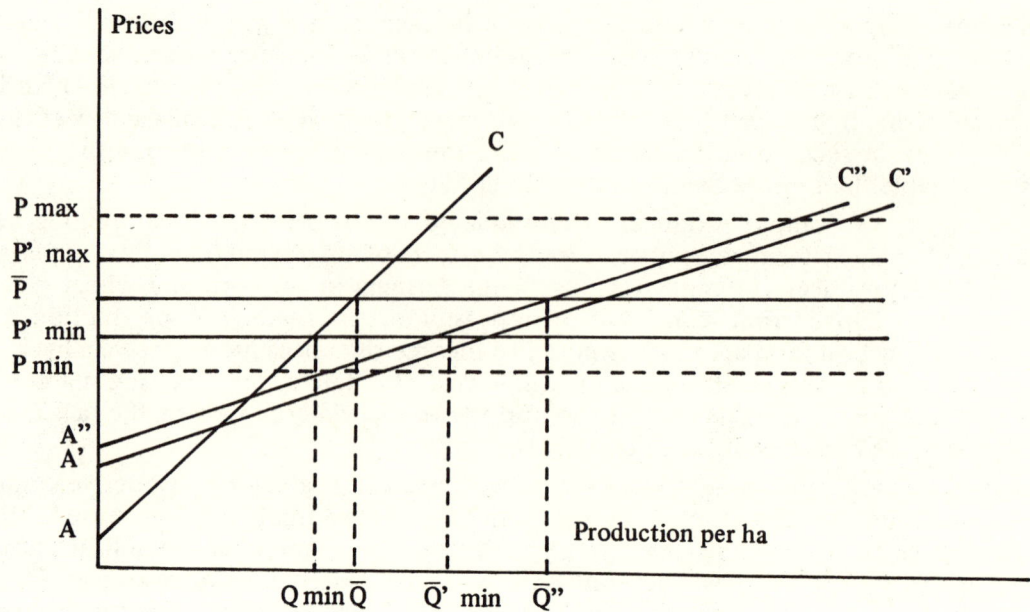

Fig. 8. Production alternatives, opportunity cost of capital, and technology adoption.

factor is generally less flexible than the land factor, this bias tends to inhibit the adoption of capital intensive technologies and favor the incorporation of land intensive technologies.

Three important consequences result from the hypothesis:
(a) In the first place, while the strategy of productive combination guarantees the short-term operation of the enterprise, it preserves conditions that limit production growth and hold back expansion of the enterprise over the medium and long terms.
(b) Equally important, the bias for land intensive technologies dictates that any overall productivity increases must be derived from the land factor. This steadily pushes up land values beyond natural historic price rises caused by limited supplies and a growing demand.[5]
(c) The third consequence is indirect, but important. Pampas producers adopt the strategy of combining activities. The criteria they use are more like those of a holder of bond portfolios than those of a productive agent. They give priority to the evaluation of alternatives and always keep in mind the opportunity costs of capital use. Thus, it is reasonable for these producers not to limit themselves to the borders of their farms, but to remain open to any opportunities that may arise off their land. This openness to the world outside the rural productive enterprise has two important implications.

In the first place, restrictions on medium and long-term enterprise growth are of little concern: producers usually look to outside alternatives for making use of any profits that cannot be beneficially reinvested in direct rural activities.

This attitude toward profits illuminates a second point that, until now, had not entered into the analysis: the size of the operations, or more generically, of the enterprises. It is logical to assume that the range of outside alternatives available to the rural producer will be greater when the enterprise is large and decisions must be made on the placement of hefty sums of capital. If in addition the capital market is flawed, as is frequently suggested for Argentina, differences in operation size are even more pivotal. The opportunity costs of capital placement are higher for the owners of large operations than for those of medium and small-scale enterprises. This produces differences in their behavior concerning investment in the operations themselves. Thus, on large operations, the demand for innovations would be biased more toward intensive land use, with a lower capital/land ratio than on the medium and small-scale operations.

Aggregate Effects of Behavior Patterns in Pampas Enterprises

The strategy of productive combination in the rural pampas enterprises extended far beyond farm borders. As it spread, it exerted strong influence on key features of the overall operation of the region. Two of these are of particular interest: the aggregate effect on technological development for pampas agriculture as a whole, and the way the strategy conditioned the actions of producer federations, particularly in their attitudes toward the government.

Aggregate effects on technological development in pampas agriculture. Theoretical models normally used for explaining technical change in capitalist enterprises are based on the assumption that the operation is under constant pressure from environmental tensions. Based on different hypotheses, these models compile historical experience to demonstrate the continuity and intensification of technical progress. In this context, the stagnation that took place in pampas agriculture from 1940 to 1960 appears as an anomaly. The models can explain it only by resorting to two arguments: (1) a hostile environment produced by faulty government policies prevented enterprises from operating as they should; (2) the fact that these enterprises—or at least those dominant enterprises that set the pace for the rest—were not entirely capitalistic.

The hypotheses introduced above provide a different interpretation. It is possible that enterprises and producers did act in accordance with the rules and rational tendencies of the capitalist behavior model; that environmental conditions were not hostile in the sense that they did not create market flaws; and that, nevertheless technical progress came to a standstill.

This suggests that the strategy for optimizing resource use adopted by pampas rural enterprises restricts the adoption of innovations that incorporate more capital. This is not because capital has a higher price than other factors; rather it is because of a two-tiered process: the broad price swings introduce market risks, and producers combine their activities to attenuate these market and production risks. To the extent that total available land is a fixed quantity and labor is scarce, the aggregate effect of a bias against technologies that require more capital to increase land productivity eventually deters the pace of technical progress. This halts production growth and the economic expansion of enterprises, a situation represented schematically in Fig. 9. The normal tendency of a region (or productive sector) in which typically capitalistic rural enterprises operate is for overall technical progress to be ever on the rise. By contrast, pampas agriculture is characterized by rural enterprises that inevitably reach an inflection point, after which technical progress in the region begins to stall as it approaches a natural ceiling. The actual situation is exceedingly complex in both cases, and neither of the two curves as drawn can be taken as a literal representation of the real world. However, the graph can use two simple lines for effectively synthesizing the proposed hypotheses. It also illustrates the profound differences between overall patterns of pampas agriculture and those of other regions to which it bore initial similarities, such as the vast North American plains.

An additional advantage of Fig. 9 is that it clearly shows the major flaw of previous hypotheses: they provided a plausible explanation of the stagnation that gripped pampas agriculture around 1940, but they had no explanation of its recovery 20 years later. This is why the proposed interpretation includes a second body of ideas on later developments in the region. As was stated, the situation presents two different but complementary facets that must be explained. Before moving into this second part, however, I will explore another important area in which prevailing business practices had consequences on the aggregate level: federated movements of producers.

Conditioning federated action: attitudes toward the government. Over a decade ago, the theoretical models of analysis began to include the crucial role of federated producer associations for galvanizing technical progress in agriculture. Hayami and Ruttan applied the classic Hicks thesis to agricultural enterprises in their study of inducement of change, and found it necessary to include another important item. As soon as rural enterprises begin to require innovations that cannot be protected by patent or monopolized by the developers, who need to obtain a return on expenditures, much research and development work for new techniques begins to occur outside of private enterprise. In other words, the work is assumed by public powers, by the governments, and is provided as a service to enterprises. However, public activity in research and dissemination of technology takes a certain orientation. It will operate as a function of the relative prices of

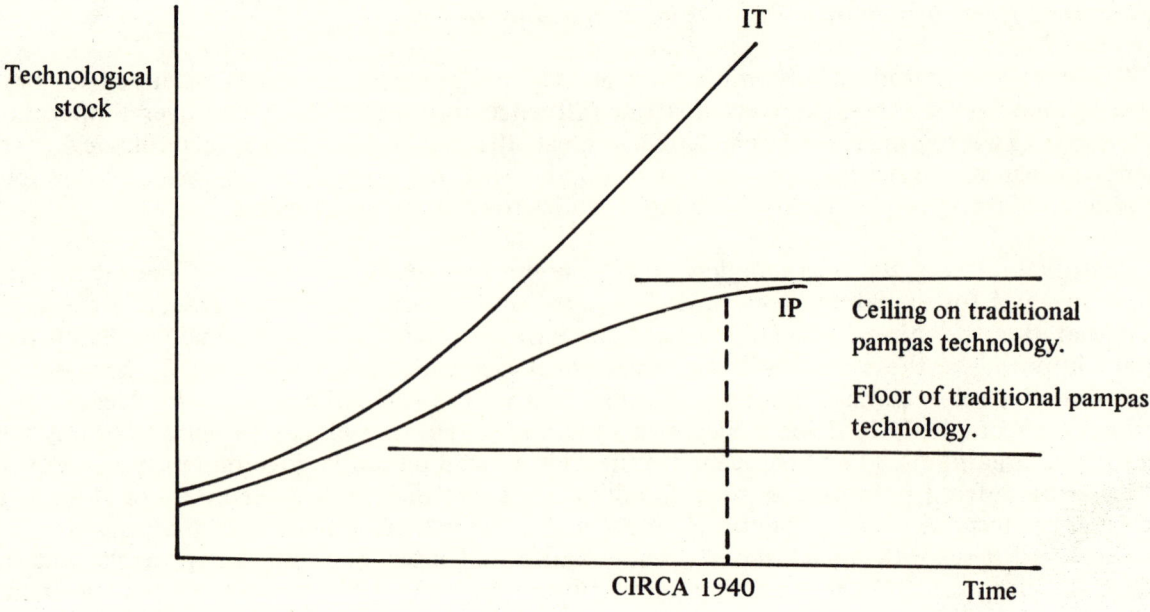

IT: Growth curve for technological stock used in the region (or sector) due to the incorporation of innovations: theoretical prediction based on North American farm operations similar to rural pampas enterprises.

IP: Growth curve for technological stock used in the pampas region, due to the real incorporation of innovations by rural pampas enterprises through combination of activities.

Fig. 9. Limits on technical progress in the pampas region.

available factors only to the extent that agricultural producers are well organized into federations to express their demands. These federations must have fluid access to public institutions operating in the field.

In their discussion of federated action, Hayami and Ruttan state "The dialectic interaction between farmers and scientific research is more effective when farmers are organized into local or regional associations with considerable political clout . . . Given active and efficient organization of farm producers, and a network of agricultural experiment stations whose activity is oriented so as to preserve regional agriculture or satisfy client demands, the competitive behavior model of the enterprise expands effectively and explains how the administrators of experiment stations and of scientific research respond to economic opportunities" (Hayami and Ruttan, 1971).

Nevertheless, despite the crucial role that Hayami and Ruttan ascribe to producer associations, they do not explore why these organizations are or are not formed, or how this happens. In short, the phenomenon is treated like an external bit of data supplied by the environment, a presumably inevitable fact, the explanation of which pertains to some other discipline.

Experience in pampas agriculture shows that producer organizations do not always perform as predicted in the induced innovation model. In fact, if they are compared with their counterparts in other regions of the world which use similar types and methods of production, such as the United States corn belt and the Canadian wheat region, the differences are striking. It is especially clear that the producer organizations in the Argentine pampas are highly heterogeneous. At times they have clashed with each other and, as a whole, appear to have exerted very little pressure for obtaining price, marketing, credit and insurance regulations such as those which are common in other countries. In exploring the technological question as such, a study of the National Agricultural Technology Institute, INTA, (Oszlak, Roulet and Sábato, 1971) showed that the federated organizations had very little participation in the direction of the research institution.

The peculiarities of pampa producer associations can be attributed largely to the risk control strategy adopted by the enterprises, and to certain basic risks that attended the historical development of the region. Let us examine first the effects of the risk control strategy.

Combining activities is not the only way to diffuse the effects of income risk. Another is the system of mutual insurance, which covers a unified group of individuals or enterprises who, at different times, may be affected by certain types of damage. By consolidating their many risks, producers transform possible damage, devastating but uncertain, into the regular, known, bearable cost of insurance.

The obvious difference between these two methods of dealing with risk is that the combination of activities is individual, while mutual insurance is collective. For this reason, the combination of activities is simpler to implement, even though it is a not always adequate to protect the producer. Only under favorable conditions, such as the climate and soil conditions of the pampas, can extensive livestock be combined with grain cultivation. Other regions of Argentina and the world are not so blessed. Producers have fewer alternatives and therefore begin to specialize, gradually incorporating some variation of the mutual insurance system to control the risks that constantly threaten their survival.

Strictly speaking, it is incorrect to call these arrangements mutual insurance systems, thus synthesizing all too briefly a historical process which has been complicated, lengthy and indirect. The only merit of using this term is that it summarizes a concept and highlights the immediate advantages available to pampas producers. By combining different activities, they individually erected a mechanism that directly, effectively protected them from income risk.

Specialized producers, on the other hand, were to undertake a long, circuitous process in order to protect themselves. Historical examples abound which demonstrate that the initial momentum for setting up federated organizations often came from the presence of risk, expressed more immediately in terms of group protection. For example, the disunity of producers was a typical rallying point when marketing was relatively concentrated. The oligopsony or monopsony enables marketing enterprises to pass the negative effects of price falls directly to the farmers, but absorb a higher margin of earning when prices rise.

Producer awareness of the power of marketing intermediaries was always an important incentive for them to join forces and organize their efforts to rectify the situation. Lipset's (1968) classic study of wheat farmers in Saskatchewan, Canada and North Dakota is particularly illustrative. There are many similar examples in other regions and for other commodities. Argentina has a comparatively high degree of federated organization among specialized producers in certain regions outside the pampas (grape growers in Cuyo, cotton growers in the northeast, *mate* growers in Misiones and Corrientes). In certain cases, federated organizations emerged for the purpose of preventing concrete, immediate market risks from being passed on to the producers. They later expanded their scope of action. In initial efforts, the government was pressured to prohibit rail companies from granting shipping privileges to marketing enterprises, but the groups then moved on to organize storage cooperatives, develop trade activities, issue demands for price regulations on the domestic market to offset the effects of international market price swings, and request or directly effect the establishment of special credit systems.

Only recently, as they reach their maturity, have the federated organizations begun to set up integrated mutual insurance systems, in the technical sense of the word, to ease losses. This is despite the fact that risk reduction was always the fundamental motive of the entire process.

The emergence and growth of federated organizations in the pampa agricultural sector shared certain features with the movements discussed above, but were also different in many ways. In Argentina, as in North America, producers tended to come together for group protection. Curiously, though, they were not worried about other, non-farmer groups, but about intra-sector conflicts. The Agrarian Federation, a group of *chacra* farmers, grew out of a production crisis in 1912 which unleashed a wave of protests by renters against the high rents they were paying to landowners and against the conditions forced upon them in farm rental contracts (Grela, 1958; Scobie, 1968). The Federation of Rural Associations of Buenos Aires and the Pampas (CARBAP) emerged during the period of abrupt, prolonged drops in livestock prices after 1922, which pitted *estancia* breeders against *estancia* growers and fatteners, the latter allied with the owners of cold storage plants that industrialized and exported meat (Pereda, 1939; Smith, 1968; Murmis and Portantiero, 1968). The only major exception to this trend was the oldest institution of all, the Argentine Rural Association. It was founded in 1867 by a small group of technical professionals and progressive *estancia* owners, and initially sought to provide a forum for spreading know-how and improving farm production, rather than defending group interests. Its identification with the interests of *estancia* owners was to emerge later, especially during clashes with *chacra* farmers, which led to the formation of the Agrarian Federation. Nevertheless, the distinctive aura surrounding its origins was to lend the Rural Association a certain ambiguity, as became evident in the 1920s, when the *estancia* breeders came into conflict with the growers and fatteners. The struggle tore the very heart of the Association, which eventually proved incapable of reconciling the opposing interests. This led to the founding of CARBAP, more clearly defined as an association for the federated defense of a specific group of *estancia* producers. These ups and downs suggest that the unified, federated nature of the organizations was and is more marked in specialized groups: *estancia* breeders (who raise few crops) and the *chacra* rental farmers (unable to raise livestock).

In any case, it is of interest to note how initial differences in the approach to risk control adopted by pampas producers and by specialized producers in other regions would profoundly influence their different federated organizations. In the first place, specialized producers had a stronger need for solidarity and were confronted with more acute, complex problems to solve. In order to respond to them, the organizations had to be unified; because of the multiplicity of problems requiring solution, they had to be sophisticated. At the same time, the greater specialization of the producers molded an approach in which production problems provided a framework for interpreting other problems.

All the demands and actions performed by these federated organizations for obtaining market, credit, price and other regulations tended to flatten income fluctuations. They also placed producers in a situation which, as suggested by the hypotheses discussed above, fostered the adoption of technical innovations to raise factor productivity and reduce production risks by stabilizing yields. It is no wonder, therefore, that their interest in technical progress, which in itself was too low to have engendered or sustained a federated movement, should have been articulated in a general strategy for protecting producer interests. Because their mechanisms of protection were more homogeneous, the major lines of conflict emerged, not from inside the rural sphere, but between producers and their environment. Consequently, the organizations had more political clout and a clearer sense of purpose for mobilizing to obtain government resources and reinforce the institutions that provided them with technical services. It is clearly unwise to omit any of these circumstances in trying to understand why the federated associations were able to play the key role ascribed to them in the induced innovation model of Hayami and Ruttan.

The concerns that drove pampas producer organizations were very different. These groups had no need to struggle for better conditions that would guarantee the survival of rural enterprises, nor were they obliged to confront large off-farm interests in order to establish complex mechanisms for regulating trade, credit and prices. Consequently, their demands were simpler, temporary, more immediate. The only issue strong enough to prevent the different federated organizations from dissolving was the desire to force succeeding government administrations to take measures for raising commodity prices, but always in a partial, immediate, point-by-point framework. Almost all other issues related to the more permanent conditions of production, such as marketing, taxes, credit systems, and land tenure systems, unleashed mutual recriminations or even open confrontations among the associations. If the pampas agricultural sector was at all capable of exerting political pressure, it was because of the key position it held in the Argentine economy, and not because of the ability of the federated movements to combine individual strengths and channel producer interests. Technical progress, which was of only secondary importance in rural enterprises, was to play an even smaller role when compared to the other issues that concerned the associations.

In sum, the federated movements were directly affected by the strategy of combining activities to cut risk and by the divisions that created conflict among the different groups. Thus, the organizations had no practical interest or political power in playing the crucial role needed by the induced innovation model for successfuly driving technical change in the agricultural sector.

THE EVOLUTION OF PAMPAS AGRICULTURE AND THE IMPACT OF PUBLIC POLICIES

So far, the analysis has covered the predominant behavior of rural pampas enterprises and the actions of the federated organizations that bind them together. It has shed light on the type of operation that was to produce stagnation. However, the history of the region and the development of the departments studied show that the condition into which the region had sunk by the early 1940s changed, and a new phase of growth began. This section will propose the thesis that the fundamental force for recovery came from outside the sector. More specifically, it was the result of certain measures taken by succeeding government administrations. While the explanation is based on simple concepts, an examination of the period beginning in 1940 gives the sensation of a chaotic jigsaw puzzle, in which successful actions were matched by frustrated ones, and predictions were generally uncertain. For this reason, it is essential to examine the following features in the development of pampas agriculture. Only in this way can we see why the new growth phase took hold and how actions from outside the pampas fell into place in the operational model described above.

The Establishment and Social Organization of the Production Model

The Argentine pampas have a long-standing tradition of combining activities to reduce risk, and enterprise behavior is a natural result. The trend formed together with the emergence of certain types of exploitation toward the end of the last century. The subsequent period ushered in a rapid sequence of changes that eventually shaped the distinctive personality of this region. Nevertheless, it was not easy to discern the features described in the preceding section. Even today, three quarters of a century later, the resulting hypotheses may prove unsettling. The mechanism of combining activities to offset risk is common everywhere and is simple to understand. However, as a mode of operation, it has been highly evasive in this particular case, perhaps because its use was sophisticated from the very beginning, instead of undergoing a deliberate, easily traceable process.

From the last decade of the nineteenth century through the first decade of the twentieth, the pampas region set up an economic model for production that combined farming and livestock and diffused risk, but worked across many enterprises instead of focussing on individual farms. In other words, the economic model of production was not directly correlated to a single social model of organizing production (the enterprise), but rather to an original combination with at least three components: livestock *estancias,* rental *chacra* farms, and temporary farm labor. The system had an internal logic and dynamism for the sector as a whole, not easy to see when each component was examined separately, as was generally done.

There were three key conditions that allowed this peculiar economic and social production mechanism to emerge, eventually distinguishing the pampas from other regions, especially the vast North American plains, to which it bore a resemblance. In the first place, the Argentine pampas enjoyed a much milder climate. This made it possible to develop the livestock industry in open, extensive fields, and still earn profits comparable to those of extensive grain farming in the same zone. However, the ecological advantage alone would have been insufficient in the absence of a second important distinguishing feature: a heavy initial concentration of land ownership. This was the fruit of political influences and economic manipulations that came to bear at the time the available lands were distributed, and it led to the establishment of vast production units, the *estancias*.[6] Finally, the pampas had a temporary labor force available from outside the region. This raised the harvest potential of several crops.

The forces that led to the establishment of this complex farming system are well known. They generally did not include the principles of income risk control, although this eventually became one of the system's most beneficial results. By 1890, much of the richest and best-placed land in the pampas, located near the port of Buenos Aires, was being used almost exclusively for livestock production. Farming had been relegated to more distant areas, a circumstance that on more than one occasion at that time, and in later years, was viewed as a maneuver by the large *estancia* owners opposed to progress. There were, however, more concrete reasons. Land in the province of Buenos Aries, dominated by livestock, was worth almost four times as much as land in Santa

Fe, where farming had spread. This was explained by the Rural Association in 1887, when it objected to provincial legislation for establishing Agricultural Centers, because "only beef breeders and fatteners (of top-grade meat) can pay rent on the land and cover the mean value that this law will expropriate and give to the farmers" (Giberti, 1964).

A massive, unexpected about-face took place shortly thereafter. In the early 1890s, live beef exports were introduced, proving to be an excellent business. However, the industry discovered it would have to convert its creole stock to English breeds that could not feed on the natural grasses of the pampas and required the introduction of artificial meadows. For ranchers, this meant vast, unwanted investments would have to be made. The method they found was described in a famous letter published in the Annals of the Rural Association in 1892. They began renting the land temporarily to farmers who, after three years of working plots of around 200 hectares, planted the area to alfalfa and departed (del Carril, 1892). Rarely had a system been developed that was so mutually beneficial to two parties. Both the owners and the renters were interested.

For renters, the system generally provided access to land through payment of a monetary rental fee which initially was less than proportional to the benefits they would obtain: according to a number of indicators and estimates, it gave them access to extraordinary earnings by comparison with other market alternatives. Later, true tenant farming was to be introduced, offering another type of advantage by turning land rental from a fixed cost into a variable cost.

For the land owners, the system tremendously reduced, and at times even eliminated, the cost of investments required for refining their cattle operations. On the surface, it may have appeared merely as compensation for lost production, through a meager monetary fee that did not cover theoretical earning potential, but it had a low opportunity cost when future benefits were added to the equation.

It is worth discussing why *chacra* farmers were initially attracted more to the farm rental system than to settlement. This helps to explain the peculiar system that finally took hold in pampas agriculture. The *chacra* farmers, like grain farmers in the United States, ran their operations by assigning an important role to personal or family labor. Therefore, their problem as agricultural producers was simply how to obtain, in the briefest period possible, the capital and land needed for making full use of their labor capacity to maximize income.

All settlement systems assigned a certain price, over the short or long term, to the land received by settlers. Occupation of the land was therefore an investment. It was a fixed cost that could be distributed over several years, depending on credit obtained for payment. As such, it competed with investments required for obtaining draft animals, plows and other tools and equipment needed for cultivating the land, as well as funds for covering maintenance costs and purchasing seeds. The enterprise was also subject to severe production and market risks, as discussed above. Finally, the plots assigned to the settlers had specific, established dimensions, and often proved too large for working with available tools. It was even more common for them to be too small to occupy the full personal or family labor force available.

Many of these dificulties were overcome after 1890, when the true tenant farming system was introduced. Farmers committed themselves to turn over a certain proportion of their final production (from 20 to 30 per cent) to landowners. The land ceased to impose a fixed investment cost and became a variable cost of operating the agricultural enterprise. Under these conditions, renting had an opportunity cost of zero, as land no longer competed with investments in draft animals and farm tools. A system soon became widespread in which small and medium-scale contracting firms took responsibility for such jobs as harvesting and threshing by hiring migrant workers available annually at the proper time of year, and by lending the use of equipment. By hiring others to do the work, tenant farmers avoided the need to acquire relatively costly equipment for the harvest. Labor, like land, was converted from a fixed investment into a variable operating cost. Expenditures could either be made or avoided, depending on the condition of the crops and the prices on the market. As a result, full accumulated purchasing power could be used for acquiring such farm tools as would maximize the amount of land that could be worked with personal labor.

The contracting firms for the harvest also enjoyed advantages: the large number of operations on which migrants could work, together with the staggered harvest periods throughout the humid pampas, provided economies of scale and risk distribution.

All these factors converged to produce a system of enterprises so vast that settlement projects could not compete with them. The plots distributed to settlers were so small that the farmers could grow more produce by renting.[7] Another important offshoot was the initial rapid spread of technical progress in pampas agriculture, quickly approaching levels current at that time in the United States. Production results were spectacular, as can be seen in Tables 20 and 21.

TABLE 20 Wheat (Thousands of Hectares Under Cultivation)

Year	Total	% annual growth	Buenos Aires	%	% annual growth	Santa Fe	%	% annual growth
1887/1888	829	13.6	221	27	8.1	402	49	14.2
1895	2,000	8.5	880	19	15.1	1,020	51	2.0
1908	5,760	2.3	2,362	41	−0.4	1,324	23	−4.73
1914	6,601		2,310	35		990	15	

Alfalfa (thousands of hectares under Cultivation)

Year	Total	% annual growth	Buenos Aires	%	% annual growth
1887/1888	390	13.7	—	—	—
1895	713	15.5	162	22.7	19.5
1908	4,657	7.6	1,633	35.1	
1914	7,236		2,280	31.5	5.7

SOURCE: National Census and Agricultural Statistics from the Ministry of Agriculture.

TABLE 21 Province of Buenos Aires: Cattle Herds, Percent of Creole Stock, Total Value and Value per Head

Year	Cattle (thousands)	% creole stock	Value (thousands % gold)	% annual growth	Per head ($ gold)	% annual growth
1895	7,746	50.0	73,923	8.06	9.54	5.7
1908	10,351	8.7	202,396	15.20	19.55	17.7
1914	9,091	3.5	472,954		52.02	

SOURCE: National census.

From 1890 to 1910, farming operations burgeoned while the livestock industry underwent profound transformation. Under this double influence, overall production swelled, outstripping growth in the plains of the United States and Canada. Pampas agriculture became the driving force of rapid, intense growth in the entire Argentine economy, providing impetus for the emergence of a society very different from that which existed previously.

During this expansion, the peculiar features of the production model used in the pampas aroused some resentments. However, the system effectively and quickly took over the agricultural frontier that had been open since 1880, when the "Conquest of the Desert" was completed, and proved capable of flexibility under adverse conditions. Economic and social tensions and conflicts were inevitable during the livestock crisis of the turn of the century (when millions of sheep were lost and England banned imports of live animals), the farm crisis of 1912 (when the entire corn crop was lost), the problems of World War I, another livestock crisis in 1922, and finally, the great world crash of 1929.

However, the imbalances were less acute than in similar zones of other countries. The production system proved its ability quickly and easily to accommodate itself to changing environmental conditions. If the crop mix were varied and the ratio of livestock to farming were alternately increased or decreased, it was possible to stabilize annual income in the face of short-term contingencies and to adjust to medium and long-term market shifts. Another favorable factor was the highly adaptable socioeconomic organization. The number and size of rental arrangements could be varied easily, according to relative price movements between farm and livestock products; similarly, the massive use of temporary labor, while perhaps increasing the cost of certain tasks in a labor-short market, effectively provided a partial response to production and market risks.

Those who originally promoted the use of productive combination had not been interested primarily in reducing income risks. However, the system repeatedly demonstrated its effectiveness for this purpose and implicity cemented the complex system of social organization that had produced it.

There were many participants in the production process, and all were not equally protected from risk. The *estancia* owners were the most sheltered because of their hold over the land, which was the most flexible factor. They were followed by the tentant farmers, who for many decades probably had more safeguards than farmers in the United States and Canada. The full weight of insecurity and loss was passed on to the seasonal workers. However, because they lived outside the region, their disadvantages were not widely perceived. In short, in exchange for the goods it produced, the pampa imported solutions and exported social problems to other areas. Inequality was the typical of the production system as established. It had the effect of sowing division and conflict among the different participants, and from it emerged the federated movements.

Stagnation

The productive system introduced in the pampas region was sound. It was flexible enough to adapt to succeeding crises, especially to the crash of 1929. This stands in surprising contrast to the seeming debility that struck it after World War II, when agricultural production plummeted (see Table 1). That situation has frequently been attributed to the mistaken policies upheld by the first Peronist administration for the agricultural sector. Theodore W. Schultz (1968) discusses the period in the following terms:

"The failure to increase productive capacity in pampas agriculture, especially during the 1940s and 50s has no parallel in Latin America, in view of the available economic potential and the earlier impressive growth rate. The fundamental cause of the failure was no doubt the absence of economic incentives. The price paid for inefficiency in farm commodities and inputs almost always obscured investment opportunities for increasing pampas farming capacity . . . The failure of pampas agriculture proved an expensive experiment. The lesson we draw from the experiment is clear . . . Economic policy is truly important. Profitability is indispensable . . . ."

Other remarks similar to that of Schultz can be quoted from several analysts and many politicians. After the war, it is said, Peronism punished pampas agriculture by paying prices below the international rates and confiscating the difference in order to finance non-productive expenditures (Martínez de Hoz, 1961; Díaz Alejandro, 1975). We have already seen that this type of criticism offered no explanation for later recovery. In addition, it is clearly difficult to corroborate with post-war data. At that time, there was a real regression in pampas agriculture, particularly for wheat and corn. However, the per hectare production value of wheat, at domestic prices, was 28 per cent higher from 1945 to 1949 than from 1935 to 1939, and while physical volume of production had fallen 24 per cent. This inverse ratio was even more dramatic for corn: production value climbed 32 per cent, while physical volume plunged 47 per cent.

These data alone are not enough to refute the "punishing the farms" theory, but they do show that the issue is not easily resolved. The same is true of contrary theories, which criticize Peronist policies for failing to eliminate the large *latifundia,* thus maintaining the root of all evil in the

Argentine pampas. However, it is also unclear that this system was the direct cause of the backsliding in agriculture, especially when problems had been minimal during previous crisis periods. Finally, the two positions appear to overestimate the presumed ability of Peronism to obtain such results as had not been achieved by two world wars and the deepest crisis ever suffered by the capitalist world.

The arguments become specious if it is also recalled that pampas agriculture had experienced only a minimum of change up to that time. In 1942, Carl Taylor visited the pampas and described the same type of *estancias* and *chacras* that Benigno del Carril had recommended half a century before, in a famous letter to the Rural Association. The system had also been vividly drawn by Jules Huret in 1911.

Carl Taylor's description suggests a reasonably simple hypothesis for explaining agricultural production falls: the abrupt loss of the temporary labor force, which placed a new ceiling on production activities. Indeed, one of the essential elements of the production model introduced in the pampas had been the availability of temporary labor from outside the region. Without it, renters would have had no access to extensive agriculture and would have been unprotected from income risk. The quantitative and qualitative importance of this situation cannot be overestimated. Gilberti (1964) calculated that by 1914, the temporary labor force in the entire Argentine rural sector numbered 573,000 persons.[8] In order to have some idea of the real significance of this figure, it should be recalled that the total population of the country was seven million. According to ECLA estimates, these temporary workers made up nearly 20 per cent of the economically active population (ECLA, 1959).

From the turn of the century to the outbreak of World War I, a considerable portion of this seasonal demand was met by the so-called yearly "swallow migrations" that swept out of poverty-stricken zones of Italy and Spain. After the harvest, the workers returned to their countries of origin. Traffic was interrupted by the war, but the problem was solved with increased mechanization on the farm[9] and, above all, through the contributions of rural workers from farther inland. The overwhelming expansion of the pampas region from 1890 to 1914 had dismantled the traditional economy in non-pampas zones, thus freeing a considerable labor force. However, these workers followed the same patterns as the "swallow migrations" and continued to reside in their areas of origin, because permanent occupation in the pampas region was not available.

However, the opportunities which did not present themselves in the rural areas soon began to arise in the city. The crash of 1929 and, particularly, the outbreak of World War II, swiftly cut off imports and produced a sudden leap in the demand for domestically produced manufactured goods. New industries arose, and existing industrial capacity was put into maximum production through the incorporation of massive contingents of workers. The demand for labor changed radically. The structure of employment was turned upside down and the face of Argentine society was permanently altered.

Statistics show that the growth of industrial employment from 1935 to 1954 took the shape of a logarithmic curve. From 1935 to 1939, growth was slow; it exploded from 1939 to 1946, at which time it abruptly fell off and, by 1954, had become imperceptible. Industrial employment doubled from 1935 to 1946, with the incorporation of over half a million workers. Even without the addition of data on urban jobs indirectly generated by industrial expansion, this figure was approximately equal to the labor force needed for the entire corn harvest in the highest planting years of the 1930s.

The sudden loss of the labor supply had a profound impact on pampas farm production because it increased costs and curtailed profitability. In any other context, it would have followed the Hicks model and engendered a search for alternative production methods to replace the newly expensive factor. The result would have been to provide pressure for a swift process of technical change.

Some of this could be glimpsed in the sudden increase in tractor purchases from 1936 to 1938. However, absolute figures were small, and the trend quickly faded for other reasons (Dagnino Pastore, 1965). It is probable that shortages of farm equipment during the war and the post-war period impeded a rapid mechanization to replace the lost labor supply.

A more decisive factor appears to have been the economic and social organization that existed in the pampas. It provided a simple, immediately available option: instead of changing methods of production, why not change the activities themselves? Rapid cuts could be made in the land surface used for farming, and the displaced lands could be turned into extensive livestock operations requiring a minimum of labor. The existing production model still appeared capable of adapting itself to changing conditions. When factors were displaced, a new balance could be sought through lower, but equally profitable, production levels. This was a normal response and had always been used to cope with changing prices and relative profitability levels for the two major activities, as can be seen in Table 22.

TABLE 22 Changes in the Land Surface Used for Livestock and for Farming in the Pampas (in Millions of Hectares), 1920–1954 (ECLA, 1959)

Five-year periods	Livestock	Farming	% of land used for farming of total estimated useable land
1920–1924	36.1	11.95	51.9
1925–1929	32.6	14.88	64.7
1930–1934	31.6	15.15	65.9
1935–1939	32.4	16.01	69.9
1940–1944	34.4	15.06	65.6
1945–1949	36.6	13.06	56.8
1950–1957	39.3	11.52	50.1

NOTE: Comparable aggregate data are not available for the pampa region prior to 1920.

In short, it appears unnecessary to ascribe the regression in pampas agriculture to any rupture or catastrophe which paralyzed production. Rather, it can be simply explained as a swift adaptive reaction, consistent with internal operating logic. The region had no campesinos to provide labor or to introduce social problems; as a result, the conflicts born of the agricultural retrenching were less acute, less widespread, and less painful to assimilate.

Public Policies and the Stagnation of Pampas Agriculture

The first paradox in the early 1940s was the conflict between the pampas region and the country as a whole. Existing problems could have been reasonably well assimilated inside the pampas agricultural sector, where they were born, but they proved intolerable for the overall society and economy of Argentina. The productive crisis was serious, yet not disastrous, where it arose, but it became critical in a broader setting, and in the end its backlash stung the pampas with full force. The process through which this occurred was characterized chiefly by confusion and contradiction. In order to understand why it took such a peculiar form, and how agriculture eventually recovered in spite of it, a number of important factors must be considered.

The decade of the 40s was not the first time the government of Argentina had attempted to intervene directly in pampas production. In earlier years, especially since 1930, faced with the shock waves of a world-wide crash, the government had adopted a broad array of measures: the establishment of a National Grain Board and National Beef Board, the introduction of a price support system for a number of farm commodities, the construction of a network of government-sponsored grain elevators, the inauguration of the Central Bank, and the introduction of an open policy for exchange rate control. All these tools were intended to protect local production from the vagaries of an international commerce system whose rules had changed.

One criticism often seen in retrospective analyses is a comparison of the success of the 30s with the failures that began during the Peron administration. The contrast is used to show that the key issue is the competence or incompetence of government leaders. However, there was an essential difference between the two periods, and it must be stressed if we hope to understand later developments. In 1930, Argentina rode out the great market crash at a time when the society and economy had become relatively stable and sound. The production crisis of 1940 was caused by sudden transformations taking place in the society and economy of the country itself. This different background caused problems of other kinds.

In the first place, the government was forced to take action, whether it wanted to or not, to counteract the agricultural recession of 1940. Since the crash of the 30s, pampas production had ceased to be the driving force of economic development for the nation. However, it continued to play an important twofold role, as indicated earlier. In the first place, it provided around 85 per cent of total Argentine exports, thus serving as the basic source of foreign exchange to pay for imports. In the second place, it supplied a substantial portion of foodstuffs for domestic consumption, and for this reason pampas produce was internally considered a wage good.

The production fall in the pampas affected the general development of the economy because it curtailed import capacity at a very critical time. During the post-war period, it was crucial to renovate industrial equipment and infrastructure that had deteriorated from overuse. The demand was rising for raw material imports and intermediate goods to keep the productive process moving. These economic needs were felt acutely by an unstable social structure in profound transformation. The relations among the different social groups were being redefined. The old rules of behavior and coexistence had been swept away, but new ones had not yet been established. Worse yet, the pampas production falls placed the supply needs of foreign trading partners in direct conflict with the demands and need for domestic supplies. This meant that one or the other had to be cut back, and the resulting tensions were increasingly acute.

The data in Table 3 cover over a quarter of a century, from the end of the war to the early 70s, during which the economy and society of Argentina were crippled by the stagnation of pampas agriculture. Directly or indirectly, all the events in the country were conditioned by the impact of this decline. The problems of the foreign sector, and the continuous need to strike a new balance, produced sudden devaluations. These caused extreme shifts in income distribution, fueled inflation, and unleashed a succession of abrupt recessions and recoveries in the economy which left astonishingly low levels of net growth in its wake. The conditions exacerbated social and political conflicts and continually threatened the stability of the government, forcing it to concentrate on stimulating pampas production growth in order to defuse tensions.

Against this background, an issue of a different nature took shape. In 1940, and for many years afterward, it was not easy to discern the basic cause of the fall in agricultural production. Obviously, no one was blind to the difficulties imposed by the sudden loss of the temporary labor force; but it was not easy to understand the close relationship between the pampas production model and the presence of structural unemployment in Argentina. The seasonal nature of the migrant force that lived outside the pampas masked the presence of a campesino problem in the region. It averted the tensions and social conflicts that became so clear in other areas, and it demanded no awareness of the conditions of economic operation that were causing them. Apparently, the Argentine pampas were free of the great agrarian social problems that continued to plague the interior of the country and the traditional non-pampas economies. Because these two interrelated problems were incorrectly viewed as separate phenomena, it was believed that the expansion of employment in industry and other urban activities would provide a solution to the suffering of the non-pampas rural population that was indirectly creating problems in pampas agriculture. For this reason, it was frequently stated that the problems of agriculture were the direct result of government policies for redistribution, which had raised wages and promoted the creation of non-productive jobs. The profound transformation that had taken place in the structure of employment in the country—and that was reflected in wage increases—was unseen, and the causal relations were inverted. The Peronist government was fingered as the cause of a social transformation (Díaz Alejandro, 1975) when, in actual fact, the social and economic transformation had created Peronism (Germani, 1962).

This is more than a mere intellectual exercise. It provides an explanation of the major stumbling-block to government action in the agricultural sphere for 20 years, before and after the first Peronist administration. The only possible solution in the context of the existing production model would have been to revive a situation of structural unemployment. It is obvious that no administration held the power to turn back history, least of all Peronism, which was itself a fruit of this process.

As this road was closed, therefore, the only alternative was to find patchwork solutions to hold back the collapse of agricultural production. Initial attempts had already been made by the conservative government during the war, when farm rental contracts were extended. Other measures were isolated, such as the establishment of an integrated agricultural insurance system (Taylor, 1948). Peronism, hemmed in by the relentless production slides of the mid-war period, was to multiply these initiatives.

A third problem was that the characteristics of the production model as a whole were simply unknown, even to the pampas producers themselves. One result of the model was that no homogeneous sector of producers and enterprises existed. On the contrary, there was a clear distinction between the land owning *estancias* and the *chacra* farmers working rented land. Each of these two groups had its own interests, which frequently conflicted. Each had developed its own view of the causes of stagnation and of the measures needed. Both the *estancia* owners and the *chacra* farmers defended their own positions to every government administration, pressuring for fast action in accordance with their own approaches and interests.

In other words, even the agricultural sector itself was unable to offer a clear idea of how solutions could be developed. Instead, their contradictory demands simply increased the confusion. Each administration tried to obtain the support of one of the two groups, but it also had to answer requests from other social sectors and overcome emergency problems by taking measures that often ran contrary to the interest of *estancia* owners or *chacra* farmers. A common example was farm prices. On such occasions, the *chacras* and *estancias* were able to join forces in opposition to government policy. One result was that prices were subject to violent swings, which intensified the negative effects of the production model.

A comprehensive look at the different levels on which problems appeared shows why the outcome was a process fraught with confusion and contradictions. In the first place, it was played out against the background of a society and economy undergoing rapid change, leaving behind it a wave of uncertainty and instability. This was compounded by cutbacks in foreign trade, due to the recession in agricultural production and the intensification of income struggles that produced additional points of conflict. At the same time, the possibility of a quick recovery was lost, and no clear alternative solutions seemed available. In short, the full array of conflictive issues that underlay Argentine society unfolded in the framework of an extremely fragile political system.

Each incoming administration felt compelled to take action on the problems in the pampas agricultural sector, but all the available measures affected one or another of the groups involved and produced considerable resistance. The precarious balance was frequently tipped, and the margin within which stable agreements could be reached was narrow. The different social sectors were continuously obsessed with short-term issues and expectations. Each succeeding administration had less freedom of movement, while the different social groups found themselves in a unique position: none was able to implement a solution, but all were quick to veto solutions proposed by the others. The situation was widely recognized throughout Argentine politics, and it inspired comments that the country had reached a social and political impasse. Clearly, however, the perpetuation of the problems ended up working in favor of those who had always benefitted in the past.

This general setting had two basic consequences for pampas agriculture. The most visible and immediate was that all the administrations were inevitably obliged to take a permanently activist position. However, their actions were overwhelmingly dominated by short-term emergency measures, and discontinuity was the hallmark of the policies of succeeding governments and even of succeeding years under each administration. Widely varying price systems were tested, running from price fixing by decree to total market freedom, sometimes for certain products, sometimes for everything. The same occurred in the areas of marketing fiscal policy, and land tenure, for which opposing attitudes were adopted, rejected, and adopted again, sometimes in ashtonishingly rapid succession.

It is impossible to trace a historical outline of this seesaw of events, trying in vain to discern a pattern. This can be demonstrated through an examination of the many measures taken by the Peron administration. The analysis clearly shows how far they were from reflecting the kind of consistency so often attributed to them, and also fails to demonstrate their alleged catastrophic effects. The same can be said of all succeeding administrations. It would be excessive to denounce everything as simply incompetence, which in fact did play a part. Simply, the process was too complex for anyone to grasp and control.

The second consequence of this situation is less obvious: by its very nature, the chaos created a high degree of uncertainty in pampas agriculture. Thus, it encouraged producers to maintain their accustomed production model. This point must be emphasized, as it shows how continuity existed through all the changes, in spite of, and at times because of, outside pressure. This continuity eventually brought about the recovery of agricultural production.

Public Policies and the Recovery of Pampas Agriculture

Because of the complexity of the situation, recovery emerged from a diverse assortment of circumstances and even accidents. However, it would be helpful to highlight the particular factors which were most important, and the basic characteristics of the process through which they exerted their influence.

The hypothesis presented here is that the decisive factor lay in initiatives taken from outside agriculture, especially from the government and its institutions, which modified the conditions of supply and demand for technical innovation. As it began to assimilate technical change, the pampas agricultural sector was able to increase its productivity. More specifically, the shifts in technology demand were induced by the appearance and survival of a policy for credits and tax cuts that subsidized the incorporation of capital. At the same time, the technology supply line was activated through the establishment of institutions, especially the National Agricultural Technology Institute (INTA).

The picture outlined above suggests a specific sequence for these two broad lines of action. The tone of the times motivated action, while at the same time restraining actions from developing. Under such conditions, a policy needed to fulfill four important conditions if it were to have continuity and produce lasting effects: (1) a particular social group, sector, governing body or government office had to take the initiative for promoting it; (2) once introduced, it had to receive support from succeeding administrations regardless of their different orientations and aspirations; (3) it must not arouse resistance or active opposition from any social group or economic sector with the power to exercise a veto; (4) it must fit in with the traditional economic model used in pampas agriculture.

This combination of requirements explains why credit, tax reductions and INTA efforts were all decisive measures. Technical change could theoretically have been stimulated in many other ways, with the major role being assigned to other factors (prices, taxes, new land tenure systems, etc.). In fact, these alternatives were repeatedly attempted, but as it happened, only credit, tax cut measures and INTA efforts were capable of fulfilling all the conditions. The restrictions described above worked like a filter, severely limiting the range of policies that could last.

A quick look at the two different lines of action that survived highlights their peculiar features. Clearly, the major initiatives for credit and tax cuts for the incorporation of capital goods did not originate in rural areas. The most pressing concerns of producers were short-term (prices, operating credits, marketing); their usual behavior did not push them toward the incorporation of many capital goods. Their demands were based on freeing imported equipment and machinery from duties, so they could be obtained for lower prices than domestic goods. Even though the introduction of lines of credit for acquiring domestic equipment and machinery did not arouse enthusiasm, neither was the measure rejected. The equipment and machinery manufacturers were clearly interested in promoting these lines of credit and tax cuts. Less visible, but more important, was the interest of groups of technical employees in certain government offices, and the influence of international financial institutions, such as the Inter-American Development Bank, which facilitated special lines of foreign credit for backing domestic credit.

The confluence of supplier interests, outside support and technical-bureaucratic interests based on the need to promote technical progress in the agricultural sector ensured adequate support for several decades by different administrations. This support was bolstered by the absence of lively opposition from other social groups or economic interests. The granting of considerable, but not excessive, credit had no direct impact on short-term, immediate conflicts, and it responded to a vague general agreement on the need for technical progress in rural areas. Clearly, inflation provided negative interest rates for producers, translating the credit into a subsidy and double income transfer from the rest of society toward equipment producers (through price protection on their goods) and toward rural enterprises. The phenomenon, however, was masked among all the other widespread, often more pressing effects of inflation.

Finally, credits and tax cuts for equipment acquisition tied in perfectly with the operating model described above. This can be seen in Fig. 10, which shows how price fluctuations tended to restrict the incorporation of innovations requiring greater capital endowments.

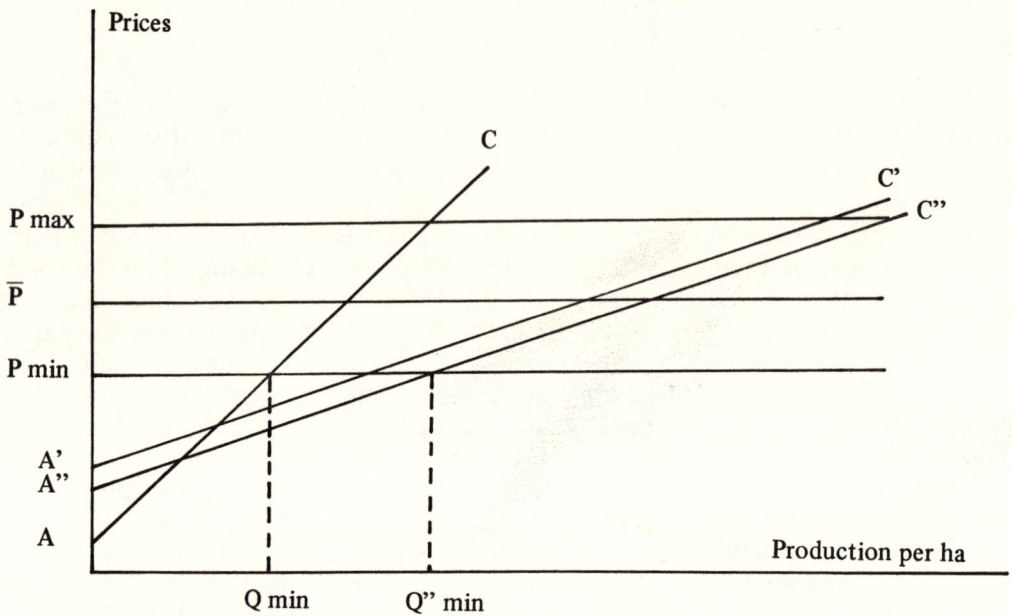

Fig. 10. Effects of implied subsidies in credit for the adoption of technology requiring increased fixed capital endowments.

The implied subsidy can be shown as a drop from the point of origin of C', caused by credits at negative interest rates and tax cuts that reduce fixed costs and transfer the function to C''. This produces an effect which is the reverse of the opportunity cost of capital, as a result of the presence of production alternatives, and favors the incorporation of a greater fixed capital endowment. An important point is that this mechanism, by its very nature, can be selective to the extent that certain goods and equipment are included or omitted from the lines of credit and tax cut provisions.

The results were impressive, even though they did not always attract the attention they deserved. At the start, a massive injection of credit made it possible to mechanize farm tasks through the acquisition of tractors to replace the lost labor force. The magnitude of the outcome is clear in Table 23, which gives estimated labor requirements for corn production from 1930 through 1968.

The first sizeable drop in labor requirements in the 1950s, according to these estimates, was attributed to the replacement of pre-harvest labor (manual weed control). Time savings after 1960 were basically ascribed to the mechanization of the harvest. Finally, post-harvest work changed with the dissemination of bulk harvest to replace bagging.

These innovations benefitted from the later incorporation of tractors, mechanical harvesters, silos and driers (all with the use of credits and tax cuts), which also had major indirect effects. For the *chacra* farmers, mechanization freed the land for livestock, by eliminating draft animals and reducing the time needed for raising crops on each plot of land. Mechanization provided *estancia* farmers with the opportunity to do the farming themselves, or through contractors, without tying themselves to tenant farmers. These possibilities further expanded with a full range of

TABLE 23 Labor Requirements for Corn Production

A. Person/hours per hectare

	1930/40	1940/50	1950/60	1960/65	1968	1979
Pre-harvest	44hs.22'	41hs.26'	11hs.44'	7hs.42'	5hs.12'	4hs.40'
Harvest and shipping	54hs.38'	53hs.54'	39hs.36'	8hs.48'	5hs.38'	2hs.05'
Total	99hs.	95hs.20'	51hs.20'	16hs.50'	10hs.50'	6hs.45'

B. Person/hours per quintal (100 kg)

	1930/40	1940/50	1950/60	1960/65	1968	1979
Pre-harvest	2hs.01'	1h.53'	32'	21'	12'	7'
Harvest and shipping	2hs.29'	2hs.27'	1hs.48'	24'	13'	3'
Total	4hs.30'	4hs.20'	2hs.20'	54'	25'	10'

SOURCE: Compiled by the author from information by Coscia and Torchelli (1968) through 1968, and by Pizarro and Carcciamani (1979) for 1979.

livestock credit for retaining breed cows, fattening calves and introducing pastures to improve land productivity, which was possible with the new tractors.

Tables 24 and 25 give two series of estimates of the magnitude of the subsidies implicit in the negative interest rate of credits. The first table contains a provisional calculation of the per cent discount obtained by paying with credit, as compared with the nominal price of tractors each year. This estimate holds only through 1974, due to the payment terms used. However, it can be seen that as of 1975, the burgeoning inflation rates in Argentina produced implied subsidies on purchases made through May, 1977 (when monetary reforms ended subsidized credits) equal to or greater than those for 1974. The importance of this subsidy does not include the relief of financial burdens by spreading payment over an average of five years. Even so, it explains why, after the credits were removed, the sale of tractors plummeted below what was required for simply maintaining current use levels through replacement of obsolete units. The second table shows the magnitude of the implied subsidy compared with the total beef slaughter rate, or the total value of annual livestock production. From this viewpoint, the subsidy functions as an increase in prices received, and in some years it totalled more than 20 per cent.

From this, it may be concluded that the system of subsidies and tax cuts had two major consequences. First, credits for mechanization and livestock induced a fundamental transformation while preserving a basic continuity. There was no longer any need for the earlier complex social organization that included *estancias, chacras* and temporary labor, in order to operate the production model of combined activities. It was now possible to operate the model inside each enterprise. The *estancias* could develop their farming activities directly, and the *chacras* could obtain livestock.

At the same time, the incorporation of greater capital endowments, which had made the change possible, set a new ground level of receptivity to technology. Without altering the former economic model of combining activities, this new level made it possible to achieve higher levels of productivity for the combinations. The aggregate effect of such a transformation in the majority of rural enterprises could be described in a graph similar to that used previously to demonstrate increasing stagnation (see Fig. 11).

Defining a new ground level of technological receptivity produces a sharp jump in the function of technology incorporation for the sector as a whole. Initially, the graph has a slope similar to, or greater than, that of the normal situations for an agricultural sector made up of capitalistic rural enterprises. This would be the case for the 1970s (perhaps beginning in the mid-60s) in the pampas. However, if the description and hypothesis are correct, the persistence of the productive combination model should eventually lead to a new inflection point after which technical progress would once again decline. In any case, the effects of the credit and tax cut policies explain why the promotion of such institutions as INTA was so important. The higher ground level of receptivity to technology in the enterprises was to create a new demand which could be met only by a higher technology supply.

However, it would be best to qualify this second phenomenon. As had occurred with the initiatives to promote credits and tax cuts, the increased incentives for opening INTA did not come from the pampas. In fact, the idea originally aroused resentment and suspicion in such organizations as the Rural Association. In this case, therefore, the decisive influence clearly came from technical groups of government employees interested in forming the institution. The initiative was able to produce results because of a specific recommendation in the report prepared by Raul Prebish in 1956 for the Argentine government, concerning the economic situation of the country. An awareness arose concerning the need to produce rapid technical change in order to boost production in the pampas and solve export problems. This coincided with a broader current, spreading throughout Latin America, and it received support from international technical and funding organizations.

TABLE 24 Tractors: Subsidy on Mechanization*

Years	Tractors produced (thousands of units)	Tractors sold (thousands of units)	Subsidy implicit in credit** (% of normal tractor price)
1956	9.8	9.8	36
1957	10.5	10.5	50
1958	11.0	11.0	56
1959	12.5	12.5	26
1960	20.2	13.1	28
1961	14.7	16.7	37
1962	11.7	11.2	36
1963	11.4	12.1	34
1964	13.1	15.0	32
1965	13.5	13.7	28
1966	11.2	9.6	25
1967	9.5	9.9	19
1968	9.8	10.9	26
1969	9.0	9.5	40
1970	10.9	11.2	51
1971	13.8	14.8	70
1972	15.4	14.8	54
1973	21.3	21.5	59
1974	24.5	24.8	78
1975	18.8	19.0	—
1976	23.9	22.8	—
1977	25.8	23.7	—
1978	5.9	9.0	—

*Preliminary estimates by Graciela Rodríguez on average price data for tractors and conditions for credits granted by the Banco de la Nación.
**Total subsidy implicit in credit was calculated as a ratio between the amount ultimately paid by the producer in real currency, and the nominal price of the tractor in the year it was sold. It does not include profits on the deferred payment of credit, nor tax exemptions granted to the producers for purchasing the tractor.

TABLE 25 Subsidy Implicit in Credit as a Percentage of Slaughter Value*

Year	Slaughter** (thousands of head)	Value (millions of pesos)	Percent subsidy/Sl. Value
1950	9,898	38,998	—
1951	8,978	39,575	—
1952	8,786	44,448	—
1953	7,896	43,483	9.5
1954	8,133	42.934	6.5
1955	10,004	45,778	5.6
1956	11,664	43,017	5,8
1957	11,961	38,108	11.4
1958	12,278	46,534	12.4
1959	9,148	52,610	18.6
1960	8,884	49,830	21.8
1961	10,212	46,128	21.5
1962	11,790	43,882	17.1
1963	12,926	52,828	8.8
1964	9,367	58,712	6.2
1965	9,134	64,742	5.5
1966	11,076	60,353	8.3
1967	12,520	63,226	10.0
1968	12,802	63,037	9.1
1969	13,821	67,046	5.7
1970	12,924	76,924	4.1
1971	9,468	73,339	4.1
1972	10,010	66,306	12.1
1973	9,818	68,392	17.0
1974	10,004	58,883	19.6

*Compiled by Lucio Reca.
**Adjusted for inflation, based on Wholesale Farm Price Index based 1960 = 100).

This conceptual foundation was the key to obtaining continued support for INTA from succeeding administrations, in spite of their opposing philosophies. Their support held more weight han the often passive and ineffective support of federated producer organizations, and lasting overnment backing was deliberately and intensely sought by INTA directors. Curiously, this placed federal government representatives on INTA's Directive Council in a position to defend he institution to political authorities, instead of serving as political emissaries to INTA. In fact, NTA policies and orientations were defined entirely by technical teams, rather than by government dictates or producer demands.

It was possible to perpetuate government support, as in the case of credit policies and tax uts, because there was no open opposition from social groups or economic interests opposed to he existence or activities of the institution. Strictly speaking, the most common attitude was ndifference or a vague sympathy, tinged with specific criticism (sometimes accurate, sometimes iot) of INTA's work. More important was the clear awareness by INTA directors of the need to void confrontation with outside social groups and economic interests. The atmosphere of political instability and uncertain support clearly meant that the institution was much too fragile to run unnecessary risks. Finally, the new technological ground level on which the pampas enterprises operated, due to credit policies and tax cuts on certain investments, paved the way toward adoption of an appreciable number of the technologies adapted and developed by INTA. Thus,

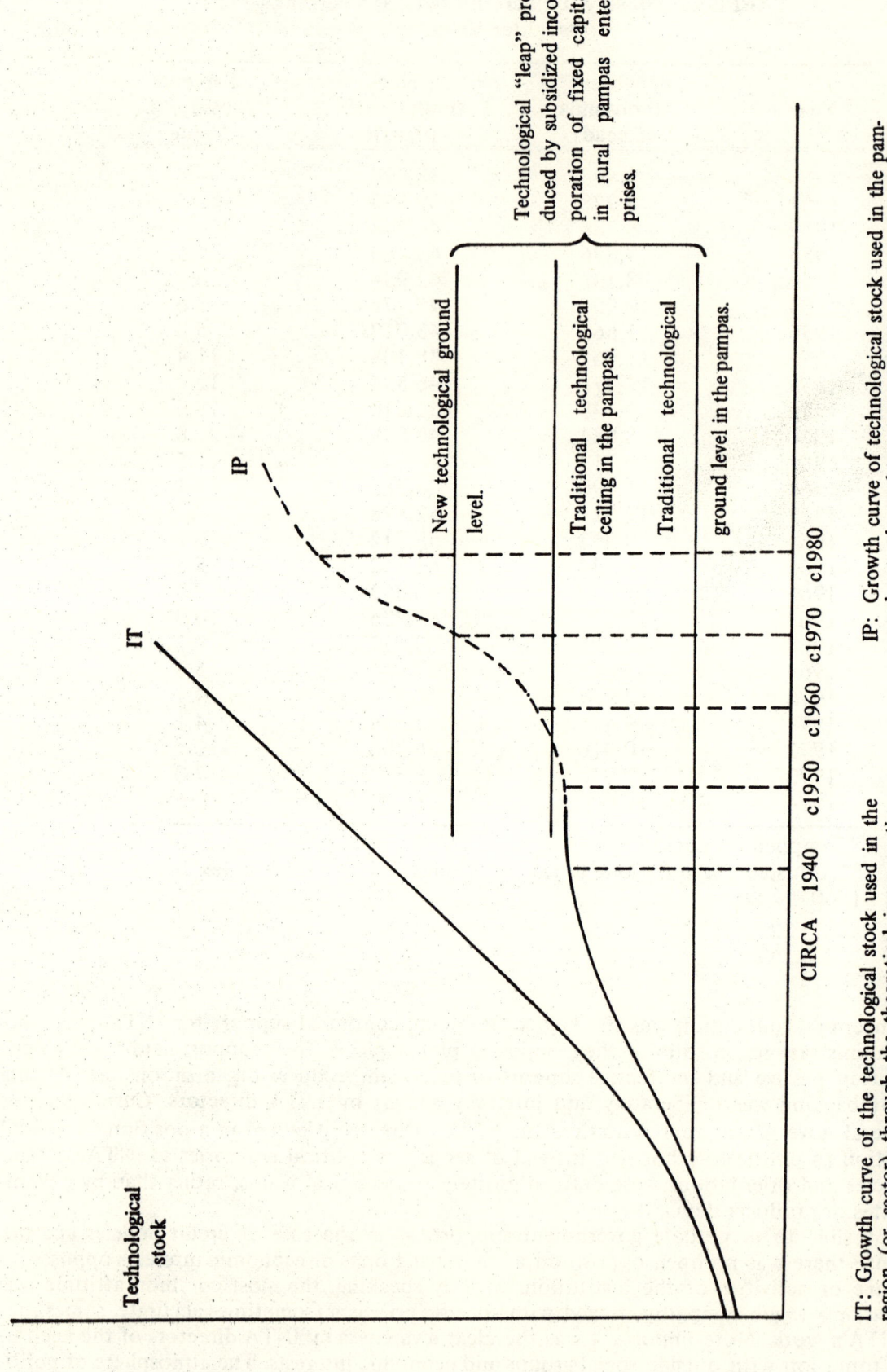

Fig. 11. The new phase of technical progress in the pampas.

the fourth and last condition was met for government action to be effective and achieve continuity. On this point, however, there are a number of interesting considerations that must qualify any judgement of the effectiveness of the technology supply, and that provide an explanation of certain severe problems that plagued INTA.

Because enterprises continued to operate on the strategy of combining production lines, their technological demand continued to have a peculiar bias and a special logic. For the farmers, productivity increases available on a specific production line were conditioned by what happened with other activities and therefore did not in themselves assume high priority. Meanwhile, INTA concentrated on producing a technology supply operating under totally different criteria. In fact, it was implicity assumed that pampas producers acted according to the same basic patterns as farmers in the United States and Canada, or, more generally, held the normal attitudes of a capitalist farm enterprise trying to maximize profits by boosting the productivity of available factors. Therefore, INTA adopted an internal institutional organization based on the successful experiences of similar institutions in other regions. Consequently, two major services were established: research and agricultural extension. The research branch worked on the development and adaptation of innovations by product, favoring technologies that would, above all, boost the productivity of factors. The extension service was seen as a conveyor belt between researchers and producers. It was a two-way system for translating and communicating information by which supply could be accommodated to demand, and vice-versa.

The system proved frustrating due to the differences between the approach used by INTA and the model used by producers. In many ways, the two were incompatible, and the attempt to accommodate demand to supply was an illusion. This produced a series of diverse consequences.

In the first place, the adoption of technology by producers struck INTA technical personnel as totally erratic and unpredictable. Unaware of the logic that lay behind producer selections, they were unable to understand why some techniques were adopted so quickly, others were rejected, and still others, after an unfathomable apathy, suddenly caught on. The producers could not understand why INTA technical specialists spent so much time developing and disseminating techniques that did not interest them. Finally, neither could see the cause of these incompatibilities, which they attributed to other factors.

In INTA, the phenomenon produced serious problems between researchers and extension agents and, in general, a growing loss of coordination among experiment stations and with central research services. The most common symptom was an ever-increasing near-sightedness by researchers, and a proliferation of bureaucratic mechanisms to attempt to coordinate disparate worlds.

A more serious result was that, in spite of INTA's unquestionable contributions to the producers, the establishment of any close, effective ties, such as those postulated by Hayami and Ruttan in their induced innovation model, was hindered. The institution lacked the kind of active political support that was provided in other countries by federated producer organizations. Thus, INTA's survival and funding from public resources continued to depend on the willingness of the government to ascribe continued importance to technology in general.

CONCLUSIONS

The ideas and hypotheses developed above are useful in offering an interpretation of *what* has happened in the agricultural sector of the pampas in recent decades, and *how* it happened. It would now be helpful to recapitulate briefly and clarify certain points.

A. Three main conclusions can be drawn from the examination of enterprises and production on the pampas:
   (1) Beginning at the turn of the century, a model of productive combination was introduced in the pampas region. It was generated and sustained by the strategy of mixing activities in order to cut income risks; over the long term, it tends to restrain technical progress, enterprise growth and production expansion.
   (2) During recent decades, this traditional productive model has been preserved, but the type of social organization in which it was practiced underwent change. The complex combination of landowning *estancias,* with their livestock operations, renting *chacra* farmers, and migrant labor gave way to *chacra* and *estancia* enterprises containing all three functions.[10]

(3) During this period, there was also a sharp rise in the "ground level" of technology used on pampas enterprises. This change was initiated by a process of mechanization, which replaced labor and allowed for the transformation in social organization discussed above. The mechanization took place through the operation of a policy of subsidized growth and tax cuts. These financial measures produced new possibilities for technical progress, which were satisfied with an increase in the technology supply.

B. These three conclusions suggest a number of facts and empirical observations that form a consistent picture:

(1) The idea of a model of productive combination that was maintained while the technological leap took place is consistent with the description of producer behavior given earlier. The swift reaction to relative price swings and to changing environmental conditions is reflected in the tremendous irregularity of annual land use decisions, which is symptomatic of the strategy of combining activities. At the same time, an examination of medium-term trends clearly shows production progress moving through distinct phases of adoption of innovations. If these phases are linked together, it is clear that the technological leap was in large part a qualitative change taking place inside the enterprises.

(2) If each enterprise individually assumed the model of productive combination, it is reasonable to assume that the dimensions of land in use were broadly revised. Accordingly, census data from 1947 to 1969 show that the mean farm size grew from 100 to 1,000 hectares (and especially from 500 to 1,000 hectares), and this cannot be attributed merely to the disappearance of traditional rental arrangements. It further suggests that these particular dimensions probably optimize the use of the strategy of productive combination, in accordance with the resources and technologies currently available in the corn growing zone and other similar areas.

(3) In any case, it is clear that the pre-existing division between *chacras* and *estancias* continues to be important. The hypotheses proposed in the second part of this paper suggest that these differences had a profound influence (through the external opportunity cost of capital) on the use of production factors and the composition of the prevailing productive basket. This is consistent with the differences observed between the more heavily *chacra* zones and the *estancia* zones in the corn belt. However, both systems practice essentially the same behavioral model and for this reason neither can be singled out as having provided the momentum for technical progress. Indeed, initiatives for change alternated among the different subzones and eventually spead throughout the region.

(4) Production progress also improved the financial standing of *chacra* enterprises, which facilitated the incorporation of innovations. One important result was that, as they continued to practice criteria similar to those of the holder of a portfolio of bonds, the *chacra* farmers found that their greater financial capacity provided opportunities for placing surplus outside the enterprise; thus, they assumed behavioral patterns similar to the traditional attitudes of *estancia* owners. The proliferation of urban industries and activities in the towns and cities of the pampas appear to support this idea. It could also explain an increasingly intense rural-urban interaction that has occurred in the region.

C. Finally, the major conclusions reached suggest a number of consequences. Two are of particular importance:

(1) The alterations in the traditional social organization of the productive model eliminated certain causes of conflict, thus attenuating the clashes and divisions among the federated producer movements. At the same time, when each enterprise incorporated the model of productive combination, and the size of the enterprises changed, it became possible for technological demands to take a more homogeneous form (even though the biases and logic inherent in the strategy of combining activities continued to exist).

(2) To the extent that the model of productive combination is preserved, it is very possible that the production growth experienced as of 1960 will reach a new "ceiling" and will gradually slow down. In other words, as occurred before, the behavior of pampas producers will prevent improvements in the use of available factors, at least from the vantage point of the society at large.

D. The study underscores three phenomena related to the interpretation of how pampas agriculture was transformed:

(1) The agricultural sector in the pampas showed very little initiative for making internal changes, and this indicated poor ability to transform its model of productive combination.
(2) The process through which changes came about was characterized by confusion and contradiction. It was part of a profound alteration of the entire society and economy of Argentina, which triggered a recession in pampas agriculture (by changing the structure of employment) and required growth in production (to sustain the irreversible transformations that were taking place). However, it stood in the way of establishing sound, stable foundations for such an expansion of production, by upsetting the pre-existing system of social relations without providing any strong alternatives for replacing it. These contradictory environmental conditions were further compounded by internal divisions and confrontations throughout the pampas agricultural sector, stemming from earlier forms of social organization. Clearly, it would have been nothing short of miraculous if the processes in agriculture (and in almost every other realm) had been anything less than confusing and inconsistent.
(3) In spite of the acute political instability generated by this situation, and as a result of it, the most energetic initiatives for inducing agricultural change in the pampas were to emerge from the different government administrations. The reason for their continued interest was simple: production growth was viewed as indispensable for political survival. Similarly, it was clearly perceived that the key to the problem lay in technological change.

E. The confluence of all these circumstances leads to three important conclusions:
(1) The confusions and contradictions that reigned during the period discussed surrounded the economy with debilitating uncertainty and high risk. In pampas agriculture, this reinforced the preservation of the model of productive combination as a means of protection.
(2) Due to the reciprocal veto power held by the different social groups and economic interests, the only policies capable of having any continuity were those that aroused a minimum of opposition. At the same time, lasting positive effects could be exerted only by policies that were consistent with the prevailing model of productive combination. This two-way, implied filter allowed the measures of subsidized credit, tax cuts and technology supply to achieve their impact.
(3) During the period, political leaders were willing to grant considerable importance and high priority to pampas agriculture chiefly because it stood as one of the great bottlenecks blocking the smooth operation and renovation of the society and economy of Argentina. This problem held more weight than the traditional power and influence enjoyed by the great land owners and, no doubt, was more significant than the activities of federated producer movements, divided and in conflict.

F. This state of events, which emerges from the preceding analysis, was decisive in events of recent years. Indeed, it follows that as production grew because of the changes analyzed, and the foreign sector was relieved of its problems and ceased to exert pressure on the governments, pampas agriculture paradoxically began losing priority. The following factors are indicative of just such a phenomenon:
(1) Pampas enterprises have suffered a severe loss of income and benefits since 1976/1977, due to the revaluation of the Argentine *peso* and the resulting alterations in exchange relations.
(2) The subsidized credit was suspended during the monetary reforms of May, 1977. After that time, equipment and machinery purchases dropped suddenly and, in general, all fixed capital investments declined.[11] Similarly, it should be stressed that tax pressures on the rural areas were sharply increased.
(3) In 1980, the special tax that had granted financial autonomy to INTA was suspended. This tax had assured the Institute a stable, growing flow of resources in a country characterized by instability and crisis.

In the final balance, all the different points recapitulated above lead to a pessimistic prognosis for the long term. Pampas agriculture continues to be governed by an operating model that hampers the capacity for growth and change, that fails to optimize society's use of available resources, and, due to technological growth promoted from outside, has ceased to be conflictive, and therefore inspires no new external incentives for continued progress.

NOTES

[1] The direction of these trends varies according to the period used as a base for calculating indices, due to variations in relative prices. The Reca series has been selected for this study because, of all those available, it uses the most consistent methodology for the entire period. Furthermore, the Reca 1935–1939 base is more useful for placing the stagnation in an *ex ante* perspective. Given subsequent relative price development, the 1935–1939 base accentuales the fall in agricultural production by comparison, for example, with the ECLA series, which use 1950 prices as a base line.

[2] Pampas livestock, because of its dual nature, cannot be studied exclusively in terms of productive economic features. This would deny its financial role, which must be included in order to explain several facts that appear paradoxical or contradictory. For example, when prices are low, sales increase, and when prices are high, sales fall. This heightens price movements and causes wide swings in stock inventories and overall production capabilities, thus accentuating the so-called livestock cycles, so intensive in Argentina.

[3] Plotting these curves, however, makes it difficult to see the possibility of function C' providing a better unit cost/benefit ratio and a better rate of earning for average price $\bar{P}$.

[4] This reasoning can also be applied to variable costs, although the effect is less intensive. Capital used for variable expenditures, which are periodic, is more flexible and lowers the opportunity cost of production alternatives more than capital invested in installations or machinery.

[5] The importance of this conclusion resides in the fact that it reverses certain causal relationships that have been used to explain events in the pampas. For example, de Janvry and Martínez (1972) believed that the bias in favor of land intensive technology was due to the presnece of large landowners for whom land was the most abundant factor. Thus, they defined a *real demand* biased toward the adoption of land intensive technologies, in spite of the *potential demand* that theoretically existed in the region and that favored more economical land use. The conclusion given above offers a contrary explanation. The bias for land intensive technologies, for reasons discussed in the hypothesis, made it possible to preserve and consolidate the enterprises of large landowners. In other words, the presence of large landowners cannot be used as a basis for explaining events in the region. Rather, it is an effect which must be explained in terms of how the region operated.

[6] During the nineteenth century, the lands in the pampas were generally measured in leagues, or around 2,500 hectares. The *estancias* typically contained from one to three leagues, and on occasion, up to ten or 20. In general, this was equal to 50 or 100 times more land than the plots turned over to farmers in the settlement projects that were unsuccessfully attempted on several occasions. The *chacras* rented to farmers by the *estancia* owners ranged from 100 to 250 hectares, or on the average, around five times as much as the *chacras* operated by land-owning settlers.

[7] In 1912, Emilio Lahitte stated that "renters were rural capitalists. Typically they had made personal investments in equipment, oxen and horses, in addition to their broad backs... Frequently they worked alone and had enough savings to buy a small farm. They were so hopeful of increasing their capital through extensive agriculture that they would sooner rent 200 hectares than own 20."

[8] Each national agricultural census different criteria for defining temporary workers. For this reason, it is difficult to give exact numbers or compile inter-census comparisons.

[9] Flichman (1977) and others have shown that the number of seeders tripled and harvesters quintupled from the 1914 census to the 1937 census. It is difficult to say whether this mechanization was a reaction to the loss of temporary workers, to technical improvements in equipment which reduced production costs, to expanded farm areas, or, most probably, to the convergence of all these factors. In any case, the changes appear not to have significantly altered the operation of the model described earlier. Fixed capital investments per hectare continued to be very small, and the forms of production had not changed significantly since the turn of the century. Technical estimates on hours of labor required for each crop in the early 1930s are scarcely less than those produced at the turn of the century.

[10] The descriptions of *chacras* and *estancias* hardly varied from 1900 to 1940, but both types of enterprises today present a striking contrast with what they were in 1940. This is clear, for example, in the recommendations for land use and rotation with livestock on *chacras* containing less than 100 hectares of land (White, 1979), or in descriptions of the operation of today's *estancias* (Capdevilla, 1978).

[11] This measure proves that the equipment and machinery suppliers (including multinational corporations) had limited influence over the establishment of subsidized lines of credit.

TECHNOLOGY AS A SOCIAL ISSUE: AGRICULTURAL RESEARCH ORGANIZATION IN LATIN AMERICA

Eduardo Trigo, Martín Piñeiro and Jorge Federico Sábato

ABSTRACT

A summary of the characteristics and evolution of the national agricultural research institutes in Latin America is presented. The role of public organizations in research and the dissemination of new production techniques in capitalist economies is discussed. The dominant institutional model is analyzed from an historical perspective, and its functionality in a Latin American context is debated.

INTRODUCTION

Technological change in the agricultural sector has an unusual characteristic: public organizations participate explicitly and decisively in research and the dissemination of new production techniques, even in those countries with a capitalistic economic organization.

There are several reasons for this. In general, the search for new knowledge and techniques requires substantial resources, without any guarantee of success; even more important, few cases have the commercial mechanisms needed to recuperate expenditures in research and development, or to ensure adequate profit to compensate for the risks incurred. This is evident in the process of developing basic knowledge which can not be used directly in the productive process, but is essential for creating new production techniques. This situation also involves a variety of techniques which the farmer may apply immediately, such as perfected cropping methods and management techniques. In these cases, it is almost impossible to charge for their use, even though they may be very costly to develop.

It also explains why a substantial amount of the research and development of new agricultural techniques is seldom taken on by private business, or even by the farmers themselves, who should be interested in having more productive techniques available. Contrary to the case of industry, where it is common for marketing and production conditions to favor the development of large enterprises, agricultural production units are seldom of the size necessary to attain the economic capability needed for developing new techniques. An additional and frequently important element is the fact that the farmer, because of the nature of his products and markets, can seldom differentiate his products on the basis of trademarks, as does industry. Thus, the farmer's possibilities for developing a "captive" market are significantly reduced, as are his chances for recuperating what he has spent on investments in developing new techniques and products.

Although these factors hinder private enterprise from researching and developing new techniques, there has been considerable social and economic pressure on the agricultural sector to increase its productivity. Experience over time in several countries has demonstrated that growing population and economic development must be based on greater amounts of inexpensive food and raw materials of agricultural origin. Given the difficulties encountered by private business in producing new agricultural technology, even in capitalistic countries, social mechanisms have emerged in response to the demand for these new technologies, thereby encouraging the active participation of public institutions in this sphere of action.

During the 19th century, Germany was the first country whose public sector became involved in developing agricultural technology. Here, the State strongly supported technical progress, creating the necessary conditions for applying knowledge and the scientific method in a broad, systematic way to improve agricultural production (Ruttan, 1982). The success of this initiative contributed importantly to the transformation of German agriculture, and it was rapidly emulated in many countries, including Japan and the United States of America, where it led to the creation of the Land Grant Colleges (Hayami and Ruttan, 1971) and other innovations.

Several attempts along these lines were made in Latin America towards the end of the 19th Century. But the larger public institutions in the countries of the area did not become seriously involved in research and dissemination of new agricultural technologies until the 1950s. These institutions have served as the main sources of technology for the agricultural sector for the past 25 years, and have been influential, directly or indirectly, in all aspects of the technical change process. At present, these agencies are showing many signs of institutional crisis. Similarly, there is a growing awareness of the need for revising the institutional model used to date, and the role which the State plays or should play in this regard.

This paper sumarizes the characteristics and evolution of the institutional model which prevails throughout the region, that is, the national agricultural research institutions. Additionally, it analyzes the model in historical perspective, commenting on its functionality within a Latin American context.

THE CREATION OF THE INSTITUTES

The institutional development of research and transference of agricultural technology throughout Latin America, particularly in South America, has been characterized by two well-defined stages which differ mainly in the quantitative effort put into research and in the degree and manner of participation by the public sector. The first dates from the beginning of research endeavors in the region, during the second half of the last century to the middle of the 1950s, a time when low levels of resources were assigned to research, largely due to the unstable, *ad hoc* institutional mechanisms that prevailed.

The first experimental stations geared to the generation and transfer of new technologies for important products began to be set up during this first stage, during the 1930s and 1940s. However, the general situation continued to be unstable, with the responsibility for station activities and the financing of their research changing hands frequently. Throughout this process of change, the agricultural universities and schools lost much of the importance they had had during the early years of the century. The process culminated in an almost total centralization of research under the auspices of the Ministries of Agriculture.

This institutional arrangement suffered from a number of problems and constraints, mostly products of the bureaucratic nature of the Ministries. The most important of these constraints were: the lack of stable budgetary support; scanty articulation with problems in the rural areas or with farming priorities; a dispersion of efforts; the lack of adequate linkage between research, technical assistance and extension; and the lack of coordination between the agency responsible for the generation of technologies and those responsible for implementing other components of agricultural policies needed to ensure the effective development of the production process, like prices, credit, and services. (Trigo, Piñeiro and Ardila, 1979; Samper, 1979).

In the mid-1950s, several elements substantially modified this situation, thereby initiating the second stage of the process, characterized mainly by a strong tendency toward descentralizing research activities and sometimes even the technology transfer efforts from the administrative structure of the Ministries. Descentralized, more autonomous institutions became responsible for these actions under plans organized along the lines of the experimental station system of the United States.

The new institutional model was rapidly adopted because: (a) it was becoming increasingly evident that technology is a major element of agricultural development; and (b) of the wide range of technologies available internationally for use by the Latin American farmer. The main problem was the transfer of this knowledge, available in developed countries, to the undeveloped nations. Adequate research infrastructures for adapting these technologies became essential, since the Research Offices of the Ministries of Agriculture were incapable of doing so efficiently.[1] This second stage had strong backing from external sources, including both technical and financial assistance which made possible the development of the physical infrastructure and human resources needed to run the new institutions.

The following institutions were set up during this stage: The National Agricultural Technology Institute (INTA) in Argentina, in 1957; the National Agricultural Research Institute (INIAP) in Ecuador, in 1959; the FONAIAP complex (National Agricultural Research Fund) in Venezuela, in 1959 and expanded in 1961; the National Agricultural Research Institute (INIA) in Mexico, 1960; the Agricultural Research and Promotion Service (SIPA) in Peru; the Colombian Agricultural Institute (ICA), in Colombia in 1963; and the Agricultural Research Institute (INIA) in Chile, in 1964. The guidelines followed in setting up these agencies were similar: the juridical-administrative character resembled that of a decentralized, autonomous public organization, with the integration of research, transfer functions, and regional projection serving as bases for the development of their activities.[2]

This tendency towards modernizing the institutional infrastructure of agricultural research also covered situations where no new agency was created. A case in point is Uruguay, where in-depth changes were introduced into the Alberto Boerger (CIAAB) Agricultural Research Center operations, although the Center was directly linked to the Ministry of Agriculture. The changes influenced the pertinent aspects of generating and transferring new technologies, relating them to post-graduate education within the Research and Training Center for the Temperate Zone, sponsored by IICA at the beginning of the 1960s. In Brazil, efforts to strengthen research and technology transfer activities were also carried out within the existing institutional structure during that time, since the Brazilian Agricultural Research Institute (EMBRAPA) was not created until 1973. Although similar to the other institutions in some aspects, EMBRAPA has a number of very important differences, among them: the exclusion of extension, a responsibility assumed by a sister institution, the Brazilian Technical Assistance and Rural Extension Institute (EMBRATER); and the implementation of a multi-organizational model with participation of the different administrative levels of the public and private sectors (national and state), to set priorities and establish objectives under the general coordination of EMBRAPA. On the basis of these differences, it may be asked if EMBRAPA followed the institutional model implemented during the 1960s, or if it represents a new approach that redefines the roles played by the State and the public and private sectors, as they affect the process of generating and transferring technology.[3]

The institutional model on which the institutes were based—in addition to the characteristics of decentralization and autonomy—includes service to a wide range of products, regions and types of farmers. In this sense, the established model represents the maximum expression of the concept of agricultural technology as a public responsibility.[4]

Budgetary Expansion and Institutional Crisis

Together with the creation of these institutes in the 1960s, a strong expansion in research and technology transfer activities backed by growing international support, and bigger budgets of national origin, can be noted. Two concrete expressions were the process of "territorial occupation" through the creation of new experiment stations and extension agency networks, and the initiation of ambitious personnel training programs including, in some cases, the development of post-graduate education infrastructure at the national level (in Argentina, Colombia, Peru, Uruguay, Mexico and Brazil). This expansion is amply confirmed upon analyzing the evolution of the human and financial resources allocated to agricultural research in the public sector.

Tables 1 and 2 describe the evolution of human and budgetary resource allocation for selected years with the 1960—1980 period, in the main sub-regions of Latin America and the Caribbean.[5] The aggregate information obtained is clear, but can also disguise institutional situations which

TABLE 1 Budgetary Resources Allocated to Agricultural Research in Latin America and the Caribbean, between 1960 and 1980, Selected Years (Constant Value of 1975; Official Money Exchange Rate: National Currency/US dollars, for Year Selected)*

SUBREGION[1]	1960	1965	1970	1974	1980
Southern zone (excluding Brazil)	31,446[2]	31,298	32,594[3]	44,702[4]	42,559[5]
Brazil	8,280[6]	15,533[7]	24,178[8]	32,879[9]	116,797
Andean Zone	15,631[10]	20,003[11]	43,056[12]	57,393[13]	60,541[14]
Panama and Central America (excluding Mexico)	4,412[15]	4,967[16]	4,904[17]	5,961[18]	10,215
Mexico	4,666[19]	5,218	9,723	14,637[20]	48,357[21]
Caribbean (excluding Dominican Republic)	1,530[22]	1,530[23]	3,280[24]	2,940[25]	2,128[26]
Dominican Republic	441[27]	496[27]	490[27]	2,278[28]	1,642
Latin America and the Caribbean (total)	66,406	79,045	118,225	160,790	282,239

*Preliminary figures, currently being adjusted (Trigo and Piñeiro, 1981: Appendix 1).

[1] Southern Zone includes Argentina, Uruguay, Paraguay, and Chile. Andean Zone includes Bolivia, Peru, Ecuador, Colombia and Venezuela. Central America includes Costa Rica, Nicaragua, Honduras, El Salvador, and Guatemala. Caribbean includes Guyana, Suriname, Jamaica, Haiti, Barbados, Grenada, Trinidad and Tobago.
[2] Information for Chile is from 1961.
[3] Information for Paraguay is from 1971.
[4] Information for Chile and Uruguay is from 1973; for Paraguay from 1972.
[5] Information for Argentina is from 1979.
[6] Information is from 1962.
[7] Authors' estimate, based on figures supplied by Boyce and Evenson.
[8] Information is from 1972.
[9] Information is from 1973.
[10] Information for Bolivia, Venezuela and Peru is from 1962; for Ecuador from 1965.
[11] Information for Bolivia is from 1962.
[12] Information for Bolivia and Venezuela is from 1972 and 1969 respectively.
[13] Information for Bolivia and Ecuador is from 1973; for Venezuela and Peru from 1976.
[14] Information for Colombia is from 1979.
[15] Information for Nicaragua and Guatemala is from 1962; for Honduras from 1963.
[16] Information for El Salvador is from 1966; for Guatemala from 1962 and Panama from 1961.
[17] Information for Honduras and Nicaragua is from 1965; for Guatemala from 1973; for Panama it was estimated as US$600,000.
[18] Information for El Salvador is from 1973; Honduras from 1976 and Panama from 1975; for Nicaragua it was estimated as US$1,000,000.
[19] Information is for 1962.
[20] Information is for 1972.
[21] Information is for 1979.
[22] Information for Barbados, Jamaica, Suriname, Grenada, Trinidad and Tobago is from 1965; for Guyana it was estimated as US$250,000.
[23] Same information as 1960.
[24] Information for Barbados, Jamaica, Suriname, Grenada, Trinidad and Tobago is from 1972; for Guyana from 1973 and for Haiti from 1976.
[25] Information for Barbados and Haiti is from 1976; for Jamaica Trinidad and Tobago from 1972.
[26] Information for Haiti is from 1978; for Suriname and Grenada from 1974, and for Guyana from 1978.
[27] Information was estimated on the basis of 10 per cent of the totals for Panama and Central America.
[28] Information is for 1977.

TABLE 2 Human Resources (Professional Personnel) in Agricultural Research in Latin America and the Caribbean, from 1960 to 1980 (Selected Years)*

SUBREGION[1]	1960	1965	1970	1974	1980
Southern Zone (excluding Brazil)	365[2]	816	1,045[3]	1,196[4]	1,364
Brazil	200[5]	500[6]	764	2,000	2,935
Andean Zone	387[7]	643	1,294	1,694	1,843[8]
Panama and Central America (excluding Mexico)	144[9]	305[10]	283[11]	333[12]	383
Mexico	190[13]	279[14]	551	1,000	1,079
Caribbean (excluding the Dominican Republic)	64[15]	96	157[16]	228[17]	198[18]
Dominican Republic	3[19]	5	12[20]	35[21]	99
Latin America and the Caribbean (total)	1,353	2,644	4,106	6,486	7,901

*Preliminary information, still being analyzed (Trigo and Piñeiro, 1981: Appendix 2).

[1] Southern Zone includes Argentina, Uruguay, Paraguay, and Chile. Andean Zone includes Bolivia, Peru, Ecuador, Colombia and Venezuela. Central America includes Costa Rica, Nicaragua, Honduras, El Salvador and Guatemala. Caribbean includes Guyana, Suriname, Jamaica, Haiti, Barbados, Grenada, Trinidad and Tobago.
[2] Information for Argentina, Chile and Paraguay is from 1959.
[3] Information for Paraguay is from 1971.
[4] Information for Chile is from 1973; for Paraguay it was estimated at 37.
[5] Information is for 1959.
[6] Information is for 1967.
[7] Information for Bolivia, Ecuador and Peru is from 1959.
[8] Information for Colombia is from 1979.
[9] Information for Honduras and Nicaragua is from 1959; for Guatemala it was estimated at 20.
[10] Information for El Salvador and Guatemala is from 1966.
[11] Information for Honduras, Nicaragua and Panama is from 1971; for Guatemala, from 1972.
[12] Information for El Salvador is from 1973; for Costa Rica and Guatemala it was estimated at 64 and 58 respectively.
[13] Information is for 1959.
[14] Information is for 1966.
[15] Information is for 1959.
[16] Information is for 1971.
[17] Information for Trinidad and Tobago is from 1971.
[18] Information for Trinidad and Tobago is from 1978.
[19] Information is for 1959.
[20] Information is for 1971.
[21] Estimated.

differ. Thus, information for several countries such as Mexico, the Dominican Republic and Brazil is given separately, as otherwise it tends to overwhelm the totals for each subregion.

In general terms, a marked increase in the human and financial resources allocated to agricultural research characterized the region. However, the results differ somewhat if the subregions are analyzed separately, and even more so if countries are taken individually. The following aspects are noteworthy at the subregional level: (a) a changing trend in the Southern Zone, where budgets for the four countries, which had peaked in 1974, declined more than 8 per cent by 1980; (b) a slowdown in the Andean Zone during the second part of the 1970s, when aggregate budgets continued to grow, but at a slower rate than that of the previous five years;[6] (c) decay and stagnation of budgetary growth over the entire period is evident for Central America, Brazil and Mexico;[7] and (d) the Caribbean closely followed the South American trend, although the variable quality of available information prevents definitive, reliable analysis.

The situation for human resources available for agricultural research is similar to that of budgetary resources, although there are some differences. For instance, subregional totals for the Southern Zone continued to increase after 1974.

When financial and human resources are analyzed jointly—resources available in terms of man/years—the picture is somewhat similar to that of each indicator taken individually. For the Southern and Andean subregions, the trend indicates stagnation or even decline. Existing information shows that the resources available per man/year dropped slightly between 1970 and 1980 in the Southern Zone, but were sustained at practically the same level throughout that time in the Andean Zone. Dramatic increases are evident for Panama, Central America, Brazil and Mexico.

When the situation is analyzed at the individual country level, the institutional instability and deterioration of the institute model appears more clearly.[8] Of the 16 countries with sufficient statistical information available, over half had budgetary peaks between 1970 and 1980, decreasing thereafter. Specific examples of this tendency in the different subregions are Costa Rica and El Salvador in Central America; Colombia, Ecuador, Venezuela and Peru in the Andean Zone; and Argentina and Uruguay in the Southern Zone. In some cases, there are differences of as much as 50 per cent between one extreme and another, as well as marked annual variations, indicating a highly unstable situation.

The human resources picture is somewhat different and, with the exception of isolated annual problems, there is little evidence of sharp trend changes. This implies that most of the countries mentioned above have experienced a reduction in resources per man/year for agricultural research.[9] It is interesting that many of the countries suffering from marked instability were among the first to develop their institutional models (Argentina, Uruguay, Peru, Ecuador, Colombia, Venezuela, among others).

This reduction in available financial and human resources indicates a change in the institutional model used since 1960 (decentralized institutes) which, while sustaining State control, was geared towards channeling greater resources to agricultural research and technology transfer, through more flexible institutional structures than those in existence at the time. Although limited, available data indicate that since the start of the 1970s the institutes have lost much of their institutional strength. It seems contradictory that this period of reduced budgetary support—perhaps the most important element in the crisis outlined above—should coincide with a marked expansion in the agricultural sectors of the region. Below, we attempt to explain this phenomenon as a result of the social implications of the national institution model used, in response to current problems and the realities of the region.

INTERPRETING THE EVOLUTION OF INSTITUTIONAL ORGANIZATION IN AGRICULTURAL RESEARCH THROUGHOUT LATIN AMERICA

In the previous section two aspects were stressed in analyzing the development of research institutions in the public sector. One was the similarity in the different organizational forms as they appeared over time in the different countries, including the development of the autonomous, decentralized institutional model. Another was the manner in which these organizations have evolved over the past two decades, with growing instability in countries like Argentina, Colombia and Uruguay, and sustained growth in others, such as Mexico and Brazil. An effort is made in this section to link these aspects with certain structural characteristics encountered in the countries under study, and to show how technology evolved as a social issue over the different periods described.

Technology as a Social Issue, and the Institutional Nature of Agricultural Research

Two aspects related to the nature and role of technology in society tend to make it a non-conflictive issue (O'Donnell, 1976). In the first place, technology is one of the bases for economic development, and is closely associated with the concept of "progress" (Oslak, 1978:48). Hence, technology is considered "beneficial" from a social point of view, and is seldom opposed by any particular sector. In the second place, the impact of technological process is sometimes scarcely perceived by those segments of society most closely related to production. The so-called "treadmill" situation is a good example of this. Here, as production increases due to the adoption of new technologies and prices drop, microeconomic behavior leads to an even faster rate of adoption, which in turn induces further drops in price and a reduction in the original margin of profit. The reason for this is that, at the micro-level, indicators of the impact of technical change are frequently overshadowed by aspects such as prices, credit availability, marketing alternatives and others, which directly effect income levels and are therefore highly visible (Cochrane, 1958).

These effects of technological change tend to hinder decision making concerning research, except when there is a sudden crisis, such as unexpected production shortfalls which cannot be attributed to climate. The "temperature" of technology, or better, its lack of "temperature" as a social issue, however, is not entirely isolated from the historical context of the agricultural sector, especially with regard to production possibilities and institutional structures. The adoption of technological change will be less important where there is unrestricted availability of natural resources to facilitate the horizontal expansion of production. Although the impact of technology may not be so evident at the production unit level, it certainly is felt at the higher institutional levels such as producer associations. Thus, as the institutions in the agricultural sector develop and are more capable of identifying and expressing technological demands as an aggregate, the more important technology will be as a social issue.

The sequence of institutional expansion and decay described above, in connection with the national research institutes, and the trends that have been observed in the different countries of the region can be explained in terms of the manner in which technology becomes a social issue; that is, how the perception of different social segments has evolved in relation to the role of technology within the development process.

Early Homogeneity in Institutional Models

During the first half of this century, most Latin American countries were concerned mainly with establishing their boundaries and consolidating their territorial possessions. Production expansion possibilities were augmented through the incorporation of new lands, and there was little pressure on food or export production. At the same time, the institutional development of the public and private sectors was just getting underway. Thus, the need for technology was not an active topic of discussion, at least up to the beginning of the Second World War, after which technological advancements were increasingly in the public eye. Institutional development responded to specific crises: in cotton, in Cañete, Peru; the sugar cane mosaic disease in Palmira, Colombia; changes in the markets of tropical crops (cacao, rubber, etc.), as a result of the war, for the Central Station in Ecuador, and others.

The lack of demand for specific technologies and the slow development of State bureaucratic structures facilitated the adoption of institutional models from other countries. These models were frequently introduced by scientists brought in from abroad to strengthen extremely weak national research and educational structures at the beginning of the century. This partly explains the rather uniform development of institutional models in the different countries of the region (Marzocca, 1967; Elgueta, 1967; Olcese, 1967).

From 1940 on, technology generation and transfer activities began to be incorporated within the organizational structure of the State. Aspects of this process were the disappearance of unlimited expanses of land that accomodated horizontal expansion for purposes of agricultural production, and the progressive change in the role played by the agricultural sector within the overall context of the national economy, as a result of the industrialization process and the trend toward urbanization throughout Latin America. Both aspects induced changes in demand and in the structures of domestic markets for agricultural products, hence the increasing importance of technology.

During the 1950s, it became absolutely essential to increase agricultural production in a number of countries. In some, such as Mexico, Colombia, Peru and Ecuador, national production was rising at a rate well below the increase in demand resulting from population growth and the urbanization process. In others, such as Argentina and Uruguay, the stagnation of the agricultural sector generated balance of payment problems, which augured the appearance of even more serious difficulties as the industrial processes started to gain headway. In still other countries, such as Brazil, the situation of the agricultural sector was inextricably linked with both foreign trade and domestic demand problems.

The crises facing the Latin America countries became a source of concern to a number of the more advanced industrial nations, particularly the United States, since they seriously limited the possibilities for expanding their markets in one of the potentially most important regions of the world. In addition, as the industrialized nations saw it, political problems were beginning to aggravate the scene. The decline of exports and the inability to meet an increasing domestic demand for food became sources of social conflict and tensions with possible international repercussions. The United States could supply food to help solve extreme problems, but only on a sporadic basis; in the long run, such aid would generate other problems by affecting customary trade patterns, such as wheat exports from Argentina to Brazil and other Latin American countries.

At the international level, it was proposed that agricultural sectoral problems could be solved through the use of available technologies. To this end, however, it would be necessary to develop appropriate institutions, something that could not be done overnight. In order to make the most of the new technologies and to transform farming techniques, the agricultural sectors of Latin America would have to modernize (Schultz, 1964). This involved three inter-related aspects: (a) modifying the existing agricultural production structures and organization; (b) promoting the use of agricultural inputs; and (c) the use of improved technological know-how.

The first implied changes in the economic and social structures of agricultural holdings, ranging from the *latifundia* to the *minifundia* situations, both justifiably subject to agrarian reform. Other less obvious changes were also considered essential, including marketing conditions (prices and marketing of products and inputs), credit and fiscal incentives. All involved major changes in existing social and economic relationships which affected the interests of politically powerful social segments.

The second aspect, promoting and facilitating the work of firms responsible for providing technological inputs, also turned out to be a source of conflict. The United States favored setting up branches of transnational businesses, an attitude not shared by many sectors of the Latin American nations, whose resistance was due to the following factors. On the one hand, changing the socio-economic relationships of the former traditional production situation seriously affected the interests of certain social groups. Moreover, these changes would inevitably imply additional cost to the national economy (increased importation, transfer abroad of benefits and royalties, etc.) and other less visible changes, such as the appearance of new interest groups within the countries and a change in orientation in the development of the agricultural sector. Although this last aspect would be least noticeable at the beginning of the modernization process, it would become in time very relevant, since it implies the need for production factors not always available locally.

The last aspect—that most material to our discussion—refers to the management of technical know-how and the crucial role of the public sector in this area. A number of interrelated elements must be defined to help comprehend how this was resolved. Latin American countries could not afford to ignore the role played by science and technology in transforming agricultural production in developed countries, especially the United States in the latter part of the 1930s. The success of public institutions like the Land Grant System in disseminating new agricultural techniques was especially impressive. Latin American countries could not, however, allocate similar amounts of human or financial resources to this process. Just to acquire the necessary information and to adapt new agricultural techniques requires a basic core of scientists and technicians (geneticists, plant pathologists, and economists, among others) and of essential material resources (laboratories, training and research centers). It was therefore becoming increasingly evident, for economic reasons if for no others, that the public sector would have to assume responsibility for this type of work.

These needs contrasted markedly with the conditions prevailing in Latin America then and even now, in some cases: insufficient (or non-existent) human and material resources in the scientific and technological fields; weak, unstable agencies working in research, adaptation and dissemination of agricultural technological innovations; and often a weak and/or inefficient public apparatus, comprised of a number of institutions submitted to daily pressures and emergencies of all sorts, highly dependent on political aspects and with very low technical and administrative capabilities.

Inevitably, these circumstances suggested the need to adopt a single organizational structure, responsible for all the research needed (disciplines, products, etc.), rather than discrete agricultural institutions in the various countries. Moreover, creating an umbrella institution would: (a) permit optimum utilization of resources; (b) hinder duplication of effort in a given area; (c) ensure that certain areas or product lines of importance are not forgotten; (d) facilitate mutual cooperation and collaboration between programs; and (e) permit the joint development of various research tasks, taking advantage of possible scale and external economies in the research process. This idea had definite implications for organization; it became essential to have centralized control over the various activities undertaken, establishing priorities, and allocating pertinent resources. Such an approach, not always explicitly presented, was implicit in a number of important social processes of the time.

Social and economic transformations which occurred during this period led to a movement to modernize the State as an institutionalized system organized to exercise its authority within society to further the development process. This resulted in significant growth and considerable differentiation of various components of the State system, and in the careful selection of the social and economic aspects in which it should participate and the manner in which this should be effected. The importance and influence of ECLA concepts in this sphere became evident, as did the need for dissemination of ideas on planning and management with regard to economic and social policy aspects.

Briefly, then, the model being considered included United States approaches and initiatives, which coincided with trends surfacing in Latin America. Fundamentally it proposed, as a basic requirement for inducing change in the agricultural production process, the need for a techno-scientific "converter" to facilitate the adoption and dissemination within the sector of the stock of technologies already available at the international level. Concurrently, there was an increasing number of international technical and financial assistance programs—particularly of North American origin—that played a very important role in diffusing and implementing the "converter" idea. Some of these efforts, based on the Point IV idea, are the research and extension bilateral agreements already described, which functioned in 18 countries of the region (SIPA in Peru, STICA in Colombia, and others); other efforts included the Rockefeller Foundation programs in Mexico, Colombia and Chile, among others, and the technical missions from the different North American universities (Nebraska in Colombia; North Carolina in Peru, etc.), which formed the basis for later USAID programs (Coto, 1967; Elgueta, 1967; Haines, 1967; Krug, 1967).

This set of circumstances set the context of post-war development of the State apparatus, with the regard to the creation and dissemination of technology. The common denominator nature of technology as part of different, sometimes contradictory, solutions to agrarian problems, facilitated its dissociation from other conflicts facing the different social sectors involved. Thus, technology was subject to special State attention, since it was the only way of ensuring necessary continuity for attaining technological modernization of the agricultural sector. This led, first, to the centralization of activities within Ministries, then to the integration of research with extension, and, finally, to the creation of decentralized, autonomous institutes to serve as bases for the generation and dissemination of technology throughout the region.[10]

Factors for Model Differentiation

The marked similarity of the research institutions, despite obvious differences between countries of the region, can be explained, in part, by strong foreign influence and participation in the developing research activities. This, in turn, facilitated the generalized acceptance of the proposed model by the social sectors involved in the process.

However, the technological institutions that were established differed from those that had served as a model. In the first place, the National Institutes were based on a set of broad objectives which encompassed a whole range of regions, products and types of farmers. In addition, their very structure and formal linkages ensured their relative independence of specific production sectors and the respective cooperative organizations, and even of government control (at least with regard to origin of resources, processes for defining policies, and selecting priorities). As a result, these institutes took on a "general interest" orientation for the setting of priorities and allocation of resources, which frequently did not coincide with those of other agricultural interest groups. These trends, together with certain changes during the 1970s in perception of different social sectors with regard to technology, partially explain the evolution of the institutional model over recent years, including the period of "crisis" faced at present by some of National Institutes.

At the beginning of the 1970s it became evident that the available technologies on which the modernization strategy was based were quite successful in given situations, but they were not neutral in their effect on production situations. Moreover, available technologies were adapted readily only for certain types of production similar to those used in developed countries, where the technology was originated. This opened a wide range of possibilities favoring commercial agriculture, but also served to widen the gap between this sector and the "campesinos" who lacked immediately applicable, successful technological alternatives. The model was certainly effective as a means for transferring available technology at international levels, but did not seem suited to generating technological response to other local Latin American problems, based on the use of regionally available resources.

Reactions to this perception included the successive reorganization of ICA Colombia, and the progressive attempts to clarify institutional objectives for specific target groups; the attempts at regionalization of INTA activities in Argentina; and several of the reasons given for creating EMBRAPA in Brazil, as mentioned earlier (Piñeiro, Trigo and Fiorentino, 1977; PROTAAL, 1978; Trigo, Piñeiro and Ardila, 1982: Chapter 4). The increasing importance of setting priorities and organizing research in response to small farmers' needs also emphasizes the change in perception of the role to be played by technology with regard to the economic organization of society (Valdés, Scobie and Dillon, 1979; Gilbert, Norman and Winch, 1980; Norman, 1980; Trigo, Piñeiro and Chapman, 1981).

Another phenomenon of a more general nature is the diminishing need for a public "converter" system, resulting in three aspects of significance: the first is the appearance and maturation of a set of institutional developments, including the establishment of associations which facilitate the direct participation of certain sectors in the process of generating and disseminating new technologies. The Sugar Cane Growers' Association and Rice Growers' Association, both in Colombia, are examples of how the responsibility for introducing technology has moved from public to private domain.[11] Nonetheless, the level and nature of participation in the process depends largely on the type of guild associations prevailing in the country. The expertise to be provided, or problems to be solved, by public agencies will vary in accordance with the type of guild associations involved. The way in which these associations evolve, however, appears to depend largely on given characteristics of the agricultural sector and on the economy of the countries, such as regional or product specialization and type of production units. In Argentina, for example, the dominance of the pampean region and its lack of specialization created producers' associations with a high degree of internal heterogeneity, only poorly able to respond to technological queries. In Colombia and Ecuador, the absence of a dominant region and a trend towards a relatively greater degree of specialization contributed to the formation of guild-type associations which responded more efficiently to specific queries. This may partially explain the differences in the relationships of the public sector with producers' associations in Colombia, as compared to Argentina (Trigo, Piñeiro and Sábato, 1981; Piñeiro and Trigo, 1983:Chapter 10).

A second factor of importance is the development of the International Agricultural Research Centers (CYMMYT in 1966; CIAT in 1967; CIP in 1971). Although the original idea was that the work at these Centers should complement that of the national institutions, in fact they turned into practical alternatives for the "converter" system. Rice in Colombia is a good example of this problem; in this case the National Institute lost much of its initiative in rice research as a result of work conducted at one of the Centers.[12]

A third factor concerns the change in prevailing technology. During the initial stages, the technologies used seldom induced the type of benefits that could be appropriated by private enterprise, and therefore State participation was essential. However, agricultural innovations of the embodied type are increasingly becoming the main sources of technological change. Benefits better suited to private appropriation have encouraged private business to participate in technology generation and transfer activities. In addition, the monopolistic or oligopolistic domination of technological inputs markets (seed, fertilizer, agrochemicals, machinery) led to the consolidation of important sectors which have participated actively in defining the nature of the technologies supplied, thereby restricting the sphere of influence of public agencies.

Within this context, then, the signs of "crisis" identified at some of the technological institutions, and the differences in their evolution, reflect mainly the different characteristics of the social conflicts generated by technical change as the modernization process advances in each country.[13]

SOME FINAL THOUGHTS: THE NATIONAL RESEARCH INSTITUTES IN THE LATIN AMERICAN CONTEXT

The institutional model used predominantly throughout Latin America was based on two main ideas: (a) the need to develop a "converter" capable of adapting to the conditions of a region the technologies already available throughout the world, as a means for increasing production in the region; and (b) the need to ensure rational, maximum utilization of scanty technical and financial resources. These two ideas led to the creation of national institutes of an all-encompassing nature, organized along similar lines, even though they would be working under very different ecological, economical and socio-political conditions. A study of these institutes has shown that, in fact, they emerged as entities somewhat different than the model on which they were supposedly based, the Land Grant Colleges and Experimental Station System of the United States.

Main differences reside in the levels of specialization of the institutions. In Latin America, the need for better science and technology and optimum utilization of available resources deflected—in terms of target group coverage and functions—the federalized organization of the Land Grant System, where the regional decentralization of its institutions facilitated a strong socio-political linkage with the farmers in each zone, permitting farmers to effectively participate in the research program development and resource allocation processes. In the United States this is a fundamental characteristic of the model, and does not depend on temporary situations such as emergencies, or on specific production structures. At the same time, it serves to ensure that farmer interest is directly linked with the survival of the organization as a whole. This local specific interest and the formal linkages with regard to the setting of priorities and resource allocation are not found in the Latin American situation.

Moreover, the adoption by most countries of ECLA's national planning concepts and models led to the further isolation of technological institutions since they were viewed as instruments for implementing agrarian policies rather than service organizations for the sector. As such, many of them entered in direct conflict with interests of the prevailing agricultural groups; this tended to reinforce the inadequacy of the model.

The broad range of action of the National Institutes was also counter-productive, mainly due to the tremendous heterogeneity of agricultural production in Latin America. The "federal" model adopted in the United States was a response to the high regional heterogeneity of the American agricultural sector. Thus, each research and extension agency in the Land Grant College and Experimental Station System was confronted with a relatively homogeneous production structure, facilitating its linkage to the different farming interest groups. On the contrary, the single agency model adopted in Latin America tended to hinder such a relationship, responding as it had to the needs of a large number of different groups, frequently with conflictive interests.

The model was also relatively inflexible, adjusting with difficulty to the new multi-organizational context within which the institutes must function in order to survive. This lack of flexibility is due, again, to the all-encompassing nature of the institutes, which view the emergence of any new institution, whether private, farmer-owned or international, as direct competition. The problem has been further exacerbated by a dramatic change in the role of the public research institutions. In general terms, they have lost the centrality they had at the initial stages of the institutional development process; for some areas the process has gone even further, as the appearance of new organizations and the changing nature of technology have together eliminated the need and social legitimacy for public participation.

The increase in the number of different types of institutions working in technology generation and transference have given rise to two other problems. The first concerns the need for establishing operative mechanisms to ensure adequate linkage between, and complementarity of, the efforts undertaken by the public, semi-public and private organizations within the system. The second concerns the need for developing the capability for coordinating and guiding the activities of the different types of agencies within the systems. This was not so essential while the National Institutes played a dominant role at the national level, when decisions regarding the allocation of resources to the different programs within the Institutes had a definitive impact on the direction of overall technological efforts.

All these factors point to the need for reevaluation of the institutional model that was adopted. Major aspects to be considered include a discussion of what the role of public institutions should and could be, and what alternatives exist for developing and disseminating new agricultural technology.

NOTES

[1] These concepts are summarized in Schultz (1964). They served as basis for the United States foreign aid policy (Point IV) developed as of 1951.

[2] INIAP in Ecuador and INIA in Mexico did not incorporate technology transfer as a formal institutional function, thereby varying the basic model.

[3] The trend to create decentralized, autonomous institutions peaked towards the end of the 1950s, and the beginning of the 1960s. However it continued into the 1970s with the creation of FONAIAP in Venezuela, IBTA in Bolivia, ICTA in Guatemala, INTA in Nicaragua and INIA in Peru.

[4] Two examples of this concept are ICA, Colombia and INTA, Argentina. In the former case, the Board of Directors excludes direct participation of representatives from associations (Piñeiro and others, 1982:Chapter 6). In the latter case, INTA tended to assume a function of expressing the social demand for technology, interpreted as being different from farmer interest (CIAP, 1973).

[5] Sub-regional grouping is based on the criteria used by the Inter-American Institute for Cooperation on Agriculture (IICA).

[6] This situation changes significantly if Bolivia is not included, since the budget changes in this country (from just under US$500,000 to over US$7,000,000) account for all sub-regional increments between 1974 and 1978.

[7] Central America and Brazil did not enter the new institutional stage until the 1970s, a process which began in the rest of South America during the 1960s.

[8] More information on the different countries may be requested from the authors.

[9] The human resources situation is also different if analyzed at the level of the more highly trained. Personnel training programs, considered a central element of the strategy for creating National Institutes, closely followed the evolutionary cycle of financing. A detailed study of the cases of Argentina, Peru and Colombia indicates that training programs expanded up to the beginning of the 1970s, then diminished, practically disappearing toward the end of the decade. This was accompanied by a growing rate of retirement of personnel from research institutes, resulting in a net reduction in the number of staff with postgraduate training. This is the case of the Alberto Boerger Agricultural Research Center of Uruguay, with a 32 per cent reduction in postgraduate trainees between 1975 and 1978; the same occurred at INTA, Argentina and the Agrarian University of La Molina, Peru, which lost 15 to 20 per cent during the same period. The situation in Colombia was analogous, although the total number changed little due to the return of graduates of an active training program initiated at the beginning of the 1970s. A detailed analysis of this trend may be found in Trigo, Piñeiro, and Ardila (1982).

[10] The trend towards decentralized autonomous institutions was late in taking hold in Brazil, and not until 1973 did the Empresa Brasileña de Pesquisas Agropecuarias (EMBRAPA) come into being, following the model of a "converter" agency. Although its functions were similar to the other institutes created during the 1960s, EMBRAPA has some significant differences: it is not a centralized agency, but rather an organization set up to coordinate the research conducted by the different public and private agencies, within a so-called holding structure. These differences in approach can be partly explained by the socio-economic and political organizational differences in Brazil, a changing process which got underway as of 1964.

[11] In both cases, the presence of an association capable of assuming control of technological aspects establishes the basis for the private sector to displace public agencies in technology generation and transfer activities.

[12] A similar process occurred with the National Rice Program of the Philippines. Although probably one of the more important national programs of the region, it has practically disappeared since IRRI opened up.

[13] For a more detailed discussion of the relationship between modernization and the organization of agricultural research, see Trigo and Piñeiro (1981:2–10).

SOCIAL ARTICULATION AND TECHNICAL CHANGE

Martín Piñeiro and Eduardo Trigo

ABSTRACT

This paper analyzes and interprets the modernization and technical change that has occurred in Latin American agriculture during the last two decades. Empirical findings from eight case studies were subjected to an analytical procedure which tied the innovation process to structural dimensions which link production, society and state. Different processes of technical change are illustrated, based on their origin and qualitative characteristics.

INTRODUCTION

Until very recently most academics and politicians considered Latin American agriculture to be in a state of general stagnation. The conviction was common that this could be overcome only with the massive transfer of the technology available from developed countries.[1]

Programs designed after the 1960s in response to this view had considerable impact, as can be seen in the number of products whose yields increased significantly after 1960 (Lazo, 1980; Cohan, 1981; CIAT, 1981) (see Table 1).

This reflects an intensification of the international technology transfer process and a gradual increase in modernization throughout the continent. This disparate and fragmented innovative process went hand in hand with considerable changes in the organization of the productive processes, in the type and relative quantities of capital and labor used in agricultural production, and in the relations of production and social structures of the agricultural sector.[2]

The literature of the social sciences has not satisfactorily explained the fragmentary nature or the qualitative features of modernization and technical change. One of the reasons is the difficulty of developing a method for tying the technological issue to the broader social processes that contain and determine it. This paper seeks to analyze and interpret the modernization and technical change that occurred in Latin America during the past two decades. It is based on empirical findings from eight case studies conducted with a formal model of analysis. It ties the innovation process to a series of structural dimensions, and examines how these dimensions forge links between the production situation and the surrounding society and State. In addition, the studies illustrate different processes of technical change, based on their origin and their qualitative characteristics.

The document is divided into five sections. Following this introduction the empirical findings from eight case studies are presented.[3] The third section gives a comparative analysis of the innovation processes studied and the fourth interprets the origin of these processes of technical change. Finally, the fifth section presents a summary and conclusions.

TABLE 1 Annual Rates of Increase in Agricultural Production and Productivity by Continent and by Groups of Products (1958–1978)

	GRAINS[a]		ROOTS AND TUBERS[b]		MILK	BEEF
	Production	Yield	Production	Yield	Production	Production
North America	2.60	2.71	1.75	2.08	−0.31	2.87
Western Europe	2.79	3.04	−2.25	1.13	1.47	4.81
Latin America	3.67	1.81	0.27	0.53	2.53	6.61
Africa	2.84	0.50	4.67	−0.85	−3.35	12.24
Far East	−1.52	0.98	4.66	2.15	−0.85	8.38
World Total	2.11	1.76	0.65	0.73	1.56	4.59

[a] "Grains" include: wheat, rice, barley, rye, corn, millet, sorghum and oats.
[b] "Roots and tubers" include: potatoes and cassava.

SOURCE: These rates were prepared on the basis of data obtained from the FAO Year Books (several editions).

EMPIRICAL FINDINGS IN EIGHT CASE STUDIES

Technological Processes with a Major Impact on Production and Yields.

The case studies on rice production in Colombia and corn production in Argentina are examples of active processes of technical change. These processes began in the mid-sixties and had considerable impact on production and per hectare yields.

The case of rice in Colombia.[4] Beginning in the early 1940s, a number of innovations were adopted in cultivating and harvesting. These include mechanized processes, the use of nitrogenated fertilizer, chemical insect control, and the use of improved varieties, of which "Bluebonnet" was especially significant. However, this combination of innovations had a negligible effect on yields, which remained virtually stagnant at under 2,000 kg per hectare. Production itself rose as a result of increases in the area under cultivation.

By the mid-sixties, improved varieties (High Yielding Varieties, HYV) had been introduced. They required a complete technological package if they were to realize their full genetic potential and this produced a true technological revolution, similar to that experienced in other areas of the world during the 1950s and 60s.[5]

In Colombia, as in other countries of the world, the key element in the process is genetic material developed by the International Rice Research Institute (IRRI). These materials were then adapted to local conditions by the rice program, jointly organized by the Colombia Agricultural Institute (ICA), the International Tropical Agriculture Center (CIAT) and the Federation of Rice Producers (FEDEARROZ) (Piñeiro and Trigo, 1983:Chapter 10).

Throughout the period studied, the process was dominated by innovations generated internationally. One by one, different national organizations assumed responsibility for rice research (National Rice Program, PNA; Agricultural Research Bureau, DIA; and finally, ICA–CIAT–FEDEARROZ). They undertook major tasks in such areas as testing and adapting new varieties and testing and determining correct doses of fertilizers and other chemical compounds. However, the key ingredients of the innovations were generated in the developed countries with the participation of international centers (IRRI and/or CIAT). The transfer of this technology was made successfully to countries whose ecological, political, economic and production conditions were appropriate.

The technological process in Colombia, especially after 1967, had a stunning impact. From 1967 to 1978, the area under cultivation increased by 50 per cent, while per hectare yields jumped by almost one hundred per cent, from around 2,200 kg per hectare of paddy rice, to around 4,200 kg per hectare. It is important to note that these yields compare favorably with those achieved in the majority of Asian countries and developed nations (IRRI, 1977). The geographic location of rice production and the type of productive enterprise engaged in this activity also underwent dramatic changes, as a result of two interrelated factors. The first is the low productivity of HYV under dry conditions. The second is the fact that in Colombia, dry rice production, which made up nearly 50 per cent of total production before 1967, was practiced essentially by small-scale producers. The result was the virtual elimination of dry rice production and, consequently, of small-scale producers, who by the second half of the 1970s were responsible for barely ten per cent of total rice production.

By comparison with these dramatic changes in yields and geographic location of production, the impact on production structures in the capitalist enterprises was relatively minor. The process of replacing labor with capital was intensified and generalized, but not in proportions comparable to those of other changes. Although no quantitative estimates are available, it is clear that the spread of mechanized harvesters and other sophisticated agricultural machinery meant an increased use of capital.

Technological inputs are another area of interest. Per hectare application of fungicides increased from 1965 to 1976 at an annual rate of 39 per cent. Herbicides rose by ten per cent. Insecticide use rose until 1968, but then slipped by almost three per cent per year. As the use of capital increased, the use of labor fell from 15.2 million person/days in 1965 to 13.2 million person/days in 1978, despite simultaneous increases in land under cultivation and in production levels.[6] In all, these lower labor requirements produced a 50 per cent drop per ton of rice produced.[7] It is important to note that these capital intensive technologies were adopted in a general economic context in which the prices of capital goods rose more quickly than wages.

Scobie and Posada (1978) and Balcázar and others (1980) have carefully documented the impact of technological processes on rice production in Colombia, examining how the different social sectors appropriated economic surplus. The strong production increases achieved after the mid-sixties took place in the presence of a closed market and this pushed real prices down by nearly 40 per cent. It was a classic treadmill process, operating under government control and characterized by production rises and planned price falls. Consequently, the new economic surplus was partially distributed toward the urban industrial sector.[8]

Inside the producing sector, the redistribution favored the capitalist farmer and worked against small-scale dry rice producers. In addition, special advantages accrued to those producers who quickly adopted the new technology, and this produced further symptoms of the treadmill process (Cochrane, 1958). In order to understand producer behavior, it is important to be aware that the irrigated rice-growing sector experienced no reductions in real income as a result of technical change, at least during the first half of the process. This is clear if price changes are compared over time.[9] Rice prices fell by 28 per cent from 1965–1969 to 1970–1974, but the costs of production per unit of output fell by 30 per cent during the same period (Scobie and Posada, 1978).

The case of corn in Argentina.[10] The case of corn production in the pampas of Argentina is very similar to the development of rice in Colombia. Corn production was introduced in the pampas at the turn of the century. Over the years, it has undergone three clearly distinguishable phases. The first lasted until the beginning of the forties, and was characterized by major production increases, especially due to an expansion of area under cultivation. During this period, per hectare yields were slightly less than those obtained in the United States, but comparable to those obtained in Australia and the European countries.

During the second phase, which began in the early 1940s and lasted until the early 60s, the amount of land planted to corn was reduced, and per hectare yields came to a standstill. The same occurred with other crops competing with corn for productive resources.

The third period began, although hesitantly, in the early 60s, taking hold more strongly after 1964. This period saw increases in the area under cultivation and in yields for almost all farm crops of the pampas region, especially corn.

It is important to stress that throughout the three periods, especially the latter two, the productive structures were undergoing profound technical changes. The second period was characterized by an explosive process of agricultural mechanization, with the resulting displacement of labor. The process was sweeping the entire country, but it was particularly strong in the pampas region. Thus, from 1947 to 1969, while the number of tractors used in the pampas rose from 25,950 to 147,680, labor use fell by 30 per cent (Martínez, Fienup and Chevallier, 1976). This advancing mechanization was a reaction to heightened migrations of labor, spurred by explicit governmental policies for industrialization. The process introduced important modifications in the relations of production, reflected in shrinking populations of the tenant farmers and renters who had been responsible for much of farm production, especially corn. These farming activities were incorporated into the productive structure of the **estancias** (productive units typified by agrarian capitalism).

In Argentina, work with hybrid seeds began in 1923, with little success. It was again taken up, this time more successfully, in 1949 (Martínez, 1973). Despite their early beginnings, hybrid seeds did not become widely used until the mid-1960s. The change began with the 1959–1960 campaign. Although it produced only 9,386 tons of hybrid seed, the figure was quickly surpassed after that year, reaching a high of 105,695 tons in 1970–1971. Production then slumped once more.

The initial efforts were made by public agencies (Ministry of Agriculture and School of Agronomy of the University of Buenos Aires) and later taken up by the National Institute of Agricultural Technology of Argentina (INTA). However, private participation began in 1946 and experienced mercurial growth, eventually becoming the dominant force.[11]

The process of incorporating hybrid seed was contingent on the use of a technological package, as had occurred in the case of rice. However, in Argentina, this package contained fewer components than that of rice, and these components were also less interdependent. The incorporation of hybrid seeds stimulated more careful cultivation practices, a greater use of chemicals for pest and weed control, and most importantly, bulk harvesting. Unlike the case of rice in Colombia, it did not include the widespread use of irrigation or chemical fertilizers. The introduction of the technological package had a profound impact on yields at the national level, and a smaller impact on production totals. From 1950–1954 to 1975–1977, yields climbed at an annual rate of 4.72 per cent, rising from around 1.5 tons per hectare to 3.00 tons per hectare. Similarly, technological inputs increased, and the process of mechanization was intensified, especially for the harvest. Nevertheless, the impact on the use of capital was limited, as the mechanization process had already progressed considerably by the end of the 1960s. Similarly, labor use in the pampas continued to fall, although much more slowly than in the preceding period.

It is important to note that the yield increases were greater in the belt (the area studied) than at the national level. By 1979–1980, almost five tons per hectare were being produced. In spite of this rapid increase, however, yields were still far below those obtained in the United States. The difference can be attributed to the use of less fertilizers and chemicals in general, and to certain agronomic practices related to livestock rotation, which are widely used as a means of protection against risk.

In terms of the distribution of resulting economic surplus, the case of corn production in Argentina is similar to the situation described for rice in Colombia, although the economic mechanisms used were different for the two cases. The price of corn in Argentina is determined jointly by the international market and the government's policies on exchange rates, export duties and market withholding. These measures for absorbing farm surplus generally include policies to protect a certain level of farm profitability.

Accordingly, production increases through higher per hectare yields benefitted the urban industrial sector because the government received and redistributed the surplus and because access to foreign exchange was eased. The producing sector benefitted by receiving more real income. This is reflected in the price rises for land observed in the corn producing zones during the period under analysis (Ras and Levis, no date; Martínez, Piñeiro and Chevallier, 1976; Flischman, 1977).

The increased profitability for agrarian production was relatively homogeneous inside the sector, due to the uniformity of land resources. However, differences did exist because the technological package was more adaptable to larger farming operations, which have greater flexibility to combine farming and livestock and thus stabilize their income. This was true in spite of steep variations in prices and yields, typical of Argentine agriculture.

Technological Change with a Heavy Impact on Production Organization.

The case studies on sugar production in Colombia and milk production in Ecuador provide a contrast to rice in Colombia and corn in Argentina. They show intensive processes of technological innovation that substantially modified relations of production and the use of labor and capital, making relatively minor increases in per hectare yields. The case of tomato production in California is somewhere between the others; the process of technological innovation triggered a dramatic rise in land productivity and substantially altered productive processes, swiftly replacing labor with capital. Because it is an intermediate case, tomato production will be discussed first.

Tomato production in California.[12] The process of technical change in tomato production in the state of California began in 1960 and was established by the middle of that decade. The key to the transformation of the productive process was the mechanical harvester, developed in the University of California and produced and marketed under license by Blackwelder Manufacturing. The adoption of this technique had profound Hirschman-type forward and backward linkages (Hirschman, 1977), and forced many modifications to be made in the overall productive process. In the first place, mechanical harvesting required the development of new varieties of tomatoes that were more resistant to the rough handling of the harvester. This genetic work was also undertaken initially by the University, but during the 1970s, it was partially assumed by private enterprise.

In the second place, the adoption of the harvester, by changing the type of product delivered to processing plants, produced a number of changes in techniques for delivery, selection and processing. If the technological package designed for production was to spread rapidly, the processing plants themselves had to accept and promote it. In other words, the full array of economic interests involved in tomato production had to be motivated, as much of the work was governed by production contracts.[13]

The incorporation of harvesters into the productive process was extraordinarily swift. In 1964, only 3 per cent of the production was harvested mechanically, but by 1970, 100 per cent of the harvest had been mechanized. During the same period (1964–1972), the number of wage-earners working in the harvest dropped from 50,000 to 18,000. This drop in labor use was firmly entrenched by the end of the 1970s, with the release of another 5,000 workers as a consequence of electronic selection processes. Technology had a heavy impact on the use of capital, and this was combined with a strong effect on productivity. From 1960 to 1978, yields and production climbed by 250 per cent. Together, these effects meant a tenfold reduction in the use of labor per ton of tomato harvested.

This dramatic process of technological transformation also prompted a considerable concentration of production, both on the farm and in the processing plants. From 1959 to 1974, the number of tomato producing farms fell by 50 per cent, while the average number of hectares per farm climbed from 57 to 173. At the same time, the use of capital intensive technologies eliminated jobs for the large contingents of temporary workers brought from Mexico. However, the permanent wage earners found themselves better able to enter into stable contractual relations.

Data available in the case study show that one of the major effects of technology adoption was that tomato production became a permanent industry in the state of California and was able to compete favorably with other regions of the United States and Mexico. Proof of this was that California tomatoes rose from 55 per cent of total United States production in 1960 to 65 per cent in 1970, and finally topped 80 per cent by the mid-seventies. Thus, technology stimulated interregional competition, and established long-term tomato production in California. This explains why the social sectors involved in production had little trouble coordinating their efforts in regard to the technology question.

Competition also began to take place among the different capital fractions, including land owners, because of the appropriation of the great surplus generated by the technological process. Evidence suggests that the benefits of technical change were distributed among the different social sectors as a function of their negotiating power. The capitalist sector as a whole, however, was the real beneficiary of the modernization process. The sudden concentration of production into large units, and the gradual vertical integration of the productive process as a result of modernization, meant the sacrifice of smaller units and the loss of opportunities for agricultural employment for unskilled laborers. By contrast, the mechanization process expanded opportunities for specialized and industrial agricultural employment.

Sugar production in the Cauca Valley of Colombia.[14] Technical change in sugar production in the Cauca Valley in Colombia began to take hold around 1960, together with the rapid expansion of the area under cultivation and increased production. These advances were triggered by a new preferential access to the market in the United States, following the blockade of Cuba. However; the process of technical change did not really become noticeable until the end of the 1960s, when the incorporation of new land reached its natural limits.

The technical change included adoption of a wide range of techniques. The most important are: the incorporation of improved varieties; the redesign of irrigation practices and land preparation; the use of fertilizers and, later, chemical methods for weed and pest control; the adoption of a more effective method for cutting cane (the Australian method). Most of these techniques were introduced from other countries. Some were adapted to local conditions by the sugar mills, which until 1973 enjoyed the cooperation of the experimental station of the Colombian Agriculture Institute (ICA) in Palmira.

These techniques had a negligible impact on yields, which increased at an annual rate of only 1.67 per cent from 1960 to 1977, for a total increase of 28 per cent for the period. However, total production climbed by approximately 150 per cent as the area under cultivation doubled. Capital intensive methods were strongly favored, although not so much as in the case of tomatoes. The use of fixed capital jumped 300 per cent, while input use soared to a 500 per cent increase. At the same time, the direct use of labor in the productive process crept up by only 50 per cent during the entire period. This combination of changes meant that the amount of labor used per hectare harvested fell by fully 40 per cent (Piñeiro and others, 1982).

The innovative process described above introduced no alterations in the capitalist relations of production, which had governed the activity from the beginning. However, it did change the organization of production, especially in terms of concentration and vertical integration, and the gradual subordination of independent farm producers to the interests of the sugar mills. These shifts explain why the technological process had such a strong impact on income distribution among the various social sectors involved in sugar production. Although consumer interests were subordinated to the needs of the producing sector, certain benefits nevertheless accrued, as reflected in the domestic consumer price reductions from 1964 to 1975. World sugar prices were high, and this price reduction took place without substantially affecting the profitability of plantations; however, the domestic prices began to rise when world prices plunged and profits had to be protected through domestic price mechanisms.

Inside the producing sector, the surplus generated by production growth and the incorporation of technology was appropriated by independent producers and by the sugar mills, to the detriment of land owning sectors and wage earners, who found their share in total income growing smaller. The bias in favor of the capital sector reached its height during the 1970s at a time when technical change was especially intense. Thus, it can be stated that the process of technical change was pushed by the mill owners as a mechanism for counteracting the increased negotiating power of wage earners and the land owners, generated by the rapid industrial expansion of the 60s.

Dairy farms in the Ecuadorian highlands. Milk production in the Ecuadorian highlands followed a pattern similar to that of sugar production in Colombia. Dairy farming became widespread early in the century, but expansion did not take place until the 1960s. The process began late in the 50s, due to the initiative of the land owning sector, which, anticipating the threat of Agrarian Reform, was interested in eliminating the *huasipungo*. However, the process did not become truly dynamic until the 1960s, when urbanization and government policies increased the demand in the cities for foodstuffs in general, and milk products in particular. From a very early date, this expansion process had included technological innovation in the form of livestock breeding programs, the recording of production for each animal in order to improve selection, and the introduction of artificial pastures. These three techniques comprised a natural package that obtained a certain following prior to 1950. Some 25 per cent of the units surveyed indicated that, during the 1940s, they had incorporated the first two techniques.

Between 1950 and 1960, more pastures were incorporated, and additional methods were adopted, including mechanization, artificial insemination, and early weaning. The process gathered strength during the 1960s, when mechanical milking also began to gain acceptance. After 1970, all these techniques had become widespread, and most are now in use by around 90 per cent of the units surveyed. The major exceptions are mechanical milking (50%) and artificial insemination (61%). Artificial pastures cover around 50 per cent of the land used for cattle and 25 per cent of total land. This process of technological adoption did not have a major impact on production. National output grew from 731 million liters in 1972 to 871 million liters in 1978, where the highlands as a whole produce 81 per cent of this total, and the valleys studied produce some 25 per cent.[15]

The incorporation of artificial pastures and mechanical milking produced a major increase in capital use. The simple expansion of milk production, with the technological pattern adopted for this purpose, replaced the labor formerly provided by the *huasipungos*, which were resettled during this time on hillside lands ceded by the haciendas.

This process of increasing technology use in Ecuadorian milk production appears backward in comparison with other parts of the world. All the techniques incorporated had been widely used in Europe, the United States and Oceania since earlier times. Therefore, it is not surprising that most had been imported and disseminated through private enterprise, as a result of international technical assistance programs. The role of the National Institute of Agricultural Research (INIAP) appears to have been important only in the testing and adaptation of pastures and the use of fertilizers.

Relative Stagnation of Technical Change

The circumstances surrounding beef production in Uruguay, potatoes in Peru and cotton, corn and bean rotation in northeastern Brazil present a contrast to the five cases discussed above. They are clear examples of technological stagnation. It is important to stress that this stagnation is only relative when compared with similar situations where the innovative process has had a major impact on the productive structure and land productivity. The definition of these cases as instances of technological stagnation is based on the absence of increases in per hectare yields (see following section).

Beef production in Uruguay. Livestock in Uruguay is of vital importance in the process of overall capital accumulation. The beef industry has been the principal productive sector since very early in the nation's history. By 1908, Uruguay already had over eight million head of beef cattle, a figure which was not to be surpassed until 1970. Because of its economic importance, the livestock industry went through a major process of technological improvement before 1930, incorporating genetic improvements, corrals, and better herd management. The innovative process increased production by around six per cent per year, while the number of animals remained relatively constant (around eight million).[16] About this time, the expansion of the sector reached the limits of available natural pastureland, and a period of stagnation began.

In 1950, the traditional production model, which had been developed several decades earlier, consisted of relatively large productive units which raised combinations of cattle and sheep of good genetic quality. They fed on natural pastures which had lost much of their original capacity after many years of overgrazing. During the past 20 years, a number of changes have taken place in the techniques used for Uruguayan livestock handling. At the same time, efforts have been made to introduce a new technological package based on experiences in New Zealand. The package calls for a combination of grasses and legumes for pastures, with the required fertilization. It also includes careful and intense pasture and corral management. The changes finally incorporated include genetic improvement, better health practices, and expanded farm infrastructure, all of which tend to reduce the labor force. Of particular interest is the increase in land surface under improved pastures. This phenomenon reflects the acceptance of one component of the technological package disseminated by public institutions.

In 1961, the Honorary Commission of the Agricultural Plan, an interesting example of the co-opting of the public apparatus by the producing sector, initiated an outreach program for pasture improvement, with the support of a loan from the World Bank. In the first phase, from 1960 to 1966, the promotional work was experimental, and each participating farm received government support. In the second phase, which was completed in 1974, the dissemination process gradually spread throughout the territory. After 1974, the process of technological adoption was reversed as a result of the livestock crash caused by conditions in the external market. From 1966 to 1975, the area under artificial pastures spread at an annual rate of nearly 20 per cent, finally covering over a million and a half hectars by 1975, or 11 per cent of total livestock area.[17] This gives an idea of the considerable capital accumulation that the process involved. The use of improved pastures was general across the different size strata of production units. although the small operations used a higher proportion of annual grasses and a lower proportion of perennial grasses than the larger farms with the New Zealand model.

This technological process had an impact on livestock production. In the first place, total estimated beef production (slaughter plus changes in herd size) increased after the end ot the 1960s, and peaked in 1973. In later years, it again declined. Nevertheless, production rose from 1,400,000 head in the final three years of the 60s to 1,700,000 head in the final three years of the period under study (1976–1978), for an annual growth rate of 6.68 per cent. In the second place, the incorporation of a technological package based on pastures meant an increase in the use of fertilizers and the development of physical infrastructure.

The increase in cattle production can be partially explained by a corresponding drop in sheep production. However, the technological package was responsible not only for effecting a real increase in livestock units, but also for providing the opportunity to make this substitution of animals possible.[18] This occurred under the protection of high beef prices received by the producing sector as world market prices soared, together with subsidized credits granted through the Agricultural Plan. The process came to a dramatic halt after 1974, when producer prices plunged.

This correlation between prices, credit and technology adoption only confirmed the microeconomic evidence of low relative profitability for the technological package. After 1974, because of greater sensitivity to climatic variations and price shifts, the rational producer decision was not to adopt the technological package. This marked the failure of the process of technological articulation launched by the Agricultural Plan. The result was that Uruguayan livestock in recent decades was labelled as a case of "dynamic stagnation". It was a stagnated because the growth of total livestock production, as well as production per unit of land used, was clearly inadequate. It was dynamic because the surface results masked major changes in modes of producing that have been translated into the adoption of new techniques and into significant processes of accumulation and disaccumulation in the sector, fundamentally linked to changing herd size, improved pastures, and expanding infrastructure in the enterprises (Alonso and Pérez, 1980).

Potato production in the Mantaro Valley of Peru. The uncertain process of technology incorporation in potato production, which began after 1960, could be divided into two stages. In the first, which lasted until the beginning of the 1970s, technological innovation was very limited, and included some crop mechanization, a gradually rising use of inputs, and the introduction of improved varieties. These technical changes were incorporated by different producer strata, including the small-scale farmers (owning from one to five hectares), who were responsible for the majority of production in the Mantaro Valley (68 per cent in 1972). However, campesino producers incorporated very few of the improved seeds, as they found the product disagreeable for their own consumption, which absorbed some 60 per cent of their output. Probably as a consequence of their continued use of older native varieties (such as the Renacimiento, produced in 1949), which were less responsive to technological processes, they also used fewer inputs than capitalist producers.

This process of differentiation was accentuated during the second stage (after 1974), when the large producers actively adopted the new, improved varieties and increased their use of technological inputs. Consequently, the capitalist producers obtained yields six times greater (1977 to 1978 campaign) than those of small-scale producers. Potato production in the Mantaro Valley was important for the entire country, and there, small-scale producers played a major role. Thus, the developments in the capitalist enterprises had little impact at the national level. Yields held steady at around 6,200 tons throughout the period, and did not begin to rise until after 1974.

By 1978, they had reached 7,500 tons. Total production stagnated, rising only slightly from 1960 through 1970, and has declined somewhat in recent years.

Because the technical change process was weak, and the nature and intensity of change varied strongly among producing strata, it is difficult to measure its effect in terms of factor use and distribution of surplus. However, some available evidence suggests that the process had been translated into a better use of technological inputs, but had little effect on the use of labor.

The case of corn-cotton-bean rotation in northeastern Brazil. The study in northeastern Brazil (cotton-corn-bean rotation) clearly illustrates a situation of production and productivity stagnation. The effects are even more striking when compared with the active process of technical change experienced by livestock production on the large enterprises, to which they are related.

During the process under study, only a few productive techniques were introduced. One included the spreading use of animal draft plows, rising from 108 units in 1960 to 8,920 in 1970. The introduction of tractor use was negligible, as was the addition of other types of machinery. Similarly, very little use was made of new varieties, limited to a few varieties of high quality fiber cotton, with lower per hectare yields, and several more productive corn varieties. These technological innovations took place after the mid-fifties and gathered strength during the period preceding 1970. The observed technological passivity corresponds to a general process of stagnation in agricultural production, low productivity, and the gradual deterioration of natural resources. This process is reflected in production and productivity figures of all three crops studied (see Table 2).

TABLE 2 Production and Yields of Cotton, Beans and Corn in Northeastern Brazil (Piñeiro and Trigo, 1983: Chapter 9)

Crop	Production (tons)		Yield (kg/ha)	
	1961	1977	1961	1977
Cotton	44,116	34,768	292	178
Beans	24,243	51,021	681	363
Corn	46,751	152,637	738	864

The dramatic stagnation in production and yields of the three crops stands in strong contrast to the situation of livestock production. From 1960 to 1975, the cattle in the northeastern region grew from 9,580,000 head to 17,880,000 head. This phenomenal expansion took place through the conversion of deforested land into natural pastures, and it was further strengthened by a process of technological incorporation of considerable magnitude. The major components of technical change were the increased cross-breeding of creole stock with the Brahma, the construction of fences and water wells, and new management practices. This combination of techniques raised birth rates and increased the average weight per animal in the corral from 144 kilograms in 1946 to 200 kilograms in 1970.

A COMPARATIVE ANALYSIS OF THE CASE STUDIES

The empirical findings of the case studies discussed above are representative of production situations throughout Latin America, in terms of productive structure, and how the product under study fits into the economy as a whole.[19] Five of these cases illustrates active processes of technical change that began during the 1960s (the key years in the five cases were: 1968 for rice in Colombia; 1964 for corn in Argentina and tomatoes in California; early 60s for sugar in Colombia and dairy farming in Ecuador). The processes of incorporating technology into large-scale potato farms in the Mantaro Valley in Peru and the livestock ranches in Uruguay, however, began in the 1970s.

Table 3 presents the annual rates of production increase and the per hectare yields of products studied, compared with the rest of the world, the four continents, and the Latin American country showing the highest overall growth rate. The figures indicate that for the eight products studied, yields at the world level, and especially in given Latin American countries, made significant progress, suggesting the international availability of technology for increasing land productivity. In the cases under study in Latin America, only rice production in Colombia and corn production in Argentina showed increases that approached those in the countries with the highest yields. In general, the empirical findings indicate that the processes of technical change involve two principal types of social phenomena. These are determined by the nature of the process of social articulation that produced them and the quality of technical change that took place.

In the first type, represented by rice production in Colombia and corn production in Argentina since the mid-sixties, the government mediates the interests of the industrial urban centers with the more specific interests of the farmers. In both examples, this was motivated by a crisis in the production levels of the commodities under study. The government then pushed negotiated solutions, which appeared in an ex-ante evaluation to respect the overall interests of the social sectors involved. The processes of social articulation are remarkably similar to those that occurred in the developed countries after the 1950s, and more recently in some Asian countries. The need of the society as a whole to increase production, and the presence of dominant social sectors capable of implementing public policies consistent with technical change, are the pivotal elements of the process.

In the two cases studied, technological articulation was based on: (a) the introduction of a technological package based on improved or hybrid varieties, which resulted from research carried out by international organizations funded and controlled independently of the productive sectors, but for which national public agencies played an important role in dissemination; and (b) the definition of an economic policy that stabilizes prices and gives high subsidies for direct investment in technology adoption or capital-embodied technology. These characteristics of the process, plus the qualitative nature of technological changes and the low concentration of supply, resulted in a relatively equitable distribution of the surplus among the different social sectors.

The impact of these processes of technical change on production and yields was significant, even by comparison with international results (see Table 2). In addition, the use of labor and land was capital intensive, with only minor effects on the organization of the productive process, the relations of production and other aspects of the productive structure, including the degree of concentration and vertical integration.[20]

A second type of technological process is illustrated by the cases of sugar production in Colombia and milk production in the Ecuadorian highlands. The social articulation was generated from inside the agricultural sector. In both cases, these sectors were able to negotiate with the government on a series of policies that served their specific sectoral interests and enabled them to enter into processes of technological innovation. The active nature of these processes, however, was firmly controlled by the productive sector, which defined the form they would take and appropriated a goodly part of the benefits of technical change. The public policies implemented were in every case specifically designed to solve particular problems obstructing the development of the dominant productive sectors. In addition, these sectors created organizational mechanisms which gave them a certain amount of control over the supply of technology. Qualitatively, the technical change had moderate effects on yields, while production expanded through the vigorous incorporation of new areas, and important changes took place in work organization. In addition, production became more concentrated and vertically integrated.

The case of tomato production in California is an intermediate situation in which the innovative process was very intensive, producing significantly higher per hectare yields and production affecting the use of labor, and increasing the processes of concentration and vertical integration. Livestock production in Uruguay was unique, in that an agrarian initiative that began in 1970 failed as a consequence of the inability to design public policies consistent with technical change.

In a third type of technical change situation, potato production in Peru and rotation in the Brazilian northeast entered technological stagnation, at least relative to cases in other parts of the world. This stagnation is due to the fact that no existing social sector linked to production was capable of mobilizing and coordinating governments action in its favor.

TABLE 3 Rates of Increase for Annual Production and Yields for Products Under Study, 1958–1978

PRODUCTS / REGIONS	RICE Prod.	RICE Yield	SUGAR Prod.	SUGAR Yield	BEEF Prod.	MILK Prod.	CORN Prod.	CORN Yield	POTATO Prod.	POTATO Yield	TOMATO[a] Prod.	TOMATO[a] Yield	BEANS Prod.	BEANS Yield
World Total	2.37	1.27	—	—	4.59	1.56	3.17	2.38	−0.03	1.37	4.84	1.09	—	—
Western Europe	0.73	0.21	−0.54	−0.58	4.81	−1.47	5.00	6.06	−2.25	1.00	2.87	1.91	−0.61	2.22
North America	4.38	1.74	3.09	−1.17	2.87	−0.31	2.57	3.09	1.75	2.03	2.93	2.95	0.29	0.12
Latin America	3.92	0.50	2.96	0.46	6.61	2.53	3.28	1.72	2.73	2.77	5.03	1.02	2.14	−0.15
Far East	1.93	1.59	3.32	0.56	8.38	0.95	3.18	1.33	5.92	0.25	3.28	0.98	2.37	0.72
Africa	3.17	0.90	1.60	−1.30	12.24	−3.35	2.58	0.62	5.82	−0.11	5.05	0.68	3.24	0.57
Country studied*	7.39	5.02	6.64[b]	1.67[b]	0.90	0.60	3.70	2.80	2.30	1.50	2.85[c]	4.16[c]	1.70	−1.60
Zone studied**	7.39	5.02	6.64[b]	1.67[b]	0.90	—	17.60[d]	—	—	—	4.91[c]	2.14[c]	1.55[b]	−4.21
Latin American country with highest growth rate[f]	14.60	5.10	10.70[f]	3.50[f]	11.70	7.30	6.20[a]	4.30	14.80	11.20	—	—	10.30	7.60

NOTES:
[a] 1964–1978
[b] 1960–1977
[c] 1950–1972
[d] 1950–1979 Calculated on the basis of indices
[e] Rice: Prod. Venezuela Yield Venezuela
   Sugar: Prod. Guatemala Yield Costa Rica
   Beef: Prod. Nicaragua
   Milk: Prod. El Salvador
   Corn: Prod. Paraguay Yield Uruguay
   Potato: Prod. Dominican Rep. Yield Honduras
   Beans: Prod. Argentina Yield Bolivia
[f] 1964–1978

*Country studied:
Rice: Colombia
Sugar: Colombia
Beef: Uruguay
Milk: Ecuador
Corn: Argentina
Potato: Peru
Tomato: United States
Beans: Brazil

**Zone studied:
Rice: Magdalena Valley
Sugar: Cauca Valley
Beef: Uruguay
Corn: "Corazón Maicero"–Pampa
Potato: Mantaro Valley
Tomato: California
Bean: State of Pernambuco
Milk: Ecuadorian Highlands

SOURCE: Based on data obtained from the FAO Yearbooks.

The following section is an attempt to interpret the technological processes that were observed, using the natural grouping of the case studies. The interpretation is based on: (a) the identification and definition of the processes of social articulation generated within each production situation under study, in response to certain structural features particular to each; and (b) an illustration of the mechanisms by which these processes were translated into various types of linkages between the productive sector and the government apparatus, and how this led to public policies on technology.

TECHNICAL CHANGE AS AN EXPRESSION OF SOCIAL ARTICULATION[21]

Introduction

A three-pronged analysis must be made in order to understand fully the different processes of social articulation that prompt situations of technical change as described above. The first component involves the relationship between the innovation process and other, broader processes stemming from the expansion of the market economies of the countries under study, and especially from the progress of consolidation of the capitalist mode of production in Latin American agriculture. This development also determines the social relations of production, the social structure in the countries and the very form the State and its institutions take (Gómez and Pérez, 1979).

The second level of analysis deals with the relationships between the innovation process in the countries and the development of science and technology as a cultural phenomenon at the international level. This international process has conditioned the development of the technological institutions of Latin America and has determined, at least to a certain degree, the real availability of technology, because it links technology development to elements foreign to national conditions (Trigo and Piñeiro, 1981).

Finally, the third level of analysis studies the close ties between the technological process and the economic and social processes occurring in each production situation. It is based on a series of structural dimensions that define both the organization of production and the position of each sector within society as a whole.

Basically, these dimensions are: the manner in which production is organized, the resulting class structure and how the production sectors interact with each other and with other areas of the government. These considerations explain the quality of the social processes observed, especially those which involve agrarian initiatives, which are particularly important because they appear to be typical of Latin America. They give rise to a form of agrarian development which significantly affects the operation of the economy and the organization of society. Following is an analysis of five structural dimensions which have been particularly useful in these case studies for explaining the nature of the technical change observed. This approach seeks to show that a number of dimensions mutually determine one another within broad historical processes.

Structural Dimensions of the Case Studies

Types of production units and control over productive resources. The most important of the dimensions is the type of production unit characteristic of each case study. This, in turn, determines the organization of the sector, the degree and form of control over the productive resources, the prevailing type of relations of production[22] and indirectly, the way the surplus is distributed within the productive sector. In addition, the sector's organization determines the behavior of the members of society who are responsible for production decisions, including their demands for technology, and the specific features of their ties with society as a whole and with the government.

In this connection, the case studies reflect diverse situations and can be categorized by type of productive unit, as follows:[23]
— Industrial capitalism: sugar in Colombia and tomatoes in California.
— Hacienda in transformation: dairy farms in Ecuador.

— Agrarian capitalism and family enterprise: corn in Argentina, rice in Colombia, livestock in Uruguay.
— *Minifundia:* (a) operating as independent units: potatoes in Peru; (b) within a large enterprise: rotation in northeastern Brazil.

Sugar production in Colombia, and to a lesser degree tomato production in California, show high degrees of concentration and vertical integration of production,[24] control over the productive resources of land and capital, and the existence of large reserves of a wage-earning labor force. The main decisions by the entrepreneurs had to do with controlling this labor force. In addition, the sugar industry's degree of economic concentration enabled it to bond itself into a cartel, and gave it considerable power to negotiate public policies with the government.

The hacienda in transformation in the Ecuadorian highlands initially constituted a typical case of a production unit with control over the land and labor force, when the non-capitalist *huasipungo* relations of production were still in effect. These conditions began to break down rapidly after 1960, although high levels of unemployment and the dependent relationship of the new "settlers" enabled the transformed hacienda to maintain control over the labor force. In addition, the hacienda owners traditionally wielded considerable social and economic power in the country. This, together with a high degree of unity, gave the combined actions of the sector tremendous influence over the society as a whole.

Livestock production in Uruguay, corn production in Argentina and rice production in Colombia represent the medium-scale family enterprise, various degrees of the use of wage-earning labor power, and an irregular control over land and capital resources. In the case of Uruguay, the productive sector controls the country's most important economic activities, and thus has gained considerable social and political power. This is also true to a lesser degree in Argentina, and in the case of rice farmers in Colombia.

Finally, the cases of the campesino potato farmers in Peru and the *minifundia* operating within the "complex" of northeastern Brazil are production situations in which family members provide the labor, and most of the production is consumed by the family itself; there is a low degree of control over productive resources, little possibility for accumulation, and a marked trend towards impoverishment and proletarization.

Uniformity of the productive sector. A second consideration is the degree of uniformity in the productive sector, in terms of production units, and the preponderant influence of one commodity in all production decisions. These elements shape the technological and economic needs common to the production units as a whole. The uniformity of economic and technological interests seems to have been fundamental in producing a high degree of federated organization. This, in turn, led to the development of producers' organizations and technical-bureaucratic structures which provided access to the media to sway public opinion, facilitated the formulation of unified, clearly stated platforms, and made it possible to negotiate with or co-opt vital parts of the public sector. At the same time, single-commodity producers' organizations began to emerge, whose demands assumed greater consistency and clarity. The cases of rice and sugar in Colombia, milk in Ecuador, tomatoes in California and, to some degree, beef in Uruguay, are good examples of this.[25]

In contrast, the productive sector in the Argentine pampas was not specialized and was structurally diverse. As a result, producers' demands were expressed through different organizations. Public policies for the product under study (corn) were negotiated in conjunction with more general policies on multiple commodities.[26] Similarly, in the cases of potatoes in Peru and campesino production in northeastern Brazil, the socioeconomic conditions associated with multiple production lines and low control over resources made it impossible to develop an organizational structure which satisfactorily expressed the type of demands that generate government actions.

Importance of the commodity in the region. The regional importance of a product determines how much power the productive sector has for mobilizing and implementing political structures and the regional state apparatus to defend its particular interests. The sugar mills in Colombia have virtually complete control over the economic and social situation in the Cauca Valley, where the country's entire sugar production takes place. The importance of tomato production in California at the regional level (80%) enabled producers to mobilize the rest of the sectors linked with production (processing plants, manufacturers of inputs, etc.), in addition to agencies of the local and state apparatus (University of California), in favor of the particular interests of the industry. In the case of milk production, the location in the highlands, and the fact that this sector provides Quito's entire milk supply, gave it certain regional characteristics.

In contrast, livestock in Uruguay, and to a lesser degree production in the Argentinean pampas, are more national in nature, because of the type of space they occupy. For this reason, they cannot defend their specific interests or negotiate sectoral policies from a regional base. In the case of the rice farmers, because of the dispersion of production over several geographic zones, the importance of the regional dimension is only relative. Similarly, potatoes are produced in Peru on more than half of the total number of farms, and although potato farmers are more influential in the highlands than on the coast, the regional dimension is of little importance. As for the minifundia in northeastern Brazil, regional considerations helped to strengthen the owners of the "complex". This strength was used, however, to defend the dominant interests of the livestock farmers in the region, which made it easier to exert control over small-scale farmers.

Relative size of the productive sector. Another dimension related to the negotiating power of the productive sector is its size relative to the rest of the agricultural sector and to the national economy. Relative size, in conjunction with the degree of concentration and vertical integration, apparently determines the level of association or conflict with other capital interests and, consequently, the greater or lesser possibility of co-opting and mobilizing specific parts of the government apparatus. Productive sectors not essential to the overall process of accumulation, like sugar in Colombia and tomatoes in California, had greater freedom and independence (that is, less conflict) for defining economic policies suiting their needs than those of greater relative importance to the economy as a whole, such a livestock in Uruguay.[27] Similarly, they were able to co-opt the institutional technology-generating apparatus without much resistance from competing sectors. Corn in Argentina, rice in Colombia, and milk in Ecuador are examples of intermediate situations.[28] Inversely, a greater economic importance for production provided political presence and power, which in some cases was translated into concrete public policies, and in every case provided certain protection against the implementation of economic policies detrimental to the vital interests of the sector.

Impact of the product on the economy as a whole. Finally, a very important dimension is the specific manner in which the product (in terms of supply, price, etc.) affects the income of other social sectors. This factor was reflected in three basic ways in these case studies: (a) the product was a wage good, and thus its price affected the reproduction cost of the labor force and had an impact on the industrial sector's possibilities of accumulation (for instance, rice in Colombia); (b) the product was a good consumed by the urban middle classes, and when its price affected the income levels of these sectors, it became a highly sensitive political issue (for instance, milk in Ecuador); and (c) the product was an important source of foreign exchange needed for the country's overall economic activity (for instance, corn in Argentina). In the first two cases, the price of the product was a potentially important social issue and consequently became the focus of social conflict in regard to the product's public policies. In the third case, potential conflicts appeared to be limited to the relationships among different capitalist sectors over the issue of how to distribute the surplus generated by export activities.

This brief examination shows that the conflicts particular to each product were determined by the type of market they targeted, and their specific function in these markets. These factors conditioned the interest of the remaining social sectors and their concern with each product. The productive sectors of an item of significance to the economy as a whole had less flexibility of action than those producing secondary goods.

In case of Uruguayan livestock, the product was central to overall accumulation because of its importance in national exports and in domestic consumption.[29] Corn in Argentina, on the other hand, was important as an export product, but was not of domestic significance. This allowed for greater leeway in domestic price policies.[30]

Comparatively speaking, rice and sugar in Colombia, milk in Ecuador, tomatoes in California and potatoes in Peru were less important to the society as a whole, as they had less effect on the overall patterns of accumulation. In each case, however, they affected the overall economy in a unique way. Thus, sugar production in Colombia affected export levels, though much less than the principal export (coffee). It also had an impact on the cost of replacing the national labor force, although less so, because of the significant role *panela* still plays as a cheaper sugar substitute.[31] The price of rice in Colombia had an important effect on the reproduction cost of the labor force, which explains the interest in increasing production.[32] Nevertheless, the fact that the supply and the price could be regulated by controlling imports provided a certain margin for maneuvering price policies. In Ecuador, the accelerated growth of the middle classes and the increased income of the urban working class considerably increased the demand for milk, causing conflicts over milk prices. Milk thus received more attention than other goods that had a more noticeable impact on the cost of reproduction of the labor force.[33] Although potatoes are still important in the national diet of Peru, they are being replaced by rice, especially in the urban areas. This has the effect of weakening the strategic role of potatoes in the national market.[34] Finally, production in the farming complexes of Pernambuco, Brazil was used mostly for home consumption.

Table 4 shows how these five structural dimensions appear in the case studies, reflecting well-defined patterns of relationships. The three dimensions directly related to the organizational and federated expression of the productive sector (type of productive unit, uniformity and regional importance) are closely tied to situations in which the processes of social articulation began in the productive sector. These dimensions lose importance towards the right-hand side of the table. The two dimensions related to the importance of the product in the process of overall accumulation carry more weight in the cases where intersectoral conflicts have not been fully resolved.

Following is a discussion which relates these structural dimensions to the qualitative characteristics of the processes of social articulation that were generated.

Processes of Social Articulation, Public Policies and Technical Change

The structural dimensions described above facilitate an interpretation of the processes of social articulation, so that their similarities and differences can be observed. This sheds light on how they were expressed in specific public policies, including those that act directly to organize institutions for supplying technology. The case studies depict four types of social processes which led to distinctive public policies.

Agrarian initiatives. The cases of sugar production in Colombia, milk production in Ecuador and tomato production in California reflect processes of social articulation which were generated within the productive sector. In all three cases, although to different degrees, the structural dimensions described above operate in favor of the development of unified sectoral actions for negotiating with other social sectors to produce conditions that further their own economic and technological interests. All three cases have very similar production units and are characterized by a preponderance of a particular product in the productive strategy. In addition, the cases of sugar and tomatoes show a rising degree of concentration and vertical integration in the latter part of the period under study. As a result, an active process of federated action emerged, with clear and decided intent to influence governmental decisions. In none of the three cases was the product essential to the overall process of accumulation, and this provided a certain margin for maneuvering the negotiation of economic policies. However, since milk is an important product to urban middle and working class consumers in Ecuador, the productive sector, despite its relatively greater strength, had more difficulties negotiating a price policy in line with its interests. This drawback was compensated by the redistribution of the substantial surpluses in the petroleum industry to the dairy sector through subsidized credit, which became the most important policy tool for accelerating agrarian transformation.

Likewise, in the cases of sugar and tomato production, the productive sector seems to have enjoyed a greater capacity for negotiating with other social sectors, due to the regional nature of production, which increased ability to mobilize regional power structures. This was particularly true in the case of sugar, and was reflected in certain qualitative characteristics of the public policies which were implemented. Both in this case and in the case of milk in Ecuador, public policies typify the ability of capitalist sectors to negotiate with other social sectors on the economic policies they need for removing barriers to further modernization. In Ecuador, the State

TABLE 4 Relationships Among the Structural Dimensions and the Nature of Social Articulation: Eight Case Studies

Structural dimensions	Agrarian Initiatives		Inter-Sectoral Conflicts			Non-Agrarian Initiatives		Subordination with Policies of Surplus Appropriation	
	SUGAR	TOMATOES	MILK	LIVESTOCK	CORN	RICE	POTATOES	ROTATION BRAZIL
Type of production unit: Control over resources, concentration and vertical integration	++++	+++	+++	+++	++	++	+	+
Uniformity of production unit and marketed products	++++	++++	+++	+++	++	+++	++	++
Regional importance	++++	++++	+++	++	++	++	+	++
Relative size of sector vis a vis the economy as a whole	++	++	+++	++++	+++	+++	++	+
Importance of the product to the non-agricultural process of capital accumulation	++	++	+++	++++	+++	+++	++	0

Degree and nature of social articulation and public policies implemented

→ INCREASING DEGREE OF MOBILIZATION AND APPROPRIATION OF THE STATE APPARATUS BY THE PRODUCTIVE SECTOR

NOTE: + Reflects an increase in the importance of the dimension as a characteristic of the case studies.

eliminated the *huasipungo* through land tenure legislation, after the process of expulsion had already begun through the initiative of the landowners. Although increased demand for both sugar and milk motivated the reactivation of production, the resulting policies differed as a consequence of the different roles each product played in the overall economy. In the case of sugar, the policy that was implemented sought to stabilize income for the mills by fixing the price of domestically consumed sugar which contrasted with the varying income received for exported sugar. The increase in the demand for milk, on the other hand, was motivated by wage policies and subsidies, and this made it necessary not only to regulate domestic prices, but also to channel financial resources from petroleum production to encourage capital formation on the haciendas. In both cases, the producers' organizations provided the arguments for setting price levels and establishing the mechanisms needed for implementing credit policies for the sector.

Research findings suggest that the public policies implemented served to strengthen the productive sectors and consolidate the original property and power structure. This boosted the sectoral importance of the mills, helping to consolidate the sugar oligopoly in the Cauca Valley, and guaranteed the predominance of the modern hacienda in the Ecuadorian highlands. This process of economic consolidation and expanding demand spurred the adoption of technology and raised the interest of pertinent social sectors in ensuring and controlling the supply of technology. The organizational means used to solve this problem were different in each situation, and were apparently linked to varying structural dimensions.

The case of sugar cane in the Cauca Valley in Colombia illustrates a complete breakdown of the model of a national institute serving multiple products, and is characterized by the total absorption of the technology generation apparatus by the productive sector. After 1960, public sector research (ICA) was gradually replaced by the mills as they began to develop their own structure for procuring technological innovations from centers abroad. This type of organization was completed and consolidated after 1977 with the creation of CENICAÑA, which was set up to conduct sugar cane research. It institutionalized the process whereby sugar cane activities were removed from the public technology generation/transfer apparatus, to be assumed formally by the productive sector. An important factor in this process was that the new entity—which was formally created as a non-profit scientific and technological corporation—was financed with public funds (National Sugar and *Panela* Fund). The private sector maintained control over the agency, however, through its hold on seven of the nine positions on the Board of Directors.

The cases of milk in Ecuador and tomatoes in California are a different type of situation. Although the public technology generation/transfer apparatus remained as the principal institutional element, the productive sectors exercised almost complete control over the orientation of the activities.

The mechanical and biological components of technology for tomatoes were developed almost exclusively by the University of California, through its experimental system and extension services.[35] The private sector—the farmers' associations and processing plants—became involved in the research process and in the development of new technology by providing funds for research activities. This enabled them to direct the supply of new technology. They made their influence felt through their participation on University advisory committees and through the legislative process that sets University budgets. In addition, further impact was exerted through the system of donations for research on specific subjects. In general, the University's budget covers only the salaries of the permanent staff, and not research expenses *per se*. For this purpose, funding had to be generated from private sources or from specific public donations. It was through the allocation of these funds, which represent a minor but strategic part of total research expenses, that the private sector was able to orient the activities to suit their specific interests.

The hacienda owners in Ecuador had a similar means of influencing the orientation of public sector activities for dairy production. The nature of the mechanisms used in this case, however, were different from those used in California, and were more closely tied to the political power of the landowners. The main component of the technology generation/transfer system for dairy production in Ecuador was the National Agricultural Research Institute (INIAP), which generated technology for the agricultural sector as a whole. Technology was formally the responsibility of the Ministry of Agriculture, although INIAP also shared in these activities through field days, training courses for farmers and staff at the agricultural establishments, and other mechanisms. In terms of organization, INIAP is a typical public sector decentralized agency directly responsible to the administrative power (Ministry of Agriculture). Its only explicit tie to the productive sectors is a representative of the Chamber of Agriculture on INIAP's Board of Directors. In fact, however, as in the case of California, the design and orientation of the institution's program of activities were conditioned by the technological requirements of the dominant productive sectors. The extensive resources devoted to dairy pastures and forage programs are evidence of this fact (Barsky and others, 1980).

Livestock in Uruguay. The case of livestock production in Uruguay presents an interesting contrast with the three cases analyzed above. The product's importance to the economy as a whole, and its role as a wage good, impose two-fold limits on economic policy. First of all, the productive sector's chances of appropriating the surplus generated by livestock production are limited, and second, the State was clearly restricted in terms of the domestic price policies it could implement to protect the earnings of the livestock sector.[36]

The interests of the livestock sector were very uniform, and active producers' organizations, combined with the sector's economic and political importance, made it possible to infiltrate the government apparatus with individuals representing their interests. This power fell short of what could have been anticipated, however, given the importance of the sector. In this connection, the State's creation of the Agricultural Plan was actually conducted by the producers, as it coincided with the pro-agrarian climate that accompanied the political change taking place in that country in the late 1950s. The system basically consisted of the Alberto Boerger Agricultural Research Center (CIAAB), in charge of the bulk of research activities, and the Agricultural Plan, devoted to dissemination. Both were responsible to the Ministry of Agriculture, and although they belonged to the same bureaucratic-administrative network, they shared no other ties. The livestock sector participated in this system to some degree by sitting on the Honorary Commission that administered the Agricultural Plan. The situation improved in the early 1960s, when certain productive sectors demonstrated their interest in incorporating technology from New Zealand, and turned to the State for the needed funding. In addition, these sectors entered the sphere of the Ministry, in an effort to control the agrarian process. Their participation at this level, however, was limited to matters of technology diffusion, and they were able to participate only indirectly in the research carried out by CIAAB.[37] Although this structure was maintained formally, it began to lose strength after 1970, when the prevailing economic conditions, determined by the foreign market, raised doubts about the viability of the proposed technological scheme.

This case, and the case of corn production in Argentina during the first part of the period under study (until the mid-sixties), illustrated situations where public policies are unsuitable for carrying out a long-term technological process. The policies were the product of unresolved inter-sectoral conflicts over the distribution of economic surplus. The two cases have common traits, but different origins. In Uruguayan livestock production, the landowners' initiative failed as a consequence of the inability to overcome, over the long-term, the objective limitations created by the foreign market, and by the importance of livestock production to the economy as a whole. In contrast, the case of corn in Argentina shows how an early strategy of modernization stimulated by non-agricultural social sectors produced positive results, once mechanization reached a certain level and a revolutionary technological innovation (the new varieties) notably increased the profitability of adopting the technology.

Processes of inter-sectoral articulation. The social articulation that took place in corn production after 1960, and affected rice production in Colombia, resulted from the negotiated action of non-agricultural sectors. The success of their action was due to the relative weakness of agrarian sector interests *vis a vis* the rest of society, which promoted the technical change process. In the case of rice, this weakness was due to the productive sector's inability to mobilize regional political interests, the sector's limited economic value, and its low degree of concentration. In the case of corn, it resulted from a lack of sectoral unity caused by the diversity of types of production units and the productive structure's focus on multiple products.[38]

In both cases, the non-agrarian social sectors mobilized in response to changes in conditions that generated new crisis situations or aggravated existing ones. In the case of rice, the deterioration of traditional agriculture and the increased demand for food products caused by accelerated urbanization resulted in food shortages and rising food costs. These factors awoke a certain degree of social interest in technology. The resulting policy sought to use technical change and increased production to contain price increases, in an attempt to harmonize the interests of urban entrepreneurs, who were struggling with the demands for wage increases by their labor force, with those of the working class sectors who demanded limits on food price increases. This strategy, and the relatively small size of the sector, made it possible to regulate import and subsidy policies so that prices could fall but remain profitable, without resorting to the appropriation of agricultural surpluses. As a consequence, credit acquired a new value, becoming a principal means for channeling the State's policy on technology. It was also tied to the adoption of technology, subsidizing the cost of the industrial components of technical change, and thus neutralizing the relatively high cost of capital goods and technological input.

In addition, the State's strategy touched on the area of the generation and diffusion of technology. An example is the rice program, promoted by the Ministry of Agriculture and the Rice Growers' Federation, through which new technology was made available to the farmers. It proved to have a strong impact on increasing yields. The division of responsibilities between private and public institutions (formalized through agreements that detail the activities assigned to each agency: Colombian Agricultural Institute, ICA; International Center for Tropical Agriculture, CIAT; and the Colombian Rice Grower's Federation, FEDEARROZ) was complemented by different types of relationships among the three participating agencies. The most important of these was the appointment of FEDEARROZ technical staff to ICA research activities (Balcázar and others, 1980). It was through this mechanism, and the participation of technical staff in setting priorities for the research programs, that the circuit of private participation in generating and transferring rice technology was completed. It remained, however, within the public domain to define the research policies to be followed.

In the case of Argentina, the State faced problems of a more general nature. The stagnation of the agricultural sector in the pampas affected exports and obstructed the process of general accumulation. The stagnation of production aggravated the traditional imbalances inherent in the Argentine economy and raised doubts about the very continuity of the political-institutional process. Policies, then, reflected the basic problems of the sector, which had lost a significant part of its labor force to industry. These workers had to be replaced by finding profitable ways to incorporate capital. An aggresive policy of subsidized credit was developed and finally resulted in the general mechanization of the sector. The State thus responded to diverse types of political pressure while at the same time attending to the need for technical change suited to the productive sector.[39]

The technology developed by the National Agricultural Technology Institute (INTA) focused on selected varieties of corn, and later on the extraordinary technological innovation of the hybrids, which could be used if suitable mechanization were introduced. The State provided the financial resources, thus making the articulation of the technological process possible, even in the face of extremely variable prices and a price policy that represented, for the most part, a partial appropriation of the sector's surplus.

This process of technical change was essentially non-agrarian in nature, as reflected both in the public sector's leading role in supplying technology and in the negligible influence exerted by the producers. The bulk of the technology generation and transfer process was in the hands of the public sector through the National Agricultural Technology Institute (INTA).

Here, the private sector had a limited degree of participation and/or integration in technology generation activities, and in how the innovations were transferred, implemented and oriented. This is particularly interesting because, more than in the other cases, INTA's organizational model specifically recognized the need to maintain ties between the productive sector and the institution in order to ensure that the activities of the latter served to improve production and productivity in the sector. Moreover, although concrete mechanisms for this purpose are included in INTA's organizational chart, few significant ties were actually made. One example is the Institution's Council of Directors. An examination of the performance of this body reveals no clear efforts by the private sector to orient technical activities according to specific sectoral objectives. Several studies on INTA, and on the activities of farmers' organizations, confirm their lack of interest in orienting technology generation and transfer activities, beyond making limited demands concerning the control of pests, deseases, etc. (CIAP, 1973; Sábato, 1980).

The productive sector's passivity in attempting to influence the activities of the institution may be explained by the fact that multiple products are produced in the Argentine pampas, and the private sector was actively supplying technology. Similarly, multiple-product agriculture explains the relatively successful technological and productive performance of the pampas when compared with livestock production in Uruguay, despite the fact that the economic policies in both cases shared certain traits, especially in regard to the appropriation of agricultural surplus and price variation. The basic difference is that Argentine farmers could distribute the risks associated with international prices among several products, with a consequent reduction of their specific technological demands for certain products. This structural advantage, in turn, was considerably increased through the special effectiveness of the technological innovations generated for corn at the international level.

Stagnation. Finally, the two cases of campesino economies included in this study illustrate situations of poor technological articulation and stagnation. This was caused by the absence of an economic and technological policy consistent with the needs of the productive sector, which in turn was unable to mobilize and influence the State. As a result, the public policies that were actually enforced were nothing but a reflection of overall, general policies whose objectives were unrelated to the modernization of the productive sectors.

In the case of potatoes, the food policy adopted for stimulating commercial production, and the fact that food was imported to supplement its supply, only served to heighten the segregation of traditional crops usually consumed in the rural areas. In the case of the Brazilian Sertão, the replacement of traditional crops by livestock was accelerated by the State's management of credit which targeted the more highly capitalized livestock holdings. Both cases reveal how public policies accentuated the unequal development between the commercial sector and the campesino economies, by promoting the expansion and consolidation of commercial endeavors.

SUMMARY AND CONCLUSIONS

This article has presented a comparative analysis of a limited number of case studies selected to represent the major processes of technical change and modernization that have taken place in Latin America since 1960. The case studies clearly show the endogenous nature of technical change, with regard to more general social processes on two levels of analysis. The first level corresponds to certain world-wide (or continent-wide) processes that have predetermined some of the overall components of the economies in Latin America. They also predetermined the development of technological institutions and the availability of technology at the international level. The second level involves the innate characteristics of each production situation, including how it related to and fit in with the national economy.

The interpretative discussion focuses on this second level of analysis. It seeks to link the development of several mutually determined structural dimensions with the appearance of certain processes of social articulation that allowed the observed processes of technical change to take place. It has already been noted that, of all the dimensions used in the analysis, the most important in determining technological behavior were the type of productive units, including the organization of production, the degree of homogeneity, and the regional importance of the product. This was especially true in those cases in which the process was the result of an agrarian initiative. In these cases, the agrarian sectors, with a certain degree of concentration and political clout, had the opportunity to develop effective institutional negotiating mechanisms from inside the government apparatus. Thus, they mobilized public power to their own benefit. This co-opting of the government produced active processes of technical change, preferentially oriented toward solving the internal problems of the producing sector. As a result, technical change had its major impact on modifying the productive process, allowing for a better control of labor.

The processes of social articulation driven by non-agricultural sectors are determined by such elements as the structure and importance of the non-agrarian sectors, the nature and operation of the government, and the availability of international technology. As a result, public policies acknowledge a higher degree of independence from the agrarian productive sector and generate technological processes which bring about major increases in production and productivity, where benefits are distributed more broadly among the different sectors of society.

Finally, the cases of social desintegration observed in campesino production (potatoes in Peru and rotation in Brazil) clearly demonstrate how the non-mobilization of producer sectors, and the unimportance of that specific production for the process of overall capital accumulation, prevent the emergence of public policies that will initiate a dynamic process of technological modernization.

NOTES

[1] See the work of T. Schultz (1964), one of the most renowned defenders of this viewpoint, and the documents on U.S. and international foreign aid programs of the 1950s and 1960s. During the 1960s, the subject of Agrarian Reform was also of great importance.

[2] It should be noted that, despite production increases in some countries for some products, overall food production barely surpassed population increases.

[3] The case studies are on: rice and sugar cane in Colombia; potatoes in the Peruvian highlands; dairy cattle in the Ecuadorian highlands; corn in the pampas of Argentina; beef cattle in Uruguay; farm rotation (beans-cotton-corn) in the State of Pernambuco in Brazil. In addition, a case study was performed in the State of California (United States) for purposes of comparison with a case in a developed country whose productive structure—for historical reasons—shares some traits with Latin America.

[4] This is a summary based primarily on Balcázar and other (1980).

[5] It is worth comparing other countries with Colombia. In Japan, the development of dwarf varieties began at the turn of the century. By 1920, these strains had spread to Taiwan (Dalrymple, 1976), which was still a Japanese colony, leading to an active plant breeding program. This early development explains the major increases that were made in production and yields in these countries after the end of World War Two. Yields approached three tons of paddy rice per hectare by the end of the 1950s (Chandler, 1979). By contrast, in other Asian countries such as South Korea, the Philippines, India, Pakistan and Indonesia, the production increase was not felt until the end of the 60s, the same time it was emerging in Colombia (Chandler, 1979; Dalrymple, 1976).

[6] For a detailed discussion of these figures, see Balcázar and others (1980). In addition, Montes and others (1980) estimate that labor used in 1975 totalled 12,937,000.

[7] This process of reduced labor use, although relatively limited, stands in contrast with developments in Asia, where the incorporation of HYVs and the implied technological package meant an increase in the total use of labor (Chandler, 1979:63; IRRI, 1977).

[8] Scobie and Posada (1978) argue that this process benefitted the urban wage-earning sector. By contrast, de Janvry and Crouch (1980) and Balcázar and others (1980) argue that the falling rice prices were internalized through wage fixing mechanisms. It is important to note that the real wages of the metropolitan population remained constant from 1969 to 1979 (FEDESARROLLO, 1980:52).

[9] This comparison is probably more significant for interpreting producer behavior than any comparison of hypothetical producer and consumer surplus estimated on the basis of the prices that would have existed had no technical change taken place. This is because producers are not aware of the income they would have received under these circumstances. See also the estimate given by Montes and others (1980).

[10] This is a summary based primarily on Piñeiro and Trigo (1983:Chapter 7).

[11] It is interesting to note that Argentina was far behind in this process. By 1949 in the United States, some 90 per cent of the area planted to corn in the major corn-producing state (Iowa) was using hybrid corn. Similarly, public sector research efforts in Mexico began to study corn hybrids and varieties of open pollenization in the early 1940s. The impact on production and yields was almost immediate (Hewitt de Alcántara, 1976). These efforts were consolidated and expanded with the establishment of the International Center for the Improvement of Corn and Wheat (CIMMYT) in 1963. However, in spite of the obvious time overlap between the establishment of CIMMYT and the initiation of the process of dissemination of hybrid corn varieties in Argentina, the two events appear not to be closely related, as hybrid seeds used in Argentina came mostly from commercial lines developed by the private sector (Cargill, Morgan, Continental, etc.).

[12] This summary is based on the work by de Janvry, LeVeen and Runsten (1980). It is useful in understanding events in developed countries under circumstances in which the historical development of the agricultural sector shows traits similar to those in Latin America.

[13] Because this type of relationship existed between production and processing, the tomato industry could be described as a situation of formal subordination of agricultural production to agribusiness capital (da Silva, 1980).

[14] This is a synthesis based on Piñeiro and others (1982).

[15] It is interesting to note that there are no major differences among the various size strata in terms of per hectare and per cow production. Cows in the Cayambe valley produce around ten liters per day per head, and per hectare production is around 2,200 liters. This was similar for all strata, except those containing over 500 hectares, which had lower yields per cow (seven liters per day) and more per hectare (3,000). The opposite occurred in the Machachi valley, where an inverse ratio was found between per cow production rates and per hectare rates for livestock land and farm size.

[16] This was possible because of increases in the coefficients of extraction.

[17] Eighty per cent of this land followed the New Zealand model.

[18] From 1970 to 1974, when expansion reached a maximum, total livestock units increased by ten per cent.

[20] In the case of irrigated rice, however, the number of farming units exceeding 30 hectares increased from 39 to 50 per cent from 1959 to 1970.

[19] A description of the method used to select the case studies can be found in Piñeiro and Trigo (1977).

[21] This section was taken from Piñeiro and Trigo (1983:Chapter 11).

[22] The organization of this sector is also determined by other elements, such as the technical characteristics of the product and the nature of the markets within which the productive units operate.

[23] This categorization of the case studies corresponds to the early 1960s, or the beginning of the technological process studied.

[24] In the case of sugar, 17 mills controlled the entire industrial process. These mills, in turn, had a vertical integration ratio in 1977 of 53 per cent with the agricultural process, and 40 per cent with the distribution process (Piñeiro and others, 1982). Tomato production in California in 1974 involved 1,493 productive units, averaging 174 acres of tomatoes, and 28 processing units. Both types of units had experienced rapid concentration. In 1954, the number of productive units was 2,896, and that of processors, 56. The gross income for a typical unit was US$225,000 in 1974 (de Janvry, LeVeen and Runsten, 1980).

[25] In the four principal rice-producing Departments of Colombia, most production takes place on units ranging from 20 to 200 hectares (Tolima 83%, Huila 65%, Meta 63%, Meta dry zone 83%, Ibagué 80%). Rice is the major product of these units. Average size of the irrigated rice farms was 91 hectares in 1972 (Montes and others, 1980). In the case of sugar, the units produce sugar only. Here, production units vary in size and nature, including independent farmers. administered farms, and mills. The 17 mills, however, were responsible for the dynamics of sectoral organization. In the case of milk in Ecuador, transformation occurred as a result of the expansion of milk production, which became the main commodity of the production strategy. Tomatoes in California and livestock in Uruguay show more diversity in terms of the size of the productive units and the combination of products. Nevertheless, in the first case, a rapid trend was observed towards concentration, and the sector became increasingly uniform. A farmer's association also played an important role in that case.

[26] These units usually produce corn, wheat and beef, and in smaller quantities, soybeans, sunflowers and milk.

[27] In 1968, the livestock complex in Uruguay represented 26.6 per cent of the GDP and provided more than 80 per cent of total exports. By contrast, in 1976, sugar in Colombia represented four per cent of agricultural production and about one per cent of the GDP. Sugar exports accounted for about three per cent of total exports. (Piñeiro and others, 1982; Kalmanovitz, 1978).

[28] In 1976, rice in Colombia represented 9.2 per cent of the value of agricultural production and 2.4 per cent of the GDP. (Kalmanovitz, 1978).

[29] In Uruguay, beef is the most important source of calories after wheat. In 1975–1977, beef provided 17.4 per cent of total calorie intake (CIAT, 1981).

[30] Corn in Argentina provided 1.3 per cent of total calorie intake (CIAT, 1981).

[31] In 1976, the value of *panela* production was twice that of sugar production (Kalmanovitz, 1978).

[32] In 1975–1977 rice contributed 13.1 per cent of total calorie intake (CIAT, 1981, based on FAO's nutritional balance tables). In turn, 71 per cent of the rice was consumed by the population receiving less than 20,000 pesos annually (this figure corresponds to 1970, which was equivalent to about $600. (Scobie and Posada, 1978).

[33] In 1977, milk contributed ten per cent of total per capita calorie intake, while 76 per cent of the total income spent on milk consumption in Quito came from the top two quartiles of the income levels (CIAT, 1981, from ECIEL data, Brookings Institute, Washington, D.C.).

[34] In 1975–1977, potatoes provided 5.6 per cent of the total calorie intake, in comparison to 11.4 per cent for rice (CIAT, 1981).

[35] For a complete description of the structure and evolution of the institutional technology transfer/generation model for tomatoes, see de Janvry, LeVeen and Runsten (1980).

[36] Between 1964 and 1974, the State collected about 20 per cent of export earnings.

[37] During the 1960s a substantial change was noted in the orientation of CIAAB's activities. It intensified its research into livestock production, and especially into the type of technology promoted by the Agricultural Plan. This change in orientation, however, does not appear to have been directly linked to the private initiative promoted by the Agricultural Plan, but rather seems to have resulted from a renewed interest in technology by the State apparatus itself. Obviously this renewed interest could not be separated from the formal incorporation of the private sector into the Ministry through the Honorary Commission of the Agricultural Plan; what should be emphasized is that there was no formal connection between the two phenomena, and that both result from the overall political climate that favored the livestock industry during that period. In addition, the livestock sector that worked with the Plan was generally more interested in the direct incorporation of technology from New Zealand than in generating and or adapting technology at the national level.

[38] In the case of corn, other elements besides intrasectoral factors came into play. These were the extensive channeling of agricultural capital into investments in the industrial and financial sectors, and vice versa, and the significant power of the industrial working class sector to resist a price policy contrary to consumer interests.

[39] Between 1956 and 1974, the amount of subsidized credit used for purchasing tractors covered from 19 per cent to 70 per cent of the tractor's price (Sábato, 1980).

PART III

AGRICULTURAL RESEARCH ORGANIZATION IN LATIN AMERICA: ISSUES FOR THE FUTURE

INTRODUCTION

A number of issues discussed in the previous chapters indicate that, over the past two decades, agricultural research and technology generation have experienced dramatic changes. These changes affected mainly the role of the public sector in the innovative process, particularly with respect to its capacity to direct the process in accordance with national resource endowments and overall development objectives.

The results of PROTAAL show quite clearly that the preëminence of public sector activities in regard to the generation of new agricultural technology is not a permanent or "natural" situation. The case studies reflect significant changes in the role of public research organizations as economic development evolves. As markets for new technologies develop and the generation and diffusion of new technological knowledge becomes a profitable activity, private organizations increase their participation as social carriers of technology. The changing nature of the institutional organization of agricultural research is a natural consequence of economic development in market economies. In this context the institutional model implemented since the mid-fifties to foster agricultural modernization with a heavy emphasis in public research is increasingly less able to determine the rate and direction of technological change.

In this model, the overwhelming importance of National Research Institutes with respect to the total research effort in each country implied that the resource allocation process within the institutes constituted a national technological policy. At present, the diversification of the institutional focus, through which new technology is made available to agricultural producers, implies that the development of new ways for the public sector to be able to guide technological change becomes one of the more pressing issues for the immediate future. In addition, this transformation of institutional character is a process not restricted to the national level. Twenty-five years ago international flows of agricultural technology existed, but were mostly confined to the exchange of scientific knowledge among countries. Today, agricultural technology has became an international commodity. There has been a dramatic increase in the activities of multinational firms now participating in many areas of the technological process, and there have been important developments in public and international organizations. In the latter, the most important event is the appearance of the International Agricultural Research Centers, three of which are actively operating in Latin America. This increasing internationalization of the research and technology diffusion processes has had important consequences at the national level, and should also be taken into account when discussing the future organization of the national research systems.

This part of the book includes three chapters. In the first by Trigo and Piñeiro, the overall issues regarding the nature of the modernization process in Latin American agriculture, the dynamics of research organization and the internationalization of the technological system are summarized as background information for the discussions of the design of a science and technology policy for the agricultural sector. In doing so, the need to recreate a role for the public sector in ways more compatible with the present day institutional environment, and the nature and limits of science and technology planning mechanisms, are given special attention. The other two chapters, by Coulter and Evenson and Evenson discuss in depth the two main institutional developments in the last decades: the development of the International Research Centers and the growing participation of the private sector. These developments are analyzed in relation to their implications for the future organization of national research activities.

FOUNDATIONS OF A SCIENCE AND TECHNOLOGY POLICY FOR LATIN AMERICAN AGRICULTURE

Eduardo Trigo and Martín Piñeiro

ABSTRACT

This paper seeks to establish guidelines for the definition of a science and technology policy for Latin American agriculture during the 1980s. It is suggested that the profound transformations undergone by the agricultural sectors in most countries have substantially altered the role of the public sector in the process of innovation. The establishment of the national research institutes is discussed, as well as the context in which technology became an important social question. The emergence and resolution of problems within the institutes is analyzed, and priorities for technological policy are considered.

INTRODUCTION

The purpose of this article is to set down general guidelines for the definition of a science and technology policy for Latin American agriculture during the 1980s. The central argument is that the profound transformations undergone by the agricultural sectors in most countries have substantially altered the role of the public sector in the process of innovation. The article is divided into three parts. The first describes the historic context in which technology became an important social question, as well as the establishment of national research institutes. The second evaluates the impact of this institutional development and describes the present backdrop against which new problems and conditions have appeared. In the third, a number of analytical and interpretative factors will be discussed as a basis for determining priorities in scientific and technological policy.

TECHNOLOGY AS A DEVELOPMENT TOOL: THE NATIONAL RESEARCH INSTITUTES

One of the most striking features of the postwar period was a new awareness of the tremendous effect that science and technology could have as tools for transforming society. At the same time, a perception began to take shape in academic circles and among those responsible for defining public policy in developed countries: the possibility of using and controlling technological advances as means for attaining pre-defined objectives, and of transferring them toward other countries with lower levels of relative development.

In the case of agriculture, this point of view was of particular significance. The most important aspect was the idea that at the international level there was a wide array of new technologies available for potential adoption in Latin America; what was needed were institutional infrastructures capable of adapting them to local conditions. This was the base for the development, from the mid-1950s on, of national research and extension institutions. They were supposed to link the receiving countries to the centers where the new technologies were being generated.

These new institutions had a significant effect on technological innovation in Latin America. They meant new and renewed efforts by the public sector to generate and transfer technology, and they mobilized public opinion concerning the need to make larger efforts in this area. In addition, they played an important role in linking national activities to events at the international level. These efforts had a considerable impact on the levels of production and yields of a broad range of products in which the economic conditions needed for the technology adoption process were also present.

Thus, contrary to what has been accepted generally, yields and production for certain products have increased significantly in a number of countries on the continent (Trigo, Fiorentino and Piñeiro, 1978). This is a result of both the expansion of international technology transfer and national efforts, which have accelerated the agricultural modernization process. This agrarian transformation process had been the key reason for the establishment of research institutes. It fit into a more general strategy to modernize the government as an agent of transformation in Latin American economies.[1] There is no doubt that, from this standpoint, the strategy was a success.

However, a later analysis of this process suggests that, like all historical processes, it led to a number of new problems that must be confronted today; future efforts to strengthen national research capabilities must be carried out under different conditions. As these problems arose, the national research institutes entered a period of reduced effectiveness (Trigo and Piñeiro, 1981; Piñeiro and Trigo, 1983: Chapter II). The three most significant problems are:
  (a) The qualitative nature of prevailing agrarian modernization processes.
  (b) The appearance of a more complex institutional model for technology generation, made up of different kinds of organization, and within which the private sector becomes increasingly important.
  (c) The growing internationalization of technology supply.

THE IMPACT OF TECHNOLOGICAL INNOVATION IN THE PAST TWO DECADES

The Qualitative Nature of the Agrarian Modernization Process

Efforts by the public sector in relation to the development and diffusion of new technology resulted in the genesis of an agrarian transformation process. However, this process has not been the same in all Latin American countries. It has focused only on those products that possessed certain specific conditions of productive structure and that are covered by appropriate public policies in the economic sphere (Piñeiro and Trigo, 1983: Chapter II). Several qualitative characteristics of the modernization process had specific impacts on productive structure. In the first place, technical change concentrated on production situations dominated by commercial agriculture, especially of two specific types. One is commercial grain production, in which major technological innovations have been made at the international level, notably in the field of genetic improvement. Illustrations of this situation include rice production in Colombia, corn and wheat in Argentina, and wheat in Mexico. A second situation is associated with items produced in highly concentrated agrarian structures, which allowed the developments of active interest groups. Examples include coffee and sugar in Colombia, and dairy production in Ecuador. In these cases, technical change increased the economic concentration of the sector by fostering vertical integration and the development of oligopolistic behavior in the product markets (Piñeiro and Trigo, 1983: Chapter II). Technical change has been highly capital intensive per unit of production. In a number of situations, despite rising total production, innovations actually produced a net displacement of labor. For example, the labor used for rice production in Colombia from 1969–1978 dropped by some 15 per cent in spite of production increases of over 100 per cent.

This capital intensive technology is introduced even when labor costs are falling by comparison with the price of capital goods, which shows that available technology is strongly biased toward promoting greater use of capital. This highlights the impact that technology generated in the developed world has on the technology supply at the national level, and it points to the lack of corrective mechanisms in the international technology transfer process.

These features of technical change and its natural concentration on commercial agriculture have had a destructive effect on small farm economies. Since they do not benefit from technical change and thus have steadily lost their ability to compete with commercial agriculture, they are eventually eliminated from the production of the crops which have more dynamic markets and greater economic promise.

The Dynamics of the Organization of Agricultural Research and the Participation of Non-Public Sectors[2]

In most countries of Latin America the institutional model adopted for technology generation, and in some cases transference activities, was that of the decentralized national institutes. This model evolves from the need to have an organizational structure adapted to the unique role that the State plays in the process of generating and transferring agricultural technology. This unique role is based on: (a) links of a theoretical nature between the research process and the economic organization of the agricultural sector, which result in a need for public participation in agricultural research activities;[3] and (b) the empirical fact that public participation has usually played a key role in technological change.

In certain situations, especially during the initial phases of the development process, these factors combined in such a way that technology assumed the nature of a public good. Thus, the government had to shoulder the activities of generating and transferring technology. In this context, the model of decentralized institutes, as proposed and implemented in the countries of Latin America, appears logical and functional. Nevertheless, the modernization process introduced a number of important changes in the economic context, particularly in terms of market incentives for investment in non-public research. These changes have substantially modified the context in which the public research institutes operate, ultimately changing the very nature of the institutional model.

In this sense, perhaps the most important issue is that agriculture in general is assuming increasingly capitalistic relations of production, and this has important consequences for the opening and growth of technological input markets. This trend is reinforced by the fact that the output of the commercial sector grows more rapidly than that of the small farm sector. Similarly, the modernization process forges new ties between the agricultural sector and the rest of economy and strengthens existing ties. This allows for the development of communications infrastructure and of services in general, which can lead to an even greater expansion of markets for new technological inputs. As changes take place in market conditions for new inputs, the availability of technological potential in the countries is also modified. This is a direct result of public efforts to develop initial basic information (soil studies, botanical inventories, etc.), to train human resources for research, and to develop and reinforce connections with the international technological pools.

In Latin America these processes led to the emergence of potentially profitable markets for technological input, and activated private interest in the generation and transfer of technology. The trend was particularly marked for technologies that can be protected by patents and other mechanisms, thus promising recovery of initial investments in research and development. To the extent that these inputs made up increasingly large proportions of the technological package involved, technology gradually lost its initial character as a public good, and tended to be driven more and more by market forces.

The process of "mercantilization," however, is not limited to physical inputs. It also extends to other areas in the technological process, including management and farming practices in general. In farming practices, the increasing institutional development of different types of producer associations plays a role of fundamental importance. Private sector indifference toward technology investment is due in part to the fact that existing farms are too small to provide sufficient income for covering investments; furthermore, those who generate new technology find it difficult to protect their property. The appearance and consolidation of producer associations reduces or completely removes these constraints. What would perhaps be impossible or unprofitable for the individual agricultural enterprise may become feasible once production units have joined together on the basis of common interests and problems.

The entry of new interests into the field of technology has caused the institutes to lose their initial central position, becoming one component in a multiorganizational complex which also includes private enterprise and different types of producer groups.

The Increasing Internationalization of the Agricultural Technology Process

The appearance and consolidation of permanent international organizations is directly related to the advancing of the modern technological process and the increasingly international nature it has acquired. In general, technological know-how is highly location specific during adoption; however, in the earlier phases of the process, this know-how is much more general and, as a result, its development is usually free of any particular geographical or ecological considerations. This has increasingly made it possible to develop a world-wide flow and exchange of information. Examples of the benefits conferred by international technology are the genetic development that has been made by the sugar industry since the turn of the centry and, more recently, the development and dissemination of new varieties of wheat and rice (Evenson, 1977:209–236).

Two types of phenomena occur in this transnational technological process. The first is directly related to the developments described above; it has to do with the role that transnational enterprises and the international trade of technological inputs play in defining the technology supply in each country. The second is the development during the last twenty years of a system of International Agricultural Research Centers and, more recently, of the cooperative programs among national research organizations.

The participation of transnational companies in producing technological inputs has tended to accelerate the swing toward the private sector in technology generation. These companies operate in several markets at the same time and thus have been able to make investments in research and development of new products that could not be justified by the level of demand for inputs in any individual country. Thus, private participation in the generation of agricultural technology is progressing more rapidly than the situation in any one country would lead one to believe. This participation, together with an increasing technology supply—the result of the growth of commercial agriculture *vis a vis* traditional farming—also means that technology is becoming increasingly dependent on patterns of international trade and on business's capital investment policies.

The international agricultural research centers have direct roots in the success achieved by the agricultural science programs of the Rockefeller and Ford Foundations, conducted in Asia and Latin America in the 1940s and 1950s. However, their history goes back to the colonial era, when research centers were active with such tropical products as sugar cane, rubber, and pineapple. The swift growth of the system, the total budget of which soared from US$1.1 million in 1965 to US$150 million today, is a result of two key factors (Coulter, 1981; Ruttan, 1982).

The first of these factors is the possibility of achieving a high rate of return on research investments, a view supported by the rapid dissemination of Mexican varieties of wheat and of the rice varieties developed by IRRI in the Philippines. The second factor is the model of the international center as a pragmatic mechanism for generating the necessary technologies and making the countries aware of their importance, so as to create political support for the centers themselves (David Bell and Lowell Hardin, personal communication). These two factors go hand in hand with the increasing interest on the part of multilateral organizations in finding alternatives to the system of technical assistance contracts, which had been used previously for channelling international contributions to research and rural development. These contracts had shown their effectiveness as mechanisms for an "institution building" strategy but they were not effective for the development of more productive technologies. The model of the international center emerged as a more efficient mechanism for conducting certain types of research, thanks to the greater stability and geographic coverage that is offered.[4] In this sense, the centers are an important contribution to the technological system of the region. However, from the viewpoint of the national organizations, they also introduce a number of problems.

As initially conceived, the international centers were set up to complement the national research institutes. They have the advantage of concentrating a critical mass of resources around specific problems common to more than one country. As a part of this concept of complementarity, the national systems are crucially important elements for the productivity of investments in the international system, as they are the natural, inevitable tie between the centers and the productive system of each country. This operational dependency between the international centers and the national organizations has inadvertently tended to convert the centers into "interest groups" which try to direct the activities of national organizations in certain directions consistent with their own programs and mandates, but not necessarily those of high priority for the various countries. Thus, the activities of the centers tend to weaken the effectiveness of the national organizations, ultimately having the same effect as the development of producer organizations and of the private sector. In addition, the centers have ended up serving, in certain situations, as alternatives to national organizations, and this has produced dropoffs in budgetary support to the national organizations.[5]

REFLECTIONS ON SCIENTIFIC AND TECHNOLOGICAL POLICY

Introduction

Scientific and technological policy for agriculture in Latin America has been dominated by two fundamental, closely interrelated concepts. The first is the role of technology in the process of agrarian modernization. In accordance with the theory put forth by Schultz (1964), the lesser developed countries can overcome their technological deficiencies by adopting the technology available in more developed countries. This view does not take into consideration the possible undesirable effects that certain types of technologies might have on income distribution or development style. The second concept is the role assigned to the state in the technological process. It is assumed that, given prevailing economic conditions, the private sector has no interest in the process of generating, adapting and diffusing new technology. Consequently, the government must take the initiative and responsibility for agricultural research.

These concepts provided a useful basis for the policies on agricultural technology which were implemented in most Latin American countries. The policies all tended to separate agriculture from the rest of the scientific and technological system and to provide for extensive participation in research by the public sector. Because the private sector was not participating in research, resource allocation within public sector organizations dictated research priorities and, indirectly, the supply of technology. The interpretative analysis of the process of modernization and technical change in the agricultural sector, as presented here, stands in contrast to the concepts that have guided technological policies in recent years. This raises a number of questions.

The first issue is that the technological process should be interpreted as a phenomenon endogenous to broader social processes that affect both the supply and the demand for technology. Consequently an effective technological policy can not be restricted to actions directed to manipulate the supply of technology. Analysis clearly shows that the presence of technology has been only one of many ingredients in the process of technical change, which also requires the presence of economic conditions that make the adoption of new techniques attractive. Thus, an effective technological policy must include policy tools to affect the supply and demand of new techniques. The second point to consider is the legitimacy of the role assigned to the public sector in generating technological knowledge. The growth and development of market economies requires the private sector to increase its capabilities and interest in participating in activities related to the creation and diffusion of new technologies (Edquist and Edquist, 1979).

This means that the role of the state in research must be redefined. Institutional mechanisms should be developed to ensure that the functions of the public sector will be carried out, and to guide and coordinate the functions of the other sectors involved. In this context, the following operational aspects of technological policy are of considerable importance: Is it possible or desirable to plan technological policy for market economies? What is the best way to organize public sector research institutions? What functions should these organizations perform? What is the role of the international organizations, and what should the countries expect and demand from these organizations for improving the efficiency of the overall system? In view of the undesirable consequences of the qualitative nature of technical change in the past, and the limitations that science imposes on planning efforts, is it possible to define an autonomous technological pattern that has different requirements for the use of factors and with different effects on productive structures? While it is impossible to discuss these questions in depth here, certain factors that emerge from the analysis can be briefly explored.

Planning Technological Policy

The traditional model that became common in Latin America after the 1950s was based on public sector manipulation of the technology supply as a means of influencing the modernization process. Experience in recent years has shown that market forces lead to major transformations in the operation of the technological process and, in particular, in the role that the state may play in guiding this process. Thus, the focal point of the discussion is the close relationship between the degree of planning of technical and scientific policies, and the degree of planning of the economy for a particular society. If the planning capabilities are present, technical and scientific policy may become a tool for direct action on sectoral decisions. When the state's mechanisms for planning and controlling the economy are weak, the state is reduced to playing a subordinate role in scientific and technical policymaking.

The traditional approach has been to manipulate the technology supply. The unit adopting the innovations has been viewed as a simple receiver of a technological pattern defined by the public sector. However, experience seems to indicate that it is not the type of technology offered that is important, but rather the ability to influence and guide the demand for new technological know-how. In this sense, policy tools of price, credit, and inputs, among others, condition the economic context in which the production unit makes its technological decisions.

Organizing Agricultural Research

The goal of the institutional model for technology generation adopted in Latin America has been the improved diffusion of technology already available in the developed countries, with the result that those countries' priorities for technological development were also adopted, within limits imposed by resource availability. In general, the resulting available technology has been capital intensive and has centered on products of temperate climates and forms of production appropriate to the natural resources of the developed countries. Thus, the processes of economic concentration, noted above, have been set into action. This trend has been further accentuated because the research organizations lack adequate mechanisms of integration into the productive sectors, often because they are part of the public sector. At this time there is a need to review the current institutional models for the generation and transfer of agricultural technology. Such a review should include both the structural features and the operational components of the models.

The question of whether or not a given institutional model is truly functional calls a number of considerations into play. These include: the overall and relative importance of the agricultural sector; the composition of the product and its concentration and regional homogeneity; target markets for production; the prevailing type of social organization (types of enterprises, presence and type of trade unions or other organizations, etc.); the type of political organization; and the historical background of each institution. Most models implemented to date have tended to imitate successful experiences instead of producing original institutional designs based on the needs, requirements and limitations of each Latin American case. In operational terms, a key area of concern is to improve coordination with the productive sectors and to develop the capability for making fuller use of native productive potential.

The Role of the Public Sector in Generating Agricultural Technology

Special importance in the new multi-organizational situation is attached to the gradual breakdown of the government's ability to guide the technological process, and the role that the government should play now and in the future. As private interests, guided by market dynamics, have increased their participation in the process of technology generation and transfer, the play of market forces has become the major force governing technological growth, and hence commodities, priorities of clients, and types of technology produced. There is no overall policy to guide these private activities, and as a result, the technological variable has lost its influence as an active tool in agrarian decision making. This phenomenon is particularly relevant in view of the importance that agricultural sectors have for national development in Latin America. These sectors help allay the balance of payments problems, while technology plays a crucial role in solving world problems of food production.

The situation today, as described, suggests the need to rethink the implementation of scientific and technical policies for agriculture. If this is done, the public sector will be better able to guide the multiplicity of public and private organizations that are involved in technology generation and transfer, so as to tap the full potential of the new organizations. Two things must be done: (a) achieve coordination between public agencies and the new institutions emerging in the private sector; this could take place within the general framework of technological policy coordination (National Science and Technology Councils), or at the sectoral level through councils or coordinating committees for agricultural science and technology; and (b) establish specific tools that will permit the government to exercise its full capability to coordinate and direct technological change. These tools include patent laws, technology imports, and the monitoring and auditing of the financial mechanisms for research investments. The governments (public sector) should continue to be direct participants in agricultural research, but in ways which compliment, not compete with, the private sector.

New, private organizations focus their attention on those types of technology which, by their very nature, allow for private appropriation of profits. Consequently, these activities cannot be expected to cover:

(a) The development of functions for generating a "technological potential;" without these functions, the ability of the rest of the system for developing new technologies would quickly be exhausted.

(b) Specific activities which, due to their generic nature (painstaking research, etc.) and to their low probability of bringing about immediate results, will not be assumed by the private sector.

(c) The development of certain types of technology which do not require inputs, such as cropping practices, pasture management, etc., in which the private sector has little interest due to the difficulty of private appropriation of their benefits.

Private sector institutions have a very specific coverage, associated with products and conditions typical of commercial agriculture and particular forms of corporate organization. This means that a broad range of users neglected by the new institutional formats can be served only by public organizations. In this new context, the participation of the public sector should be selective, giving special attention to the groups bypassed by the private sector and based on the needs of a comprehensive technological policy. Similarly, it becomes more important to wield sector-wide mechanisms to guide the activities of the other components of the new institutional model. The institutional formats, as well as the type of mechanisms that will be used, depend on the nature and background of each situation.

The International Nature of the Technological Process

World-wide developments also have much influence on scientific and technological policy tools and on the role of the public sector in the technological process. The importance of technological innovations has grown steadily, and international trade is responsible for their distribution. The most important issue for the International Centers is the development of liaison mechanisms between the national and international levels. This will make it possible to improve the use of available resources and to define international priorities consistent with the needs of the national programs. This topic cannot be discussed without taking into consideration the comparative advantages and functional limitations that each of the institutional components presents for the various types of research activities that must be developed.

Several authors have suggested that the research process can be oriented toward the following four types of activities (CGIAR, 1981): (a) *basic research* for the discovery of new knowledge; (b) *strategic research* for solving specific and predetermined scientific problems; (c) *applied research* for the creation of new technology; and (d) *adaptive research* for adapting technology to the specific conditions of a given locale or production system. The International Centers have a comparative advantage in applied research and, to a lesser extent, in strategic research for specific areas such as genetics. These advantages have been demonstrated by the actual outcome of research efforts, and the logical consequence is that the national agencies concentrate their efforts on adaptive research, on particular crops specified in the mandates of the Centers, and in general on products not studied by the Centers. National priorities must be reoriented, and organizational structures altered.

The Limits of Autonomous Technology

Many studies have analyzed the effects of technical change on factor use, income distribution and other economic variables directly related to development style. In response to the concerns generated by these studies, one school of thought suggests that the technological pattern can be guided by certain parameters determined by qualitative measurements. In its most extreme form, this school of thought finds theoretical backing in the concept of appropriate technology as proposed by Schumacher (1975).

This viewpoint suffers from two important theoretical problems related to the implied assumptions on which the hypothesis is constructed. The first problem has to do with the limitations of planning efforts for science and with the growing role of the private sector in the technological process. These issues have already been discussed. The second problem is related to the concept of appropriate technology. The thrust of this concept is the search for a technological pattern adapted to the relative availability of factors in the lesser developed countries, which are characterized by abundant labor, scarce capital, and small production units. However, there are two important points to keep in mind. In the first place, the new technologies must also be efficient for market economies. This means that they must be capable of generating an average factor productivity equal to that of capital intensive technologies, so that the production units can remain competitive on the market. In the second place, the technologies must also be efficient in open economies so that the production will be competitive in international markets.

This problem must be analyzed while keeping in mind that the fundamental purpose of technical change is to develop more efficient productive processes for energy transformation (Boulding, 1978). Therefore, any restrictions on how much capital can be used or on how to use it, serve to impose restrictions on the range of possible scientific discoveries. This argument suggests that capital intensive technologies are easier to invent than labor intensive technologies, as has been clearly demonstrated throughout the long history of technological innovation.

A related problem is the fact that the technology available to developing countries does not constitute the entire universe of theoretically possible technologies, but rather a sub-set of them, and they have been developed in the industrialized countries in accordance with the conditions of relative factor prices. As a result, the imposition of restrictions on the type of technology can also reduce the utilization of scientific discovery.

It is worthwhile to note that, historically, the processes of technology adoption in Latin America have been associated with price and credit policies that use capital subsidies to distort relative factor prices and place them on a par with the prices available in the developed countries where the technology was created. This leads Sábato (personal communication) to suggest the idea of perverse, but inevitable technology. If production is to increase, technology must be adopted; but the types of technology available imply shifts in relative factor prices. These shifts, in turn, lead to overuse of capital and the perpetuation of structural unemployment.

NOTES

[1] This strategy received most of its momentum from ECLA; IICA played an important role in the specific case of the agrarian sector.

[2] The key factors of the dynamics of research organization have already been examined by the authors (Trigo and Piñeiro, 1981).

[3] This is because of the nature of the resources required by the research process: a high level of investment and risk associated with research; the relatively low possibility of private appropriation of the resulting technological know-how; the small size of the agricultural enterprise; and the limited or non-existent possibility of market differentiation for primary agricultural produce.

[4] This new system found concrete expression in Latin America with the establishment of the International Center for Corn and Wheat Breeding (CIMMYT) in 1966, as an outgrowth of the Rockefeller Foundation Program in Mexico; the International Tropical Agriculture Center (CIAT) in Colombia in 1967, with a mandate for the American humed tropics; and with the International Potato Center (CIP) in Peru in 1971.

[5] A clear, although extra-regional, example of this problem is the decision by the government of the Philippines to cut back the size of the National Rice Program, a decision expressly based on the presence of the IRRI.

INTERNATIONAL TECHNOLOGY: THE INTERNATIONAL AGRICULTURAL RESEARCH CENTERS

John K. Coulter[1]

ABSTRACT

A brief history of the international agricultural research centers is presented, noting how emphasis shifted, from the 1950s thorugh the 1960s, from extension services to research, basically as a result of the first provable yield increases and methods of disease prevention—the famous "green revolution." An attempt is made to chart the future of agricultural research centers, taking into account the role of the Consultative Group for International Agriculture Research (CGIAR), formed in 1971. The need for coordination between the centers and individual national policy makers and priorities is stressed.

INTRODUCTION

Tropical agricultural research, long the Cinderella of the agricultural support services, has been receiving well deserved, increased attention during the last two decades. This is not to say that it is a new phenomenon in tropical agriculture; research saved the cotton industry in the Gezira in the 1930s (Gaitskell, 1959) and it allowed the natural rubber industry to compete with synthetics from the 1950s; it has also brought highly significant yield increases in such diverse crops as tea, oil palm and cacao. There has been relatively little impact on food crops, however, though some discoveries like those of mosaic virus in cassava and streak virus in maize formed the basis of successful breeding programs by the international centers.

During the 1950s and early 1960s, there was a widely held view that the application of existing technology, either generated locally or imported from more developed countries, would move agriculture forward, and that extension rather than research should be strongly supported. However, yield increases were small and it was not until the end of the 1960s that the combination of high yielding varieties, fertilizers, irrigation and the right price to attract farmers started the so-called "green revolution" in rice and wheat. The early international research centers thus set the pattern—and the expectations—of the series of international research institutions that followed International Rice Research Institute (IRRI) and the *Centro Internacional para el Mejoramiento de Maíz y Trigo* (CIMMYT). Since the Consultative Group for International Agriculture Research (CGIAR) was set up in 1971 to finance international centers, much has been written about the international agricultural research system, its origins, its achievements and its uniqueness. It has strong supporters, and equally strong critics. In this paper I have not set out to defend or criticize the past, but rather to try to gauge the future of the system, based on the achievements and pro-

---

[1] This paper represents the views of the author and not those of the World Bank.

jections of the international centers and the plans of the CGIAR for the future. In doing so, I think it essential to emphasize that the international center system is only a relatively small part (about 10%) of all the research systems operating in the developing countries.

THE MANAGEMENT OF THE CGIAR SYSTEM

The philosophy behind the founding of the first center, IRRI, in 1960 is well expressed by the following quote from Hill, 1964:

> Experience to date strongly suggests that an organization of this kind (IRRI) can perform the following important functions: (1) By bringing together a competent staff on indefinite tenure, well balanced among the relevant disciplines, and by providing them with the facilities required for high quality research and experimentation, it is possible to increase materially the speed with which higher yielding varieties of crops and improved cultural practices are developed. (2) An important contribution can be made towards training professional personnel within the region in which they expect to work: the high level manpower required to staff agricultural colleges, experiment stations, extension services, and administrative posts in ministries of agriculture. (3) An institution such as the IRRI can serve as catalyst and pace setter—an instrument for helping increase the efficiency and effectiveness of other research, training and extension organizations in the region it serves. Through its library facilities, its germ plasm banks, its regional seminars and conferences at the working level, and its grants for research work of promise, and by other means, it can stimulate and help improve the quality of research within a large region. (4) Such an institution can demonstrate to visiting administrators and other laymen, in a manner that is not otherwise possible, the kind of balanced, sustained attack that is necessary if the foundations are to be laid for steady and reasonably rapid progress in increasing agricultural production. This can be important in countries where administrators and the public generally have little real understanding of what is involved in shifting a traditional agriculture to an agriculture based on the application of modern science and technology.

It will be noted that the center was thus intended to speed up the generation of new technology, to increase the numbers of well trained scientists and to demonstrate the impact of good research on agricultural production to policy makers. This philosophy has been followed by each of the succeeding international centers, though there have been considerable changes in the concepts of what the centers should be doing and the ways in which they should do it. The founding of the Consultative Group in 1971 could be thought of as the start of the system in its present form, though the evolution towards an integrated system, both conceptually and practically, has been slow and the process has revealed both advantages and problems.

The managements of the international centers have generally supported the strongly held belief that, to be successful, research organizations must be allowed a great deal of freedom to organize their research programs. As independent centers with independent boards of management, each has been free to follow this philosophy, but the degree of freedom allowed to individual scientists varies considerably. On the other hand, many research organizations exert a fair degree of control over their individual research institutes, and strong financial and program constraints are often in effect. The amount of freedom varies with the cultural and historical background of research organizations, but in general research institutions working on technology generation or applied research are more closely controlled than those involved in basic research. While the directors of the international centers have justified fears of bureaucracy, particularly of the international kind, one of the unique features of the international agricultural research system is that the individual centers have been able to retain a great deal of freedom within the system. Nevertheless, the independence of the centers has been eroded somewhat in the past two or three years because donors are beginning to develop a collective sense of priorities; they are asking for an increased degree of "accountability," and perhaps most important, a decrease in the rapid growth of funding that characterized the first seven or eight years of the CGIAR. Worldwide inflation and rapid currency fluctuations, particularly severe in some of the host countries of the centers, have added to the problems of financial management of the system. The question of "accountability" is viewed with mixed feelings by research directors. Schultz (1980) has said "those who provide the funds establish the accounting rules; they are rarely aware of the sharply diminishing returns due to the burdensome accounting they demand." This could impose a burden on the system that should be avoided if at all possible. The growth in financial resources as indicated by Fig. 1 could be construed as denoting a strong faith in the system, but the general consensus among donors at present appears to be that, in the near term at least, there will be relatively little growth in real resources. This suggests that major new activities, particularly in the existing centers, will need to be financed by reordering priorities within the present resources of the centers and the system. This consideration will have a very important impact on the centers' future programs.

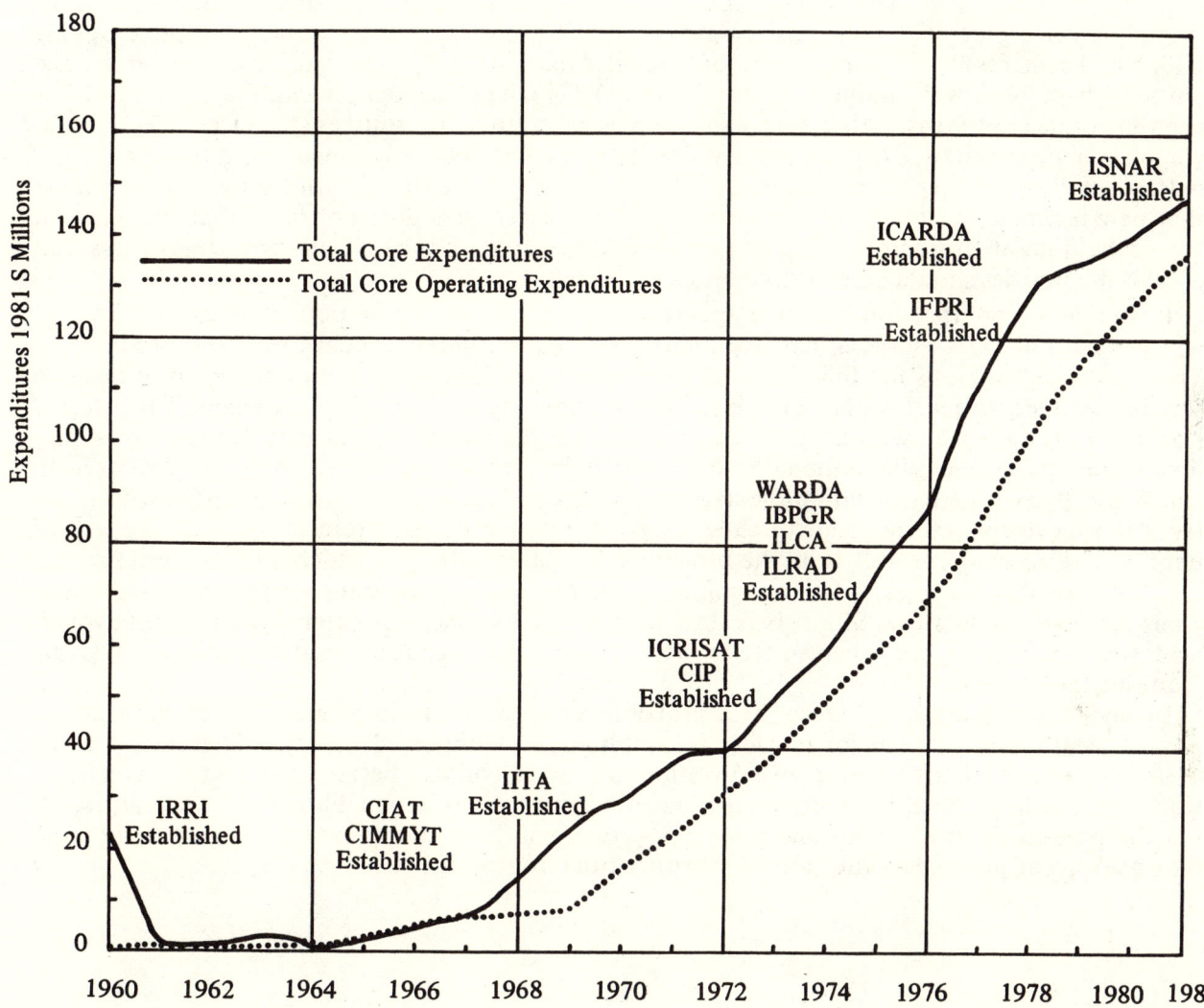

Fig. 1. International Agricultural Research Centers. Annual Core Expenditures and Core Operating Expenditures 1960–80. (In Terms of Constant 1981 Dollars) (CGIAR Secretariat, 1980).

THE ACCOMPLISHMENTS OF THE INTERNATIONAL CENTERS

Successful agricultural research has a high payoff, but it operates within a relatively long time frame. Much has been written about the impact of the new varieties of wheat and rice (CGIAR Secretariat, 1978; Dalrymple, 1978). One of the more interesting aspects of these increases in yield is the rapidity with which they occurred. In some states in India, for example, production growth rates in wheat reached nearly 10 per cent per annum. Annual increases of 5 per cent per annum occurred in many areas, a sustained rate seldom reached in developed countries. It is perhaps worth emphasizing that these achievements illustrate the impact of good technology even where extension services are not especially strong.

Much of the discussion about high yielding varieties has not dwelt on their role in the development of international and national research and extension programs, but rather on their social and economic impacts, particularly their effect on income distribution. Undoubtedly, farmers with access to adequate land, irrigation and other inputs have benefitted very substantially. Other have not, but the technology itself is inherently scale neutral. Some factors, such as improvement of irrigation, have generated additional employment; on the other hand, double or triple cropping with rice has necessitated introduction of machines like threshers. The impacts of the "green revolution" have been well summarized by Ruttan (1977); he states that, where the technology has been introduced in areas with a reasonable degree of equity in resources, the effect has been favorable on both productivity and equity. Where there has been great inequity in resources, the pattern of inequity has been reinforced. The conclusion to be drawn from these experiences is that new technology cannot solve inequalities in resources; this does not mean that less research should be done in prosperous areas, but rather that research on the difficult problems of less well endowed countries should be increased and accelerated.

It should be pointed out that the spectacular increases in production of irrigated rice and wheat have not been repeated, and are unlikely to be, in rainfed annual food crop production. There are many reasons for this. Little research has been done on many of the major crops like sorghum, millet, tropical roots and tubers, and legumes such as cowpeas and beans. The international centers have been working on these crops for 10 years at most, and their technology is only now beginning to enter the national research programs. There are many other problems as well, and World Bank experience with projects in rainfed crops suggests that it will be difficult to make rapid improvements in the yields of these crops. A number of our projects focus on the lack of improved technology, as well as on the problems of dealing with risk, which increases proportionally with augmented inputs. Others emphasize market dynamics, since crops like cassava and yams are difficult to store. Still others deal with the replacement of extensive systems, in which land was not limited but where there is little or no animal or mechanical power, with systems requiring more labor.

In many of these rainfed areas, it is probable that no one technological aspect, such as improved seeds, will have a major impact. A combination of improved practices is needed, among them better seeds, better timing of cultivation and weed control, better water and soil conservation, improved soil fertility, and perhaps improved mixtures of crops. Farmers in these areas, often the poorest, least educated and most risk-prone, will have to put into practice a fairly complex package of practices if they are to get substantial improvements.

Research Output: Improved Genetics and Resource Management

While all of the centers would like to be able to measure their success in terms of the impact of their technologies on agricultural production, they recognize that the outputs of a research organization are new knowledge and technology, and that their clients are the national research and extension services of the developing countries. The contributions of the centers to technological development are published annually and were summarized in the 1979 Integrative Report (CGIAR Secretariat, 1979). As more than 50 per cent of the resources of the system go into plant breeding and related activities, this is the single most important activity in most of the crop-oriented centers. The general consensus, as demonstrated by statements of national agricultural researchers, is that the collection, conservation and incorporation of useful genes in the mandate crops of the centers is a major activity that will continue to be valuable in the forseeable future.

The centers distribute their breeding materials at various stages, sometimes as a finished variety, often as segregating materials. As will be discussed later, the centers expect a considerable change over time; as the national programs develop strength, much of the material will be distributed at earlier stages in the breeding cycle. Nevertheless, the international centers may have to come to the rescue when serious outbreaks of pests or diseases occur. This has happened in the case of rice in Indonesia, when a new biotype of brown plant-hopper caused severe damage over large areas. Fortunately IRRI had a back-up variety which was resistant and which could be multiplied quickly and distributed to replace the susceptible variety. Still, this example demonstrates some of the problems that confront modern agriculture in developing countries. It has been estimated, for example, that 25 per cent of the 130 million hectares of the world's rice is in short strawed, photo-period insensitive cultivars. This area of about 30 million hectares is planted with some 100 cultivars which have replaced thousands of traditional varieties. Although many of these new varieties have resistance to many pests and diseases, it is often based on a single major gene for each disease or pest. It would be desirable to have more of the so called horizontal or field-type resistances, but they have been difficult to obtain. The use of multilines in wheat is one way of approaching this problem. In sum, the international centers may do some of the defensive research needed to maintain recent gains, but it is likely that the burden will fall increasingly on the national programs as modern agriculture spreads.

In addition to plant breeding and genetic improvement, the international centers have worked to develop better systems of land and water management. As indicated earlier, much of the research has been in irrigated agriculture, but it is already obvious that improvements in production of rainfed annual crops are much more difficult to achieve. Improved water conservation in the semi-arid areas and better control of excess water in the humid areas are two of the several improved practices needed to obtain meaningful increases in production.

Land conservation is also of great importance. Increasing concern is being expressed about the ravages of soil erosion, not only the destruction of agricultural land in the catchment areas, but also the rapid siltation of reservoirs and irrigation systems in lowland areas. Soil erosion has, of course, been an emotive topic for many decades, and the crusades undertaken by people like Bennett in the U.S. are well known. Soil conservation was also a major objective in many of the colonial agricultural systems, and was often achieved through coercion. It is significant that many of these projects have broken down since independence; this, together with the spread of population to land not suitable for annual crop farming (but which did support relatively stable swidden systems), has made soil erosion a widespread phenomenon.

There is no doubt that destruction of land, in some cases in a spectacular manner, in others in an insidious way, will be increasingly serious in the next decades. We may find ourselves in the paradoxical situation of having given a great deal of attention to the conservation of genetic resources and too little to the land resources on which improved crops can be grown. However, while the destructive impacts of soil erosion are obvious, the benefits of soil conservation are not always easy to demonstrate and are delayed well into the future. Thus, they are not only intra-farm, but also inter-farm and inter-generational.

Many developing countries are neither in the position to subsidize soil conservation projects (up to 80% in the United States) nor to enforce soil conservation by coercive practices. This makes it all the more important to develop farming systems that will be economically viable, socially acceptable, and that will confer immediate benefits on the farmer as well as conserving the soil. Important work is being done at IITA, whereby the combination of minimum tillage and conservation of crop residues is used to improve yields *and* conserve the soil. In one large scale experiment, it was shown that the construction of graded contour banks for conventional tillage costs about $430/hectare. Even with such banks, blocks with conventional tillage had 250 mm of runoff (carrying 42 kg/hectare of plant nutrients) compared with 16 mm of runoff (carrying 13 kg of plant nutrients) in blocks with minimum tillage, during a four month maize crop (Hartmans, 1981).

When considering research priorities, another important variable is transferability. The system has given considerable attention to the potential adaptability of plant breeding technology, making it a high priority. The new varieties of wheat and rice can be grown in widely varying conditions, thus showing high transferability. On the other hand, research focused on farming systems and land conservation is regarded as location specific, with limited transferability. Ideally, concepts and methods in both kinds of research should be able to cross national and environmental frontiers.

Training

Every country gives training a high priority, and the centers have recognized this by making the formation of trained human capital one of their major efforts. Different approaches have been used. The crop oriented centers have had production training courses which concentrate on developing national scientists in the context of a crop research program, usually lasting for the period of one crop. They have also had a variety of in-service training programs at the master, doctoral and post-doctoral levels, though numbers have been limited and the capacity of the centers in relation to needs is small. Many countries do not have the required educational levels. In Africa, for example, even the more advanced countries had only about 10 per cent of the population of secondary school age enrolled in secondary schools, compared with 56 per cent in the Philippines (Lele, 1981). Regardless of the efforts of the international centers and other organizations involved in training agricultural scientists, these countries will need a very long time to reach a broad educational base.

Work with National Programs

One of the changes in the international agriculture research system involves the attitude towards national research. The early international centers were thought to offset the weaknesses in national programs, and indeed there continue to be suggestions that the centers could disappear, absorbed by national or regional programs as these become stronger. Center directors have realized, however, that in the absence of strong national programs their impact would be limited, and so work to strengthen them. One way has been through training, but some of the centers have had substantial technical assistance programs in which they were the implementing agency for a commodity-based research project in a specific country. By and large, this kind of activity, which is normally supported by special funding from a specific donor, has not had the universal support of the Consultative Group; the centers were being drawn into areas which could detract from their major concern, doing research of transnational importance, and it was considered that they had little comparative advantage in this area.

Nevertheless, national program involvement poses a dilemma for a worldwide system that includes countries and regions with highly variable resources. National programs in Africa are usually much weaker than those in Asia or Latin America, and the centers operating in Africa face a difficult problem. ILCA, for example, is concerned with livestock improvement in Africa, where national livestock research is particularly weak. Yet by the very nature of the problem, ILCA can only make substantial progress by working with national programs. Overall technological progress in livestock husbandry is likely to be slow, even slower than is normal in livestock research.

Involvement in national programs is only one aspect of what have come to be called off-campus activities. There are many others: regional programs, off-campus core programs (such as that of the IRRI research on deep water rice in Thailand, where the center itself does not have a suitable environment on the campus), international testing programs and networks which have been described and summarized in the stripe review by the TAC (TAC Secretariat, 1980). The centers now have a major effort focused outside the main campuses, amounting to about 25 per cent of the total resources of the system. Indeed, strengthening national research institutions may be one of the major accomplishments of the CGIAR system. Intangible benefits occur when administrators and governments are convinced that agricultural research is a worthwhile investment. They occur when the research systems of developing and developed countries are brought closer together, where their synergistic effect will be of incalculable value. Agricultural science does not recognize political or ecological boundaries, and the international research system is one to which all may contribute and from which all may benefit.

FUTURE DEVELOPMENTS IN THE CGIAR SYSTEM

At present, the CGIAR is considering its directions for the future, especially those issues that the system will be faced with in the next decade or so. The system's role as a research organization dealing mainly with food crops wll be played out in a world context dominated by three main projections: (1) population growth, particularly in the urban areas; (2) increased demand for food due to rising incomes; (3) increasing costs of energy.

Population growth and the impact on food needs has been widely discussed; Latin America, for example, will have a population of about 600 million by the year 2000, compared with 350 million in 1980. Eight of the major oil exporting Third World countries, with approximately 360 million people, have had an average increase in food imports of 19 per cent per annum (IFPRI, 1980). Indeed, food imports into the developing countries doubled in real terms between 1970 and 1977. Furthermore, the demand for poultry and pork livestock, which consume grain, has been growing at a rate of 4.5 per cent per annum. Major increases are expected in the centrally planned economies. The People's Republic of China is expected to import about 15 million tons of grain in 1980–81, compared with an average 4 to 5 million tons in the 1961–77 period. The combination of population growth rates and rapid increases in per capita income seems likely to promote an unprecedented growth in demand for food in the next few decades.

Increasing costs of energy and their impact on food production are also the subject of much discussion. Modern agriculture is often accused of being a profligate user of energy. Yet the facts seem to point the other way. For example, from 1945 to 1975 the primary energy input for the average hectare of U.S. corn production increased from about three barrels of oil equivalent to about five. Of this, the energy for fertilizers increased from near zero to about two barrels, but yields increased from 2 to 5 tons/hectare (Leach, 1979), so it took an average of two and a half barrels less to produce the five tons of corn. Similar kinds of calculations have been used to show that if agriculture in the Sahel was modernized (to include irrigation of 600,000 hectares), energy consumption would be of the order of one million tons of oil equivalent per annum, which would be only a small proportion of the Sahel's total energy equivalent. Dispassionate analyses of energy use related to food production, then, seem to show that the energy used in actual production of food is not likely to be so critical as is sometimes suggested. However, many people, including vast numbers of rural people, depend on renewable sources of energy for most of their other needs, particularly heating and cooking, and agriculture supplies a considerable proportion of this in the form of crop residues and manure. The increasing emphasis on agro-forestry is a recognition that wood will supply an increasing amount of these energy needs.

Alcohol production is another aspect of the energy equation in agriculture. Brazil, for example, has a target of 4.8 billion litres of alcohol from 72 million tons of cane in 1980–81, rising to 10.7 billion litres from 160 million tons of cane by 1985 (Silva, 1981). The country also has plans to build 11 plants for production of alcohol from cassava, giving 345 million litres per annum. Plans such as these could, of course, have very serious impacts on other aspects of food crop development, particularly on food for export.

The Determination of Priorities

One can thus look at the future research plans of the CGIAR system against a background of food and fuel needs. What are the priorities as determined by the system and by the centers, considering that there is unlikely to be a substantial increase in resources in the near term? It is important to distinguish between short and long term plans. Recent program and budget documents show that the centers regard the present financial stringency as temporary, so cuts have been made in such programs as training and germ plasm collection, easily accelerated once additional funding is available. Posts have been left unfilled where the financial situation is especially severe; obviously such solutions can only be of a short term nature.

Ordering priorities for research is a process fraught with difficulties even at the national level. At the international level it is even more so; the international system represents only a small part, perhaps less than 10 per cent, of the total resources going into agricultural research in the developing countries. It exerts more influence than this proportion of the resources would indicate, and has a degree of stability of funding, resources per scientist and freedom that is the envy of most national programs. Nevertheless, these advantages carry with them the obligation to show that the resources are well used.

The overall priorities of the system are determined by the Technical Advisory Committee (TAC), a committee comprising 13 agricultural scientists from developed and developing countries that was set up when the CGIAR Group was formed. The broad objectives of international support to agriculture are stated by the TAC as follows: (a) increasing the amount, quality and stability of food supplies in the LDCs and meeting the total food needs; (b) meeting the nutritional requirements of the less advantaged groups in the LDCs.

Some of the quantitative parameters for use in priority assessment include planted areas, production, numbers of countries and people involved in the production, contribution to rural and urban diets, and resources allocated to the commodity by national research programs. By taking these and other factors into account, the TAC advised the Group that its resources be directed first towards assuring the continued support of the IARCs and other related activities already established by the Group. Given that support, the TAC suggested that additional resources be directed to five priority areas: tropical vegetable research, water management research, plant pest and disease physiology and ecology research, food policy research, and aquaculture research. Since priorities were established by the CGIAR in 1979, the International Food Policy Research Institute (IFPRI) was accepted by the Group for funding, but the International Center for Insect Physiology and Ecology was not. An acceptable mechanism for carrying out international research on water management is being sought at the moment.

Using a matrix system, the TAC (TAC Secretariat, 1979) summarized its recommendations on priorities under commodities, production systems and production factors as illustrated in Tables 1, 2 and 3. Table 1 shows that there are some areas of research in the first priority group on which no CGIAR supported research is at present taking place (tropical vegetables and aquaculture), and some second priority areas where there is research (barley, triticale, cowpea, sweet potato, broad beans). In terms of production systems (Table 2), some work is under way in all of the first priority areas and in some of the second priority areas. For production factors (Table 3), most of the existing research at the centers is concerned with improved seeds and pest and disease management. There is some work on plant nutrition, mainly nitrogen fixation, and on mechanization, but relatively little is being done on post-harvest technologies and soil and water management.

Priorities in the International Centers and National Programs

The above summarizes the Group's views on priorities for international research and the extent to which these are being presently followed in the system. However, while the TAC can advise the Group on priorities for the system as a whole, the international centers themselves have to develop priorities in terms of resource allocations for the commodities included in their mandates. Even the single commodity centers have to make allocations; for example, the IRRI has to decide how much of its resources should be devoted to irrigated and to upland rice, and the CIP has to decide how much of its work should go to potatoes in the highland areas of Latin America and the tropical and subtropical areas there and elsewhere. Although it has a global mandate for beans and cassava (excluding Africa), the CIAT has devoted much of its resources to work in Latin America; even there it has to make difficult decisions on how to allocate resources to the different commodities.

Several of the international centers, including CIP, CIAT and CIMMYT, have published or are in the course of developing long range (5–10 year) plans. By and large, the centers have not suggested any radical changes in the programs they have pursued since they started; there are no proposals, for example, to drop one mandate crop and adopt another. The CIP (1980) has probably been the most adventurous in projecting its path through to the year 2000. Thus it expects to terminate its collection efforts for both cultivated and wild potato species by 1982. It expects to have populations with good resistance to late blight, nematodes and viruses by 1990. Beyond that, it will focus increasingly on long term needs such as maintenance, exploitation and utilization of the world germ plasm collection, global communication networks for the transfer of potato information and training in new techniques. Direct help to national programs is expected to decrease, but regional programs will become increasingly important.

Although the centers do not envisage radical changes for the 1980s, a recurring emphasis is work with national programs. This is demonstrated by stated intentions to put less emphasis on

TABLE 1 Research Priorities by Commodities

COMMODITIES	1st PRIORITY	2nd PRIORITY	OBSERVATIONS
Cereals	Rice, wheat, maize, sorghum, and millet	Barley, triticale	
Roots, tubers and other starchy foods	Cassava, potato	Sweet potato, plantains, yams	
Pulses	Dry beans	Cow pea, chick pea, pigeon pea, broad beans, and lentils	Importance of 2nd priority crops is high in certain regions
Vegetables	Selected tropical vegetables which, of highest priority, include amaranths, winged beans, peppers and okra	Other tropical vegetables	Nine vegetables have been selected for consideration in the research program of the proposed IVRIT
Oilseeds	Groundnut, soyabean	Coconut, cotton, several annual oilseeds	Importance of 2nd priority crops is high in certain regions
Animals and animal products	Nutrition & reproductive efficiency of ruminants Trypanosomiasis Theileriosis Aquaculture (low-cost systems with plankton, herbivorous and omnivorous feeders)	Non-ruminant livestock African Swine Fever and other important animal diseases Aquaculture (other systems)	

TABLE 2 Research Priorities by Production System

	1st PRIORITY	2nd PRIORITY	OBSERVATIONS
	Rainfed and irrigated croplands of arid and semi-arid tropics and sub-tropics Arid rangelands Infertile lands of sub-humid and humid tropics	High rainfall tropical lowlands Tropical and sub-tropical highlands Agroforestry	Includes related aspects of socio-economic research at micro-level and macro-level and regional assessments of soil and climate conditions Includes also related animal production systems
	Food policy research related to 1st priority commodities and production systems	Broader aspects of food policy research	

production of varieties for testing and release by national agencies and more on the production of elite germ plasm for use in national breeding programs. Furthermore, several of the centers have consulted the leaders of national programs on their long range plans. It is also of interest to note that several of the centers, especially those working on the lesser known crops (lesser known in the sense that there had been little research on them prior to the formation of the IARCs), expect to have solved the major disease problems and to be able to switch more of their resources to work in increasing yield potential. This has important implications for the future when it is recalled that, even in crops like rice and wheat, there has been little increase in yield ceilings in the last decade. Even in these crops, more and more attention has had to be given to fighting a situation whereby minor pests and diseases became major ones through changes in cultural practices or varieties, or the evolution of new strains of pests and diseases. In some instances, completely new pests (green spider mite and mealy bug on cassava in Africa) have appeared (IITA, 1981). Downy mildew, a serious disease of maize in India, has now appeared in Africa. Situations like these may prevent the centers from moving beyond research on pests and diseases into other areas.

To summarize, I suggest that we shall see gradual changes in emphasis rather than in kinds of research or changes in commodities over the next decade, unless there are fairly major additions in funding. A greater involvement of the national programs in breeding work and a greater involvement of the centers in back-up research, particularly in pest and disease control, drought tolerance and some aspects of more basic research (tissue culture), would describe the picture as I see it.

SOME CHALLENGES OF THE 80s

This seminar has focused on the prospects for technological progress in agriculture in Latin America in the 1980s. As indicated above, the international centers expect the national programs

TABLE 3 Research Priorities by Production Factors

PRODUCTION FACTORS	1st PRIORITY	2nd PRIORITY	OBSERVATIONS
Improved seeds and breeds	Improved seeds of 1st priority crops	Improved seeds of 2nd priority crops Conservation, evaluation and use of special animal breeds	
Plant nutrition	1st priority crops	2nd priority crops and efficiency in fertilizer use	Includes nitrogen fixation
Plant pest and disease management and control	Genetic resistance to pests and disease of 1st priority crops and integrated pest management in 1st priority production systems	Genetic resistance to pests and disease and integrated management in 2nd priority crops and production systems	Includes basic research aspects of pest and disease physiology and ecology
Soil and water management	1st priority crops and production systems Design parameters and low-cost technologies for irrigation and drainage	2nd priority crops and production systems Irrigation and drainage of special problem soils	Includes related aspects of soil fertility management Mainly at farm and village level
Post-harvest technologies	1st priority crops, especially roots and tubers	2nd priority crops	Mainly at farm and village level
Mechanization	1st priority crops and production systems	2nd priority crops and production systems	Some qualifications

to take over an increasing share of the work in the 1960s and 1970s. This can occur only if the national programs grow in strength and stability, but will always be a very uneven process, both in Africa and Latin America. Unfortunately, the information on the quantitative and qualitative research resources of many countries is very meager. Furthermore, the instability of funding and of government support for research is well documented, as is the lack of planned designs for agricultural development. A good illustration of this is the case of the "resource poor" farmers. Donors to the CGIAR have stressed repeatedly that they wish the work of the centers to focus on increasing the prosperity of these farmers. Implementation of this goal is illustrated by many centers' programs on the development of technologies that will require low amounts of purchased inputs for millets, sorghums, and cassava, common crops of marginal area farmers; emphasis is also placed on grain legumes that will improve the quality of nutrition among poor people. However, this goal is not always congruent with national strategies, where priority may be given to quick pay-offs in more productive areas, and to cash crops for foreign exchange. The situation may be exacerbated by the predominance of rich farmers and landowners at the policy making levels of governments.

Historical aspects suggest that many national programs will continue to develop in an irregular fashion, with periods of strength followed by periods of weakness, but with an overall trend toward continuing improvement. It is also likely that once a national program reaches a certain "take off" point, it will not be so subject to fluctuations in support and quality; there are indications that several countries have reached this stage. The whole process will be encouraged by enhanced training programs in both the international centers and the many other institutions involved in training, and by a better knowledge of the capacities and capacities and capabilities of the national programs at both the center and the international system level.

Assuming that the current plans of the centers and the Group are successfully implemented, will other problem areas appear? It is my opinion that at least four of them will. The first is non-food or cash crops. In many countries most of the poorer farmers derive the major part of their cash income from non-food or industrial crops. Examples are coconuts and cotton. Cotton has had a good deal of research, but much more, especially against pests, is necessary to maintain higher yields. Coconuts have had only sporadic research, and there are still problems with serious diseases and improving yields. Without some cash income, farmers in poor areas will be unable to purchase even the modest inputs that will enable them to increase food supply. Of course, there is the possibility that some national programs, recognizing the attention being given to food crops by the international centers, will divert a larger proportion of their scarce research resources to non-food areas.

The second problem is the management of difficult lands (steep lands, shallow soils, deeply flooded areas). The international centers can supply aid in the form of improved varieties of mandate crops for these areas. The work on farming systems can supply ideas on better ways of conserving soil and water. Work on agro-forestry could help in the development of more productive species for fuel and energy production, although transferability of such technology is still a problem that must be solved.

A third problem is that of increasing the efficiency of inputs like fertilizers and water, although much is already being done. Better varieties make more efficient use of these inputs. International institutes outside the CGIAR (such as the International Fertilizer Development Center) are researching better fertilizers, including such materials as local phosphate rocks. However, methods of advising small farmers on the use of fertilizers are too often based on techniques from the developed countries. How to maximize the benefits of fertilizers in mixed cropping systems, in areas where two or three crops a year are grown, and in areas where climatic risk is high, are all challenges which seem to be receiving little attention from either national or international programs.

Irrigation is a great stabilizing factor in agricultural production in many countries, yet there seems to be universal agreement that water is used very inefficiently. There is also agreement that social factors play a major role in this inefficiency, but there is no consensus on how to attack the problem in terms of agricultural research.

Again, the major constraint seems to hinge on the question of transferability of results, since it is often suggested that most of the problems are location-specific; nevertheless, I feel this is an area where ideas and methodology would be transferable.

A fourth problem concerns the transfer of technology to small farmers. There are many examples of success where small farmers have improved their production significantly, in spite of weak extension services. Several of the international centers (for example the IRRI rice constraints program) have made major contributions to understanding why small farmers behave as they do. The Training and Visit system in India has shown that a well organized, single purpose system, armed with suitable technology, can reach large numbers of small farmers and persuade them to take up new technologies. Nevertheless, millions of small farmers remain untouched by improved technology. Perhaps greater efforts in the social sciences could lead to new ideas in communication and novel ways of disseminating improved agriculture.

This is not, of course, an all inclusive list; there are many many areas where additional efforts are needed, but they must be considered in the context of the resources, international and national, that will be available. We all recognize that spending on agricultural research is well below the figure of 1 to 2 per cent of the value of agriculture production, considered the optimum range, and we must also recognize that many countries will not approach this range in the next decade.

However, I would conclude on a reassuring note, for I do think that in spite of all the problem areas, agricultural production has made remarkable progress in the last two decades. There seems to be increasing recognition at every political level that good agricultural research has had, and will continue to have, an important role in improving production.

LEGAL SYSTEMS AND PRIVATE SECTOR INCENTIVES FOR THE INVENTION OF AGRICULTURAL TECHNOLOGY IN LATIN AMERICA

Donald D. Evenson and Robert E. Evenson

ABSTRACT

This paper discusses the role of private sector research in agriculture, where, until recently, research sponsored by public institutions was of primary importance. This is especially true in under-developed countries, where the evolution of appropriate patent and other legal systems has failed to keep pace with the competition in international invention. A series of alternative legal systems is reviewed, then compared to Latin American legal systems. The problem of agricultural patenting is discussed, with emphasis on two recent developments that will have a considerable effect: The USDA Plant Variety Protection Act and advances in genetic engineering. Three policy issues are addressed: the welfare implications of alternative invention stimulation, the specialized requirements of developing countries, and the implications for both public and private sectors. An extensive appendix cites the appropriate patent laws for several Latin American countries and the United States.

INTRODUCTION

Societies have devised several types of direct and indirect policy instruments to stimulate the invention of improved technology. They have: (1) encouraged monopolistic industrial organization; (2) offered prizes and rewards for inventions; (3) developed contractual mechanisms by which research is indirectly purchased; (4) devised and administered legal instruments, chiefly patents and trade secrecy protection laws; (5) enabled private associations to cooperate to support research and development centers; and (6) directly invested in research and development in public research institutions. All but the last two entail research by private firms and individuals with a self-interest in income and other rewards.

This paper discusses, in five parts, the role of the private sector in agricultural research. It is a peculiarity of research and invention relevant to agriculture that the private rather than the public sector requires special attention. In most industries, the public sector plays a minor role in technology development. Legal systems of patent protection and trade secrecy protection, along with certain industrial organizational characteristics, are relied upon to stimulate invention. The role of the private sector in agricultural research is acknowledged to be large in developed countries, but since agricultural supply industries are of minor importance in low-wage countries, the private sector in these economies is seen as an almost negligible source of invention.

Most developing countries have failed to develop legal systems appropriate to their competitive position in international invention. Even if firms in developing countries have adequate financing and access to scientists, they often function in "laboratory settings" which place them at a competitive disadvantage *vis a vis* U.S., European and Japanese firms. On the other hand, they do have a laboratory ideally suited to localized "adaptive" invention. Very few countries have shown imagination in tailoring their legal systems to their economic situation. In the first part of this paper we provide a review of alternative legal systems designed to do this, while the second part discusses Latin American legal systems.

In the discussion of legal systems, we note that in the past biogenetic inventions, which tend to dominate the improvements in agricultural technology, have had very limited patentability. Until about 1970, only asexually reproduced plants were offered patent protection because of the restrictive view of the "disclosure" requirement in patent laws. Two recent developments have changed this situation and it is likely that much more private invention activity in plant (and even animal) improvement will be forthcoming in the future. The first is that alternatives to the classic patent have been devised. The Plant Variety Protection Act administered by the USDA is typical of these alternative systems, which provide limited protection to breeders of new varieties on the basis of identifiability rather than full disclosure. The second and possibly more significant development is the extended protection possible through advances in genetic engineering, particularly in recombinant DNA technology, where disclosure is actually possible. The third part of this paper is devoted to these two changes and their possible consequences.

A summary of available data on private sector research and development in agriculture and on agricultural patenting is also presented. It shows that in the poorest developing countries the private sector devotes little attention to agricultural invention. This activity becomes significant only when developing countries reach a fairly advanced stage. However, data for countries with "petty" patent systems suggest that these systems stimulate more private activity at earlier stages.

The final parts of the paper address three policy issues: (1) the basic welfare implications of alternative invention stimulation systems; (2) specialized requirements of systems in less developed countries which take into account the prospects for borrowing technology and adapting it to local conditions; (3) the implications for the public as well as the private sector. It appears that many public sector scientists fear that an expansion of private inventive activity will have a deleterious effect on public systems and even on the welfare of the population affected. This concern is expressed even in cases where effective public research systems are not in place. We conclude in this paper that optimal investment in agricultural research in developing countries requires a major expansion of both public and private sector activity.

LEGAL SYSTEMS FOR STIMULATING INVENTIONS

Types of Invention

The term "invention" broadly covers any new process, device, chemical composition, or thing which has been developed, devised, or discovered. Although every invention has different economic effects and origins, inventions stemming from agricultural research can be generally grouped as follows: (a) mechanical/electrical; (b) chemical; (c) biogenetic; and (d) other.

Mechanical/electrical inventions include new machinery for planting, cultivating, and harvesting agricultural crops, as well as machinery and equipment used in animal husbandry. These "mechanization of agriculture" inventions, as they apply to developing countries, are usually quite limited in their geographic adaptability because of varying local wage rates and differing soil and climatic conditions. Mechanical/electrical inventions also include processes for making and/or using such machinery.[1]

Chemical inventions include fertilizers, herbicides, insecticides, and pharmaceuticals for animals. These inventions are usually more expansive in their geographic adaptability than are mechanical/electrical inventions.[2]

Biogenetic inventions include improved plant varieties and animal breeds resulting from selective breeding programs, as well as results of the new bio-engineering processes such as recombinant DNA technology, and such processes themselves. Biogenetic research end products, especially plants, are often limited in their geographic adaptability due to varying soil and climate, as shown by the large, increasing number of corn and soybean varieties in the United States. However, procedures for carrying out biogenetic research may have much broader adaptability.

Other inventions include computer programs and general business systems for improving agricultural research procedures, as well as the organizational aspects of agricultural production, that is, accounting systems for allocating research/production resources. These "other" inventions are usually not encompassed in patent or other legal systems for preserving private rights, but they illustrate the breadth of activities properly considered as inventive agricultural research.

Legal Systems for Protecting Inventions

Legal systems for securing private rights to inventions can be grouped as follows: (a) seed and breed certification systems; (b) copyright systems; (c) trade secret contract enforcement systems; (d) utility patent systems; (e) utility model or "petty patent" systems; and (f) plant patent and variety protection systems. All of these systems provide some type of legally enforceable right to limit the use of inventions by others than the inventor and licensees/assignees of the inventor.

The seed and breed certification systems normally provide that seed and animals be marketed with labeling sufficient to identify the origin of the seed or animal, as well as a true indication of its genetic heritage. Such certification systems operate like trademark systems to prevent others from trading on the goodwill that a breeder has established by introducing new plant and animal varieties. These certification systems do not usually prevent others from using and selling the same plant and animal varieties as long as they do not misrepresent the source, thus their effectiveness in securing private rights is dependent in large part on the capability of the breeder to market (including advertize) his products. For small breeders, such systems provide little protection to inventions unless they include provisions proscribing the sale of crops obtained from certified seed for further seeding purposes.

Copyright systems are designed to prevent unlicensed copying of works of art and authors' writings. The "copyright" is limited to "copying" of the writing and does not preclude use of the information contained in the writing. For example, an article describing a new plant breeding technique could be copyrighted, but this would only limit the "copying" of the article and would not prevent anyone from using the technique. However, there is one area of invention where copyright protection may be meaningful to an agricultural researcher, namely computer software. Since practical utilization of a computer software invention many times requires actual "copying" of the software itself, copyright protection may be significant.

Trade secret contract enforcement systems are designed to prevent persons, primarily ex-employees and ex-collaborators, from disclosing secrets of manufacture and the like to competitors. This type of protection is meaningful only for inventions that can be maintained in secrecy, such as manufacturing processes that are not readily apparent in the marketed products exposed to the public. For the most part, trade secret protection is of little use for mechanical/electrical inventions but of potentially great use for chemical and biogenetic inventions. For example, a new fertilizer-producing technique involving critical temperatures at critical phases might reduce production costs substantially and be quite valuable. With enforceable trade secret contracts, employees having knowledge of the critical parameters (as well as competitors knowingly accepting trade secrets) are liable for losses resulting from their unauthorized disclosure to others. In the biogenetic area, certain strains of microorganisms may be maintained in secrecy indefinitely as "parent" stock from which commercial stocks are developed.[3]

A utility patent system gives the inventor a right to exclude others from practicing the invention for a certain time period (usually 15–20 years). Utility patent systems traditionally include a requirement that applications for patents include an "enabling disclosure" that sufficiently describes or illustrates the invention so that those in the technical field of the invention can make or use it. Thus early publication of the invention is encouraged in return for the limited monopoly given.

Inventions covered by utility patents generally must be novel, useful, and exhibit a difference over prior art greater than would be expected from the average endeavors of those working in the art ("level of invention" requirement). For example, minor modifications to farm implements, like making some parts detachable for ease of repair, would be insufficient to support a utility patent under most systems. A utility patent is not an exclusive right to use a particular invention, but rather a right to exclude others from using it, with the scope of the exclusion being defined by claims describing the novel contributions made by the inventor.[4]

Utility *model* or "petty patent" systems are similar to utility patent systems in that they give the inventors rights to exclude others for a term of years, but differ in that they usually require only novelty and utility and do not require any "level of invention" over the prior art. Petty patent protection is usually granted for a shorter period of time than that for utility patents. Also, since "level of invention" need not be determined, such systems are usually much cheaper to administer than most utility patent systems. The petty patent systems are most effective for preserving rights to minor variations of known devices, rather than major technical innovations having broad adaptability. Nationals of developing countries, where minor adaptations of machinery to accomodate local conditions are very helpful to the local economy, but may not be themselves widely adaptable, are more likely to utilize petty patents rather than more costly and difficult to obtain utility patents.[5]

Plant patent and variety protection systems are analogous to utility patent systems in that they provide breeders and developers with limited rights to exclude others, for a term of years, from commercializing new plant varieties they develop. These systems usually require that the plant varieties be novel by exhibiting uniformity, stability, and distinctness from all known varieties. Typically some type of control depository is provided to preserve seed samples of plant varieties being protected, in addition to description requirements, to the extent that varieties can be described with words, drawings, and pictures. To encourage use of the protected varieties in further breeding/research programs, such usage is typically exempted from an infringement claim. Also, other provisions may be included, such as in the United States system, which permits farmers to save seed from their crop for use on their farm and for sale to other farmers for other than reproduction purposes.[6]

There are international conventions relating to patent and variety protection systems which provide for equal treatment of applicants from any convention country and for priority rights based on applications filed in the convention countries, provided applications are made within a certain time period (usually one year) of the first filed application. These conventions also include certain substantive requirements that the convention countries must include in their national systems, such as minimum protection and disclosure requirements. However, these conventions do permit many substantive differences in implementation of the national system. The final part of this paper discusses the policy implications of international convention membership by developing countries.

LATIN AMERICAN LEGAL SYSTEMS FOR PROTECTING INVENTIONS

Although Latin American countries have various forms of copyright and seed and breed certifying systems, this section will concentrate only on the patent and plant patent variety protection systems, as these are the primary modes for protecting the results of agricultural research. Although all patent systems encompass at least some agricultural research results, all Latin American legal systems considered exclude patent or similar protection to much agricultural research activity. There are many differences in the various countries systems, such as administrative procedures, examination requirements, chronological terms of protection, government fee structures, requirements to actually employ inventions to keep patents in force, and others. This paper will focus only on the differences in the scope of protectable "subject matter" in the various selected countries. To depict the differing approaches, we will assume a hypothetical "Latin American Agricultural Research Corporation" (LAARC). LAARC is presumed to be involved in an extensive agricultural research program covering the entire spectrum of invention and development activities directed toward improving agricultural productivity. Following is a list of "tangible" results of LAARC research activities. For the purpose of this discussion, it is assumed that all of these research results were developed exclusively by LAARC using LAARC funds, were never previously extant or described anywhere in the world, and were in fact better than the state of the art.

Mechanical/Electrical Inventions

(1) LAARC Plough I: a basic innovation in ploughs which is adaptable to a wide range of soil and climatic conditions.

(2) LAARC Plough II: a minor modification for adapting a plough to be utilized in a specific localized soil condition.

Chemical Inventions

(3) LAARC Chemical Fertilizer: a new chemical fertilizer compound which optimizes tolerance to extremes in soil moisture conditions.
(4) LAARC Insecticide: a new and useful chemical compound insecticide.
(5) LAARC Herbicide: a new and useful chemical compound herbicide.
(6) LAARC Pharmaceutical for Animals: a chemical compound pharmaceutical for controlling disease in farm animals.

Biogenetic Inventions

(7) LAARC Soybean: a new, improved soybean variety developed in a plant breeding program.
(8) LAARC Corn: a new, hybrid corn seed variety developed in a plant breeding program, with LAARC retaining control over the hybrid parents.
(9) LAARC Rose: a new variety of asexually reproducible ornamental rose.
(10) LAARC Beef Cattle: a new pure breed of beef cattle developed in a selected breeding program.
(11) LAARC Bacteria: a new and improved nitrogen fixing bacteria strain developed using recombinant DNA techniques.
(12) LAARC Vaccine: a live virus vaccine.

Other Inventions

(13) LAARC Computer Program: a new and improved computer program utilized in the determination of the optimal mix of chemicals used in LAARC Chemical Fertilizer (invention 3).
(14) LAARC Accounting System: a new and improved accounting system used for optimally allocating research personnel and facilities in the various research projects, adaptable to any large agricultural research organization.

Table 1 depicts the availability of patent and plant variety protection for the various LAARC inventions in several Latin American countries and the United States.[7] The United States has been treated in this table for comparison purposes, since it has a comprehensive scheme of protection.

For mechanical/electrical inventions, such as the LAARC Plough I, all of the countries provide some type of utility patent protection. For the minor adaptation of the LAARC Plough II, only Brazil has a utility model or petty patent system which would provide protection. Unless the LAARC Plough II was sufficiently innovative to satisfy utility patent statute requirements for "level of invention," no protection would be available for it.

As to the chemical inventions, only Argentina and the United States provide patent protection for fertilizer, insecticide and herbicide. Most countries' patent statutes specifically preclude patents on certain types of invention related to food production and health. In Argentina pharmaceutical drugs are expressly excluded from protection.[8]

For the biogenetic type inventions, none of the countries provides patent or variety patent protection for the beef cattle breed. Only Argentina and the U.S. provide protection for the new plant varieties. The U.S. does not provide plant variety protection for the hybrid corn—presumably since such protection is not needed because corn breeders can effectively maintain control over the hybrid parents and thus already have so-called "genetic" protection. Although there are legal systems in some countries (Hungary, Romania) which provide breeder's rights to new animal breeds, none of the countries in Table 1 has such a provision.

The possibility of patent protection for the nitrogen fixing bacteria and the live virus vaccine was only recently ruled upon by the U.S. Supreme Court, the matter having been an open question as to whether a "utility" patent on living organisms could be obtained.

TABLE 1 Comparison of "LAARC" Research Results and Availability of Patent/Variety Protection for Various Countries

	MECH/ELECT INVENTIONS		CHEMICAL				BIOGENIC					OTHER	
	Plough I Major	Plough II Minor	Fertilizer	Insecticide	Herbicide	Vaccine Chem.	Soybean	Corn Hybrid	Rose	Beef	Bacteria Nitrogen Fixing	Vaccine (Live Virus)	Computer Program	Accounting System
ARGENTINA	yes Art. 1 Law No. 111	no	yes Rule No. 27/74	yes Rule No. 27/74	yes Rule No. 27/74	no Art. 4 Law No. 111	yes Plant Protection Act	yes Plant Protection Act	yes Art. 2B Law of Seed	?	yes Rule No. 27/74	no	maybe Rule 15/75	no Art. 4
BRAZIL	yes ?	yes Art. 10	no Art. 9b (Process)	no Art. 9b (Process)	no Art. 9b (Process)	no Art. 9C (No Process)	no Art. 9C	no Art. 9C	no Art. 9J	no Art. 9J Product of Nature	no Art. 9J	no	see Art. 9h	no Art. 9h
CHILE	yes Art. 4a	no	no Art. 9a (Process)	no Art. 5a (Process)	no Art. 5a (Process)	no Art. 5a (Process)	no Art. 5a	no Art. 5a	no Art. 5c	no Art. 5a	no	no	old law (1935)	no Art. 5b
COSTA RICA	yes ? Art. 4	no	no Art. 10 §V	no Art. 10 §IV	no Art. 10 §IV	no Art. 10 §IV	no Art. 10 §I	no Art. 10 §I	no Art. 10 §I	no Art. 10 §I	no Art. 10 §I	no	no Art. 10 §V Art. 9 §III	no Art. 9 §III
MEXICO	yes Art. 1	no	no Art. 5C	maybe	maybe	no Art. 5C	no Art. 5b	no Art. 5b	no Art. 5b	no Art. 5b	no Art. 5b	no	maybe Art. 3 Art. 4C	no Art. 4C
PERU	yes Art. 5 & Art. 14.2	no	no Art. 15.1	no Art. 15.1	no Art. 15.1	no Art. 15.1	no Art. 15	no Art. 15	no Art. 15.3	no Art. 15.1	no	no	no Art. 14.5	no Art. 15.2
VENEZUELA	yes	no	yes	yes	yes	yes	yes	no	yes	no	no	no	no	no
U.S.A.	yes	no	yes	yes	yes	yes	yes	no	yes	no	no	no	no	no

The "other" inventions, namely the computer program and the accounting system are not provided with patent or variety protection under any of the legal systems. However, a copyright system which includes computer programs may provide some limited protection.

BIOGENETIC INVENTIONS: SPECIAL CONSIDERATIONS

Historically, inventions and discoveries involving living organisms have contributed greatly to increases in agricultural productivity. Until recently, most advances stemmed from domestication and selective breeding of plants and animals, a technology which could be termed the application of "classical genetics." In the recent past, much has been learned about the genetic make-up of living things and new techniques have been developed for manipulating the genetic structure to speed development of new strains of microorganisms, new plant varieties, and new animal breeds, an application of technology which could be termed "molecular genetics." In addition, there have been technological innovations which speed up the process of selective breeding, including sperm storage, artificial insemination, estrus synchronization, superovulation, embryo recovery, embryo transfer, embryo storage, sex selection, twinning, *in vitro* fertilization, and others.[9]

To date, patent/variety protection for living organisms has been much more restrictive than patent systems for devices and chemical compositions. Even in the United States, it was not until 1930 that a very limited Plant Patent Act was enacted, the current version of which provides:

Whoever invents or discovers and *asexually reproduces* any distinct and new variety of plant, including cultivated sprouts, mutants, hybrids, and newly found seedlings, *other than a tub-propagated plant or a plant found in an uncultivated state,* may obtain a patent therefor, subject to the conditions and requirements of title.

In the case of a plant patent the grant shall be of the right to exclude others from *asexually reproducing* the plant or selling or using the plant so reproduced. (Emphasis added).

To this day, this 1930 U.S. Plant Patent Act is limited to orchard fruits and ornamental flowers (primarily roses), with most of the plant patents being for roses, apples, peaches and chrysanthemums. The U.S. Plant Patent Act provides no meaningful protection for new varieties of major farm crops such as corn, wheat, soybeans, and oats since they cannot be commercially exploited without utilizing seeds, thereby falling outside of the "asexual" limitation. The requirement for asexual reproduction was apparently based on the belief that sexually reproduced varieties could not be reproduced true-to-type and that it would the senseless to try to protect a variety that would change in the next generation.

It was not until 1970 that the United States enacted a Plant Variety Protection Act which covered new varieties of sexually reproduced plants, the timing of this enactment supposedly being related to the advance of plant breeding technology which enabled new, stable and uniform varieties to be sexually reproduced. Fungi, bacteria, first generation hybrids, and six soup vegetables (okra, celery, peppers, tomatoes, carrots and cucumbers) were excluded. The hybrids were excluded since it was believed that the breeders had sufficient natural protection by controlling inbred parental stocks. In 1980 the U.S. Plant Variety Protection Act was amended to include the originally excluded vegetables.[10]

As to living organisms other than plants, namely microorganisms, it was not until the Chakrabarty decision of the U.S. Supreme Court in 1980 that the U.S. utility patent law was construed to cover genetically engineered bacteria.[11] Although *processes* for the development of new living organisms had been held to be patentable subject matter, this was the first instance where the bacteria product itself was deemed appropriate for patent protection, a very important expansion of the patent protection in view of the way in which the bacteria would have to be exploited commercially. Once the bacteria strain was introduced to the public, even by a single sale, it would be difficult, if not impossible, to establish infringements of the process patent claims, thus effectively limiting the value of such claims. Most other countries, especially developing countries, have not gone as far in providing private protection for living organism inventions.[12]

Neither have developing countries participated in the industrial biogenetic research boom currently underway. The largest part of current private biogenetic research is directed toward health and pharmaceutical products, but significant research directed toward agriculture has already begun (probably at least 100 million dollars in the past year). A 50 to 100 billion dollar market was projected by a recent assessment in the agricultural sector within a 10 to 20 year time span;

if only a small part is realized, it will have important implications for developing countries. It is possible, but highly unlikely, that the new biogenetic technology emerging from this work will be less location-specific than current technology. It is clear that developing countries will have to be concerned about their capacity to undertake this research in both the public and private sectors. At present, they have virtually no capacity to do so.

INVENTION: A DESCRIPTIVE SUMMARY OF INTERNATIONAL DATA

We now turn to available data, primarily on patenting, to develop a quantitative picture of technology invention internationally. Our focus will be on agricultural developing countries generally and Latin American countries specifically. For perspective, we first turn to a comparison of invention patents in developed, semi-industrialized and developing countries. Table 2 summarizes these data for 49 countries with roughly comparable invention patent systems.[13] The developing countries, broadly defined, encompass both semi-industrialized and developing countries in the table. With the exception of Brazil, the Latin American countries fall in the slow-growing, semi-industrialized countries or the developing countries categories.

The table shows that in 1967 Mexico and Argentina granted relatively large numbers of patents to national inventors. By 1979, however, Brazil had surpassed Mexico on this score. On the whole, it is clear that developing countries generally have not exhibited a capacity to generate a large number of invention patents nationally.

Table 3 provides a regional summary of these data, plus data on R & D spending and numbers of scientists and engineers for the late 1960s and late 1970s.[14] The data show that the developing world generally had roughly the same share of world patenting in the late 1970s as in the late 1960s, but that it took the form of a rise in the share of the rapidly growing or newly industrialized countries. Other developing countries are losing ground. The ratio of patents to GDP is a measure of national invention intensity. This ratio has fallen in all groups of countries over the past decade (see line 4). For semi-industrialized countries, this ratio is roughly comparable to the ratio for industrialized economies (at least in the 1960s). However, it is much lower for the poorer countries, a fact which also holds for agriculturally related patenting.

Lines 5 and 6 in Table 3 show that developing countries have very low ratios of R & D spending to industrial GDP, but that they are able to purchase more scientists and engineers per dollar of R & D. The "productivity" of scientists and engineers as reflected in line 7 appears to be higher in the semi-industrialized countries, and lower in the developing countries, than it is in the industrialized countries.

Line 8 shows the pattern of change in these ratios from the late 1960s to the late 1970s. The data of most concern from a policy perspective are those showing a decline in invention productivity (patents per scientist and engineer) in every region, being most severe in the slow-growing, semi-industrialized (primarily Latin American) economies. It is important that we understand the reason for this decline. The fact that the decline has occurred in virtually every country supports the contention that a general exhaustion of invention potential has occurred. For a number of developing and semi-industrialized economies, it also appears that policies to reduce dependency on foreign technology are responsible for the decline. Latin American countries, for example, have pursued policies which have reduced the number of patents granted to foreigners by half over the past decade. These policies include the exclusion of certain basic needs-oriented technology fields from protection, as well as other administrative requirements. Unfortunately, the same policies were probably responsible for a large part of the 40 per cent decline in patenting by nationals and the 50 per cent decline in national inventive productivity.

Table 4 shows that there is a large difference between regions in the ratio of patents granted to nationals in other countries to patents granted to nationals at home. Low income countries have very low ratios, reflecting the fact that many of the national inventions are adaptive in nature. Table 4 provides further data for selected countries showing that Latin American patenting abroad is primarily in developed countries (chiefly the U.S.). There is evidence, however, of some "stagewise" patenting in Latin America. Some of the adaptive inventions in semi-industrialized Latin American countries appear to be relevant in other Latin American countries.

Table 5 provides a distribution of patents granted (to both nationals and foreigners) by broad industrial groups. This classification is by industry of origin and does not fully reflect all inven-

TABLE 2 Invention Patents Granted by Country: Selected Years

	PATENTS GRANTED TO NATIONALS				PATENTS GRANTED TO FOREIGNERS				PATENTS GRANTED TO NATIONALS IN FOREIGN COUNTRIES			
	1967	1971	1976	1979	1967	1971	1976	1979	1967	1971	1976	1979
I. Industrialized Market Economies												
a) Moderate to Rapid Growth												
Japan	13,877	24,795	32,465	34,863	6,896	11,652	7,582	9,241	6,843	15,832	20,246	16,406
Austria	1,188	1,230	1,177	1,163	6,896	7,460	5,235	5,337	1,913	2,399	1,065	1,051
France	15,246	13,696	8,420	6,846	31,749	37,760	21,334	17,772	14,393	17,150	12,677	9,942
Denmark	338	252	208	250	2,002	2,212	2,068	1,217	1,165	1,650	1,217	922
Germany	5,126	8,295	10,395	10,895	8,300	9,854	10,570	11,639	41,775	44,862	37,316	28,188
Belgium	1,586	1,345	1,034	837	15,041	15,004	12,110	5,081	2,701	2,894	1,905	1,305
Norway	225	386	210	250	1,831	2,343	1,883	1,687	618	658	617	548
Netherlands	322	318	370	455	1,913	2,396	3,219	3,003	7,283	8,745	5,901	4,879
b) Slow Growth												
Canada	1,263	1,587	1,301	1,408	24,573	27,655	20,449	22,138	2,789	3,201	2,661	1,855
Italy	9,076	4,320	—	1,810	26,180	13,180	—	6,190	5,621	6,749	5,416	4,591
Ireland	28	16	27	16	635	788	1,055	1,255	113	151	146	89
Switzerland	5,388	4,165	3,482	1,638	16,462	11,914	8,818	4,976	12,452	15,409	10,954	8,748
Sweden	1,776	2,245	1,888	1,514	7,532	7,748	6,956	4,320	5,031	6,327	5,719	4,024
U.S.A.	51,274	55,988	44,162	30,605	14,378	22,328	26,074	18,248	73,960	87,589	90,273	47,633
Australia	752	979	910	—	10,371	9,662	10,074	—	905	986	1,065	980
U.K.	9,807	10,376	8,855	4,182	28,983	31,178	30,942	16,618	17,579	21,179	14,072	19,348
Finland	231	350	291	394	739	1,312	921	1,325	345	559	650	690
New Zealand	—	—	211	221	—	—	1,314	1,314	135	1,420	91	—
II. Semi-Industrialized Market Economies												
a) Rapid Growth												
Spain	2,758	2,042	2,000	1,569	6,827	7,764	7,500	8,643	627	933	766	543
Israel	178	202	200	207	935	1,225	1,200	1,135	219	231	146	210
Greece	975	1,227	1,343	1,607	2,302	698	1,285	988	61	70	81	99
Singapore	5	2	—	1	26	334	—	548	—	—	5	5
Portugal	84	214	46	66	1,045	3,238	1,319	1,242	53	57	50	50
Brazil	262	429	450	349	684	1,543	1,500	3,494	63	85	88	81
Korea (South)	207	200	1,593	1,358	152	117	1,727	1,161	20	20	50	50

TABLE 2 Invention Patents Granted by Country: Selected Years (continued)

	PATENTS GRANTED TO NATIONALS			PATENTS GRANTED TO FOREIGNERS			PATENTS GRANTED TO NATIONALS IN FOREIGN COUNTRIES		
	1967	1971	1976	1979	1967	1971	1976	1979	1967	1971	1976	1979
b) Moderate to Slow Growth												
Chile	80	58	60	60	1,237	1,115	514	514	—	—	—	—
Venezuela	41	237	50	39	954	1,599	514	614	—	—	—	—
Argentina	1,244	1,346	1,300	1,264	4,488	3,484	2,800	2,843	81	152	102	131
Mexico	1,981	412	300	236	7,922	5,199	3,000	1,790	149	148	181	158
Turkey	30	52	35	34	438	357	588	424	—	—	—	—
Uruguay	165	88	46	15	351	161	110	93	—	—	—	—
III. Developing Economies												
Ecuador	5	8	7	7	126	180	103	103	—	—	—	—
Iraq	22	5	12	9	146	67	150	73	—	—	—	—
Morrocco	28	24	23	31	391	313	334	341	—	—	—	—
U.A.R.	48	13	16	6	873	236	511	370	—	—	—	—
Colombia	49	62	30	36	851	651	600	808	—	—	—	—
Philippines	16	46	108	82	498	946	767	755	—	—	—	—
Kenya	0	1	5	5	104	121	98	98	—	—	—	—
India	428	661	433	500	3,343	3,256	2,062	2,000	72	70	73	74
Sri Lanka	1	10	4	5	4	148	156	36	—	—	—	—
O.A.M.P.I.	1	15	3	26	573	455	545	545	—	—	—	—
IV. Planned Economies												
Germany (East)	11,520	8,295	3,755	4,318	8,351	9,854	2,735	1,629	976	2,240	1,652	1,051
Czechoslovakia	3,613	2,824	4,880	5,928	787	1,276	2,220	2,196	1,718	1,735	927	552
U.S.S.R.	24,008	33,534	40,259	41,259	662	2,098	1,883	1,883	1,379	2,973	3,309	1,683
Hungary	414	559	594	644	663	1,054	1,155	845	596	1,020	1,116	1,067
Poland	1,564	2,331	5,619	5,225	485	543	2,380	1,531	447	538	347	358
Bulgaria	423	674	750	1,364	90	240	393	568	78	164	167	158
Yugoslavia	173	143	58	58	650	706	355	355	95	90	87	87
Romania	2,955	1,075	1,123	1,428	1,283	1,246	572	571	224	313	106	94

TABLE 3 Regional Summary

	Industrialized Economies			Semi-Industrialized Economies		Developing Economies	Planned Economies
	Rapid Growth	USA	Other	Rapid-Moderate	Slow Growth		
1. Share of World's Invention Patents by National Inventors (N)							
Late 60s	.251	.316	.110	.026	.017	.004	.277
Late 70s	.310	.233	.075	.033	.010	.004	.336
2. Ratio: Patents Granted to Nationals Abroad to Patents (N)							
Late 60s	1.94	1.51	2.28	.28	.092	.10	.155
Late 70s	1.31	1.69	2.65	.20	.165	.09	.109
3. Ratio: Patents (N) to Total Patents							
Late 60s	.39	.75	.10	.25	.17	.11	.76
Late 70s	.51	.62	.18	.27	.20	.12	.84
4. Ratio: Patents (N) to Industrial GDP							
Late 60s	.144	.125	.158	.116	.081	.028	.076
Late 70s	.097	.084	.089	.060	.025	.016	.080
5. Ratio: Industrial R&D to Industrial GDP							
Late 60s	.0352	.067	.042	.015	.0040	.0074	.044
Late 70s	.0345	.062	.034	.008	.0027	.0048	.045

TABLE 3 Regional Summary (continued)

		Industrialized Economies			Semi-Industrialized Economies		Developing Economies	Planned Economies
		Rapid Growth	USA	Other	Rapid-Moderate	Slow Growth		
6.	Ratio: Industrial S&E to Industrial R&D							
	Late 60s	21.93	19.90	16.22	18.29	58.11	79.81	12.93
	Late 70s	16.10	11.88	14.52	21.60	57.11	96.61	15.62
7.	Ratio: Patents (N) to Industrial S&E							
	Late 60s	.1865	.144	.231	.432	.328	.048	.133
	Late 70s	.1745	.113	.146	.353	.164	.034	.115
8.	Ratios: Late 70s/Late 60s							
	P (N)	1.26	.76	.70	1.29	.61	.91	1.25
	P (F)	.77	1.38	.76	1.21	.49	.75	.73
	P (NA)	.85	.85	.81	.91	1.07	.82	.87
	R&D	1.84	1.05	1.23	1.57	1.23	1.06	1.19
	S&E	1.35	.97	1.11	1.35	1.21	1.28	1.44
	P (N)/Ind GDP	.67	.67	.50	.52	.31	.57	1.05
	R&D/Ind GDP	.98	.93	.88	.53	.68	.65	1.03
	S&E/R&D	.73	.92	.90	1.18	.98	1.21	1.20
	P (N)/S&E	.94	.79	.63	.82	.50	.71	.87
9.	Share of World's Design Patents							
	Late 70s	.852	.025	.041	.046	.017	.010	.029
10.	Share of World's Trademarks							
	Late 70s	.412	.080	.064	.309	.092	.036	.007

TABLE 4 Patenting in Foreign Countries: Selected Origin Countries – 1976

Total Patents	France	West Germany	Italy	Switzerland	U.K.	U.S.A.	U.S.S.R.	Eastern Europe	Argentina	Brazil	Mexico	India
% Developed	86.0	87.5	78.4	82.4	86.8	86.2	87.0	72.7	67.1	68.0	77.5	91.9
% Planned	4.5	6.9	7.3	7.7	3.7	2.1	9.4	22.0	1.4	.9	1.5	1.4
% Asian Developing	3.7	2.2	6.5	3.9	4.7	5.1	2.2	3.1	.7	3.7	6.0	1.4
% Latin American Developing	3.9	2.7	6.3	4.8	3.9	6.0	.4	1.2	30.8	27.4	11.0	1.4
% African Developing	1.9	.7	1.5	1.2	1.1	.6	1.0	1.0	0.0	0.0	4.0	3.9

TABLE 5 Distribution of Patents Granted by Class According to the International Patent Classification: Average for the Period 1970–74 (WIPO, *Industrial Property*, several issues; for U.S.A., National Science Board, 1977)

	Agri-culture	Food-stuffs & Tobacco	Personal and Domestic Articles	Printing	Trans-porting	Metal-lurgy	Textiles	Paper	Building	Mining	Instru-ments	Electricity Related[a]	Chemical Related[b]	Machinery Related[c]	Total (%)
Canada	1.5	1.5	2.3	2.2	8.8	3.3	4.2	1.3	2.6	.9	10.3	14.0	25.2	21.6	100
France	2.1	1.3	3.5	2.3	10.8	1.8	2.7	.3	4.9	.4	13.5	14.5	18.9	22.9	100
Germany (F.R.)	1.5	1.0	2.6	2.4	8.9	2.8	3.6	.6	4.5	1.0	13.3	19.8	15.6	22.2	100
Switzerland	2.3	1.6	3.7	2.0	8.4	1.7	5.1	.2	5.1	.1	12.1	13.9	26.4	17.2	100
U.S.A.[d]		1.3			6.3	1.0	1.0		2.3		10.4	19.6	19.5	38.5	100
United Kingdom	0.8	0.9	2.4	1.3	7.2	2.2	3.1	.03	3.6	e	11.1	15.4	19.2	20.0	100
Chile	8.2	3.9	2.4	1.0	7.4	4.7	3.2	1.5	3.6	.5	5.1	6.9	37.5	13.9	100
India	.8	1.6	2.1	1.0	6.0	4.0	4.0	.4	2.7	.5	4.8	16.4	32.9	22.6	100
Kenya	6.4	8.0	1.4	1.3	4.0	2.2	1.0	.2	2.2	.8	1.3	3.7	58.0	9.4	100
Philippines	5.2	5.7	3.0	.9	2.8	3.5	3.7	1.0	2.2	.5	.7	2.8	58.2	9.7	100
Venezuela[f]	4.0	.7	10.7	.04	1.7	4.2	2.1	.2	.9	.3	3.1	.05	70.7	1.3	100
Korea (Rep.)[g]	5.6	12.8	6.6	3.6	1.4	7.2	3.8	4.4	5.9	2.5	3.8	8.0	20.8	13.5	100

[a] Includes the following subsections of IPC: electricity, lighting and heating, and neuclonics.
[b] Includes medicine and hygiene, and chemistry.
[c] Includes separating, shaping, engines and pumps.
[d] Grouped patents according to the industrial classification.
[e] Included under machinery.
[f] Up to 1972.
[g] Some patents are unclassified.

TABLE 9 Patenting Activity Related to Agricultural Production, Selected Countries (WIPO, 1965–72)

REGION	ANNUAL NUMBER OF PATENTS	
	1965–1968	1969–1972
Northern Europe		
Denmark	186.5	179.3
Finland	44.2	49.7
Ireland	48.7	45.5
Norway	n.a.	70.8
Sweden	206.2	185.2
United Kingdom	476.7	427.0
Central Europe		
Austria	304.0	327.0
Belgium	415.3	362.7
France	1178.5	836.0
Germany, West	473.7	299.2
Netherlands	76.5	65.0
Switzerland	497.0	417.5
Southern Europe		
Italy	n.a.	792.5
Portugal	35.0	n.a.
Spain	448.0	393.7
Eastern Europe		
Bulgaria	48.0	61.7
Czechoslovakia	95.7	162.5
Hungary	34.5	115.5
Poland	n.a.	49.5
U.S.S.R.	828.0	1231.3
North America		
Canada	576.7	444.7
U.S.A.	785.0	830.0
Oceania		
Australia	116.5	133.5
Latin America		
Chile	53.0	79.7
Colombia	26.0	30.5
Costa Rica	22.0	10.0
Cuba	11.0	4.0
Trinidad & Tobago	7.2	5.5
Uruguay	32.0	24.0
Venezuela	25.5	111.7

TABLE 9 Patenting Activity Related to Agricultural Production, Selected Countries (WIPO, 1965–72) (continued)

REGION	ANNUAL NUMBER OF PATENTS	
	1965–1968	1969–1972
Africa		
Kenya	8.0	10.2
Malawi	22.0	15.0
Mauritius	n.a.	6.5
Morocco	29.3	19.0
Rwanda	0	0
Sierra Leone	0	.5
South Africa	326.5	n.a.
S. Rhodesia	55.0	86.5
Tunisia	5.7	5.7
Uganda	7.0	6.7
Zaire	n.a.	6.0
Zambia	26.7	8.0
Asia		
India	32.5	33.7
Israel	87.2	118.7
Malaysia	8.7	14.0
Philippines	19.2	43.7
South Korea	7.4	11.0
Sri Lanka	8.0	n.a.
Syria	7.7	n.a.

TABLE 10 International Agricultural Patenting Activity by Per Capita Income Group, 1965–72 (Boyce and Evenson, 1975)

Income Group	Annual Number of Patents		Patents per Billion Dollars Agricultural Product	
	1965–68	1969–72	1965–68	1969–72
I (over $1750)	5574	4745	65	51
II (1001–1750)	1998	2792	50	64
III (401–1000)	1089	1171	80	80
IV (150–400)	111	113	21	20
V (under 150)	116	97	10	8

TABLE 11 Origin of Agricultural Patents Granted by Selected Countries and Ratio of Total Patents Granted Domestically to Total Patents Granted to Nationals Domestically and Abroad, 1972–77

COUNTRY	PLOWS				HARROWS				TRACTOR–RELATED IMPLEMENTS			
	Total Patents Granted	Domestic Origin	LDC Origin	Ratio	Total Patents Granted	Domestic Origin	LDC Origin	Ratio	Total Patents Granted	Domestic Origin	LDC Origin	Ratio
Latin America												
Argentina	3	1	0	3/1	2	2	0	2/2	1	1	0	1/3
Brazil	47	39	0	47/39	55	42	1	55/42	46	28	0	46/28
Cuba	0	0	0	0	1	1	0	1/1	1	1	0	1/3
Other LDCs												
India	2	2	0	2/2	1	1	0	1/1	0	0	0	0
Philippines	1	1	0	1/1	9	5	0	9/5	3	0	0	3/0
Developed Countries												
U.S.	232	200	0	232/285	291	182	0	291/274	347	304	0	347/786
U.K.	80	37	2	80/98	249	73	4	249/108	229	52	0	229/151
France	330	172	0	330/211	642	240	1	642/265	837	203	0	837/266
Japan	77	70	0	77/74	302	261	0	302/264	408	371	0	408/384
USSR	322	322	0	322/322	301	298	0	301/298	462	452	0	462/456

COUNTRY	PLANTING–SOWING				DIGGING				GRAIN HARVESTING				THRESHING			
	Total Patents Granted	Domestic Origin	LDC Origin	Ratio	Total Patents Granted	Domestic Origin	LDC Origin	Ratio	Total Patents Granted	Domestic Origin	LDC Origin	Ratio	Total Patents Granted	Domestic Origin	LDC Origin	Ratio
Latin America																
Argentina	1	1	0	1/1	3	2	0	3/2	5	3	0	5/3	0	0	0	0
Brazil	61	38	1	61/38	23	18	3	23/18	147	85	0	147/91	8	8	1	8/5
Cuba	5	3	0	5/3	0	0	0	0	21	11	0	21/14	0	0	0	0
Other LDCs																
India	3	1	0	3/1	1	0	0	1/0	5	2	0	5/2	0	0	0	0
Philippines	2	2	0	2/2	0	0	0	0	4	4	0	4/4	0	0	0	0
Developed Countries																
U.S.	206	155	1	206/323	63	56	2	63/136	744	643	1	744/1474	206	173	0	206/231
U.K.	112	52	0	112/178	54	24	3	54/55	424	152	4	424/472	80	29	0	80/70
France	411	199	0	411/258	264	153	0	264/170	1215	357	3	1215/469	138	45	1	138/71
Japan	1075	1028	0	1075/1042	104	102	0	104/103	1465	1397	1	1465/1428	64	32	0	64/37
USSR	637	628	1	637/628	194	190	0	194/191	779	761	1	779/761	179	176	0	179/182

modification to the standard invention utility model system undertaken by developing countries has been some form of exclusion of certain fields from patentability. This exclusion usually affects foods and drugs. Only two or three developing countries have designed petty patent systems. None to date has attempted to use plant patents that include plant variety protection systems. The developing countries are not encouraging biogenetic inventions.

The reluctance to provide stimulus to indigenous private sector invention would be understandable if public sector agencies were actively investing in developing countries. This, however, is not the case, even in agricultural invention fields which have been the province of the public sector the world over. Public investment in agricultural research is perhaps one-fifth as intensive as it has been in developed countries for the past half century (Boyce and Evenson, 1975). Public investment in mechanical/electrical and most chemical fields is of minor importance in most developing countries.[17]

Why is it that developing countries, particularly Latin American countries, show great reluctance to use legal systems to stimulate indigenous inventions? Why, at the same time, do most join international patent conventions which result in automatic and mandatory recognition of intellectual property rights of foreign inventors, thus cutting off their option to imitate at low cost? These same countries intervene regularly in commodity markets through tariffs and quotas to limit the payment of rents to foreigners.

We do not propose to answer these questions in this final section, although several factors can be indicated. These include a belief on the part of policymakers that there is little scope for adaptive invention in developing countries and that most technology must be imported. General industrial policy toward transnationals and local firms affects attitudes about legal systems. In a number of countries, existing industrial firms see legal systems which broaden the invention base as a threat because they will produce more competition. We believe, however, that there is a fundamental misconception regarding monopoly rights in patents, which has led to a reluctance to use them.

Below, we address three topics. The first is the issue of monopoly rights in patent systems. We then discuss some of the possibilities for the design of more effective legal systems to induce agricultural invention in developing countries. Finally, the compatibility of public and private invention in agriculture is considered.

Monopoly Rights, Rents and Patent Systems

Perhaps the first point to be made regarding monopoly rights and the ability of an inventor to collect monopoly rents is that patent systems are not the only means by which such rents accrue. In the absence of formal patent systems, a few inventions freely and easily imitated will be made. Most inventions, however, will tend to be held in trade secrecy, enabling the inventor to collect monopoly rents through the sale of products embodying the invention. The legal literature views the granting of limited monopoly rights in the form of a patent as a "bargain" in which inventions are disclosed, that is, brought out of secrecy, in return for the granting of monopoly rights for a limited period of time. This is in contrast to the economics literature on patents, which views the bargain as one of incentives versus monopoly rights. This literature stresses the fact that the opportunity to patent inventions provides an incentive for inventors who are not in a position to collect rents through trade secrecy mechanisms. Without patent protection, only oligopolistic firms have a strong incentive to engage in invention.

Secondly, monopoly rights are limited in two ways. The first, the legal time span over which monopoly rights are granted (usually 18 years or so), is relatively unimportant. The real threat to monopoly rights is the erosion of monopoly due to subsequent invention. The point here is that an effective legal system not only grants monopoly incentives, but provides a disclosure mechanism which enables subsequent invention, which in turn erodes the original monopoly rents.

Inventions, whether biogenetic, chemical or mechanical, have a "public good" character. Since they are fundamentally conceptual, even if embodied in plants, animals, machines or chemicals, they are not scarce in an economic sense once they are discovered. They may be employed in a new way without sacrifice to other employment effectiveness. Thus the marginal cost of employing an invention in another way, say to produce an added bushel of wheat, is simply the cost of imitation.

From the perspective of standard welfare economics, at the current moment a known invention will be of maximum advantage to society if employed in all ways where its value exceeds the costs of imitation. We earlier discussed six types of policy instruments utilized to stimulate invention: monopoly, prizes, contracting, legal systems, private associations and public systems. Of these, three tend to meet this standard. When public sector research systems produce inventions, they usually make them available at little or no cost to all parties. Similarly, if a public sector agency induces inventions by a prize system or via contracting with a research organization, and then makes the invention freely available to the public, the efficiency requirement is met. In the case of the private growers' association, the requirement is often met as well, since inventions are usually freely available to all members.

In the other two cases, monopoly and legal systems, the requirement is usually not met. Monopolistic firms do not produce where marginal costs equal the social value of the goods produced and do not employ inventions efficiently. Trade secret protection and patent protection allow the inventory to charge a price for use of an invention. This price will be higher than the cost of imitation. As a consequence, inventions are not used optimally. A social or welfare loss is thus incurred with monopoly, trade secrets and patents. Yet most Western countries justify some monopoly on the grounds that it increases invention and usually give patent protection, trade secrecy protection, or both to private firms.

It is argued that, for some inventions, private self-interested firms and individuals are superior to public sector research systems in producing inventions, for the following reasons: (a) many more private individuals can be stimulated to work on a problem than could be stimulated in a public sector system (this improves the probability of discovery); (2) the private sector knows what is required in the way of an invention, and inventions are thus tailored to the market, while many public systems do not have an effective clientele; (3) private systems can be of many sizes and types, utilizing informal and formal institutional arrangements; and (4) private, self-interested individuals have more "animal spirit" than public employees, who are too security-oriented and lack boldness.

We have no extensive evidence to support these arguments. Most major inventions have, in fact, been generated under the protection of legal systems. When public sector agricultural research programs compete directly with the private sector (as in hybrid corn, for example), they come out second best. We believe that the arguments of this section will show, however, that if an effective patent system is in place, there is reason *not* to provide trade secrecy protection and little reason to encourage monopoly. It is suggested that combinations of public sector research systems, public sector contracts with private firms, and imaginative patent systems are optimal.

Consider the basic analytics of the case where a single invention is made. Fig. 1 portrays the cost curves (average and marginal) for a monopoly firm (panel A) and a competitive firm (panel B). It also portrays the supply curve for a competitive industry, which is the sum of the marginal cost curves of a large number of small competitive firms. The initial effect of an invention is to lower the costs of production,[18] as depicted by the downward shift in average cost curves and in the supply curve of the industry.

If a monopolist makes an invention, it may matter little whether patent protection is available or not. The monopolist may hold the invention in secrecy or, if he has no close competitors to imitate the invention, secrecy may be of little importance. The average costs of the monopolist fall from C_0 to C_1 and, if the monopolist is maximizing profits, the price charged will fall from P_0 to P_1. Note that the monopolist has an incentive to invent without a patent system and without trade secrecy since he has no competitors. This is because his costs fall more than the price. Note also that consumers benefit from the invention because prices fall.

Now consider the competitive firms (panel B). First suppose that there is no patent protection, but that the invention can be held in trade secrecy; costs will fall. The competitive firm will expand output (to q_1^1), but this will have little effect on industry supply, and industry price will not fall. In this situation, the small, competitive firms will enjoy increased, although small, profits relative to the potential value of the invention.

Let us suppose the invention could not be held in trade secrecy, allowing other firms in the industry to imitate the invention. All would enjoy lowered cost curves and the industry supply curve would fall to S_1. A new equilibrium would be reached where prices would fall by the full amount of the fall in costs. No firms in the industry, including the inventing firm, would realize increased profits, hence no competitive firms would have a significant incentive to invest.

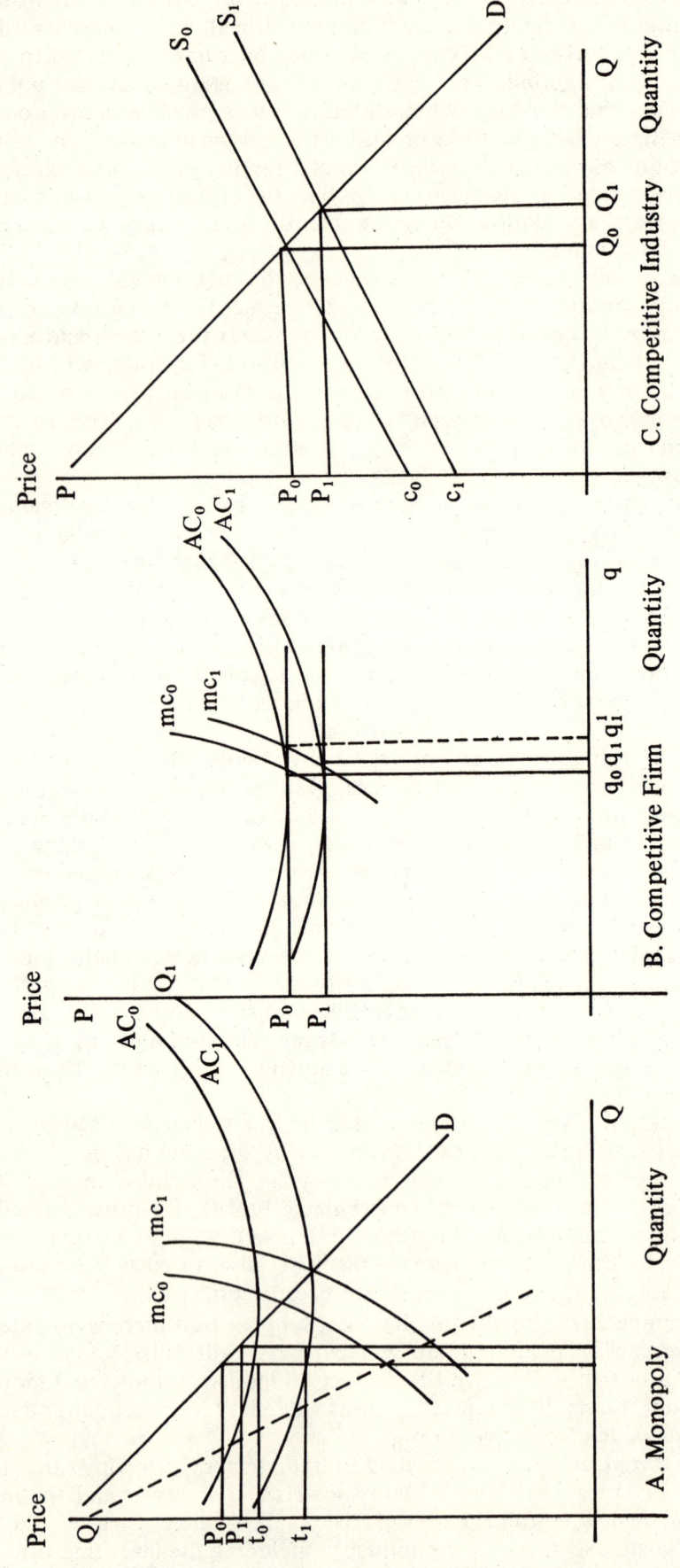

Fig. 1. Cost Curves for Monopoly and Competitive Firms

This is the substance of the analysis put forth by Schumpeter a number of years ago. It stresses that, in the absence of specific legal systems, only monopolistic firms have an incentive to invest. Now, however, consider the introduction of an inventing patent system. The monopolist will be interested in patenting only if he can sell rights to his invention, and if these outweigh the benefits of trade secrecy. He may, for example, be able to sell these rights in other countries which grant him patent protection. He could license his invention to a similar monopolist in another country and charge a fee approximately equal to the added profits the invention would provide.

The introduction of a patent system where competition exists would enable the inventor to obtain a patent and then license it to all other firms in the industry for the full amount of the cost reduction which the invention afforded each firm. In this case, industry supply would not change because the license fee would be added to costs. Industry prices would not fall and the inventor would capture, in the form of license fees, the full amount of the value of the invention.

A single invention case is unrealistic, however, since inventions occur in sequences, with each invention providing information that increases the probability that a subsequent invention will be made. When one invention stimulates another, it causes its own rents to be eroded and to flow into public welfare in the form of consumer surpluses. Consider the patented inventions just discussed. If another invention were to be made which lowered costs below the original AC_0, but not as much as AC_1, this second invention would reduce the royalties that the first inventor could charge. The second inventor could sell his invention at a lower price and take the first inventor's market from him. Thus the second inventor forces a transfer of rents from the first inventor to the consumer of the product.

This process reduces the incentive for invention, as well as reducing the efficiency loss associated with invention. It also increases the extent of copying and imitation. In rapidly advancing technology, such as semi-conductors, firms often consider the period of erosion to be so short as to eliminate the incentive to patent at all.

Patent disclosure is not the only form by which invention potential is increased. In fact, public research systems often concentrate on providing genetic materials, technical concepts, instrumentation, etc. freely to inventors so as to stimulate invention by making discovery more probable. Much technology is transferred in this form. In biogenetic technology, new varieties or lines produced in one station add to the gene pools of public and private breeders in other locations.

Many public station scientists and administrators fear that the provision of publicly produced genetic material to private firms will produce monopoly rents for the firms. This is simply not the case where the private firms in question are competitive. In that situation, the full benefits of research findings in the public sector will be passed on to the consumer, and thus into the public domain. Private firms will capture rents only on the "value-added" component that they provide, even if the value-added component is patentable. As long as competition prevails, the public sector genetic materials will enable a competitor to produce another variety which limits the rents available to the first.

The situation is different if the public material is made available to a monopolistic firm. In that case, it will result in private monopoly gains, whether the monopolist is able to patent it or not! In any situation where full monopoly power is allowed to exist, the monopolist will benefit from any publicly produced technology which lowers costs. The availability of a patent system enables the public system to limit the transfer of rents to monopolists. It could, for example, force the monopolist to pay royalties or licensing fees. We will argue below, however, that in situations where monopolists have control over final technology, the public system has a responsibility to provide competition.

Legal Systems Suited to Developing Countries

We have noted that most developing countries simply join international conventions regarding patenting, and then modify their legal system by precluding certain technology fields and administering the systems in such a way as to make foreign patenting more costly. Few are seriously attempting to encourage indigenous invention. This appears to be the worst of all options. Many countries obligate themselves to pay rents for technology from abroad and receive little in return.

The typical international convention does not stimulate national invention. These systems are oriented to only one part of a differentiated international demand for technology and most developing countries are at a serious disadvantage in serving that particular demand. The large "primary" market for most types of technology is in the developed, high wage economies serving high-income customers. The demand in developing countries, on the other hand, is for differentiated technology suited to the lower wage conditions and the commodities produced in these countries. Firms and individuals in developing countries are at a comparative disadvantage in competing with firms from Japan, Germany, France and the U.S. to produce new, primary type inventions. They do not have the financing or the access to skills, but most importantly, they do not have the economic "laboratories" that firms in the high wage countries have. On the other hand, they have a comparative advantage in the production of "adapted" technology because they can modify primary market technology to suit it to the smaller secondary markets in their own economies.

The standard legal systems of many developing and semi-industrial countries provide little stimulus to secondary adaptive invention. A good deal still takes place, but if anything, the legal systems discourage it by awarding a broad range of blocking patents to foreign and multinational firms. Further, it would seem that while the provision of patent protection may be required to attract certain types of foreign capital and goods, an across-the-board provision of patent protection to foreign firms may not be optimal. Certainly developing countries obtain little in the way of a *quid pro quo;* inventors in developing countries obtain very few patents in the developed countries.

The experience of the Philippines, Brazil, Japan and other countries with "petty" patent systems seems to have been generally good. By providing a weaker and narrower standard of patentability and by administering the system efficiently, local adaptive inventive activity has been stimulated. This is important in agriculture, where a great deal of minor modification to farm machines for different types of soil, crops, etc., is desirable.

We also think that some countries would be wise to review their general industrial strategy and to question the unlimited granting of foreign patents. During input substitution phases of development, many governments have pursued policies which favor selected large national and multinational firms over smaller firms. The standard legal systems may reinforce this bias, providing mechanisms for purchase of foreign technology to large firms but not to small ones. An effective patent system enables the small firm and individuals to invent actively, and to receive rewards in competition with larger firms.

On the matter of the extension of patent protection to plants and animals, a system which adequately safeguards public programs and stimulates private activity makes good sense. This type of invention is generally so location-specific that the issue of international patent recognition is not too important. It would be unwise to allow a legal system to hamper the work of public research systems. These systems should play two roles in the private-public mix of research systems: they should provide raw materials to private breeders and they should provide competition.

Laws have to be carefully drafted to insure that a public station can serve both roles. There is always a question of adequacy of sensible incentives for private activity. It is critical for both public and private research institutions that competition be monitored. It would be a serious mistake to franchise the right to produce technology to any private or public institution.

There is probably scope for more research contracting from private firms to public agencies, and it may become more feasible in the future. The International Centers could lead the way by more experimentation with contracts to produce materials meeting certain standards. Fundamentally, however, the fact that indigenous invention effort is at very low levels should be borne in mind. Almost any program which stimulates more indigenous inventive activity is likely to have a beneficial effect.

Public-Private Research Compatibility

As private research activities expand, they have several effects on existing public research systems. These can sometimes be unproductive as both public and private systems work toward a compatible relationship. By private systems we mean research programs conducted by private, profit-making firms, not research by semipublic commodity associations, such as sugar cane growers associations, which present rather different problems.

The experience of the public system in the United States has shown that it usually finds two roles compatible with its responsibilities. The first is to undertake a range of intermediate or basic research and to make findings available to all private firms. The second is to serve as an alternative source of technology and to undertake various testing and verification activities designed to curb abuses of the private sector. In some areas of technology, this works quite effectively. In hybrid corn production, for example, public experiment stations have produced inbred lines and other types of basic or general research on corn production, particularly on agronomic practices. The inbred lines have been made freely available to private breeders. At the same time public stations have produced hybrids in direct competition with private breeders. They have not achieved large sales, but have provided important competition. It appears to be the case that, in situations where private firms have adequate incentives to produce technology, they are more effective in doing so than are public bureaucracies.

On the other hand, the public stations have been successful at producing inbred lines in competition with private breeders. Thirty years ago most inbred lines were produced by private breeders. Today almost all important inbred lines in use have been produced by public experimental stations. (It might be noted that the economic value of hybrid corn varietal improvement since 1957, the date for which Zvi Griliches calculated benefits in his classic study, have been several times as great as the early improvements. Much of this is attributable to inbred line improvement).

In terms of engineering research, public sector agricultural experiment stations have not made valuable contributions to the private sector, except in new machines where the potential market is small. In fact, if one looks at the history of invention of farm machinery in the U.S., several interesting patterns emerge. The first is that all "new" items of farm machinery were brought to a commercial stage by small independent firms engaging in what might be called "wildcat" invention. This holds true for machines and implements developed after c. 1890, when large manufacturing firms with sophisticated engineering staffs had formed (most of them in fact originated as small wildcat invention firms). The large firms have tended to dominate the production of new farm implements because of superior capacity to produce, market and repair them.

Thus we have something of a parallel to oil exploration. The truly high risk, innovative invention is left to the independent wildcat inventors who, on balance, have lower economic returns for their effort. Through sales, mergers and bankruptcy, large firms with the capacity to maintain parts and repair systems and to produce better engineered models acquire most production.

Public sector experiment stations have contributed little to the work of the large firms, except some wildcat prototype inventions and advances in testing and standards. They have not produced any appreciable competition in mechanical invention, however.

The situation for chemical invention is like the hybrid corn case, with some development of herbicides, insecticides, etc., but little in the way of competition.

Public systems thus have reason for concern at the prospect of more private activity and more competition. All public institutions have reason to fear competition because of the basic nature of bureaucracies. The tendency of bureaucracies without competitive alternatives to become inefficient and to serve their own, rather than the clients' interest, is great. Bureaucracies with monopoly power are inefficient in many cases. Yet competition has been vital to the success of public sector agriculture research systems. In the U.S., this takes the form of competition between states. Plant breeders in Indiana are pressured to match competitors in Iowa by clientele farmers with political clout. International centers are providing some competition for national programs in the developing world.

If more effective legal systems are devised for developing countries, encouraging inventions, research centers should advance in biogenetics. Mechanical and chemical inventions would have an impact on developing economies, but the public role is not large in these. It is possible that more hybrid corn situations will develop (in fact, it is hoped that more private work in hybrid corn will take place). If private soybean breeders, for example, are successful in developing new varieties protected by PVPCs in the U.S., this will create interest in developing countries.

Public research facilities such as the International Centers have a responsibility to provide their own material with broad protection and than make it available to other breeders. They also have a responsibility to encourage competition among users of the material and to provide some competition themselves. We have noted that providing technology to monopolists is non-optimal, but the private sector breeding activities can be quite competitive if encouraged (there are large numbers of hybrid corn producers, for example).

The use of private activity will create problems of another sort. These firms are likely to be able to hire scientists away from public sector experiment stations, a serious problem in many countries at present. Agricultural scientists are trained at very high cost (usually supported by aid funds). The institutions employing them then provide low salaries and poor working conditions, losing them to the private sector.

Finally, the increase in research by private associations such as sugar cane growers or coconut growers raises very different questions. This is not really private research, but public research performed for a specific client group. It may not differ from public research also conducted for client groups. Many of these groups have the option of providing support to the public system or creating their own research centers. The decision depends on their assessment as to which institutional arrangement is most efficient. Public sector research which does not serve a clientele probably should not be supported.

APPENDIX 1

Following are selected excerpts from English language translations of statutory provisions for various Latin American countries as well as the United States. These excerpts include provisions specifically precluding certain subjects from patenting and were used in developing Table 1.

ARGENTINA

Law No. 111 on Patents of Invention of October 11, 1964, Title 1: General-Provisions*

Article 1. New discoveries or inventions in all classes of industry confer upon their authors the exclusive right of working, for the time and on the conditions hereinafter expressed, in accordance with the provisions of Article 17 of the Constitution. This right shall be established by documents called "Patents of Invention," issued in the form which will be determined by this law.

Article 2. The preceding article applies not only to discoveries and inventions made in the country, but also to those made and patented abroad, provided the applicant be the inventor or the legitimate successor to his rights and privileges, and in the cases and with the formalities which will be hereinafter set forth.

Article 3. Discoveries or new inventions are: new industrial products, new means, and new application of known means for obtaining an industrial result or product.

Article 4. The following are not patentable: pharmaceutical compositions, financial schemes, discoveries or inventions which have prior to the application been sufficiently made public, in this country or outside it, in works, pamphlets or printed periodicals, to enable them to be carried out; those which are purely theoretical, without their industrial application being indicated and those which are contrary to morality or the laws of the Republic.

Article 5. Patents shall be granted for 5, 10, and 15 years, according to the merit of the invention and the wish of the applicant. The ratification of foreign patents shall be limited to 10 years, but in no case shall it exceed the term granted for the original patent, with which it shall lapse.

Argentine Patent Office Rules 27/74 and 15/75

Rule 27/74. (3) Patent applications involving the production of chemical species by fermentation or biosynthesis must also conform to the provisions of point (1). When the novelty of these patent applications resides in the microorganism or strain utilized, the Culture Collection where one or the other has been deposited shall be mentioned in the Specification and the Main Claim, with an identifying indication. (Ministry of Education, Department of Industrial Development, Buenos Aires, 16 October 1974).

Rule 15/75. Having examined the norm proposed by the Patent Office in order to facilitate the consideration of applications related to an arrangement of means which include processor, a computer or a logic circuit, or to a process using said means, and
Whereas:
The norm proposed will facilitate the study of the aforementioned applications, by promoting a better understanding between the applicants and the Patent Office;
In many industrial processes it is becoming ever more important to include data processing machines in the devices controlling the process, the reasons for their use being as varied as the processes themselves, it being possible to use such means to avoid or minimize human error, or for reasons of ecology, product purity, economic aspects, etc.;
In each and every case patent application must show an "inventive creation" which cannot consist only in the program for controlling the operative sequence of a computer, a data proces-

---

*English translations of these Rules were prepared and made available through the courtesy of Hausheer Belgrano & Fernandez, Patents, Trade Marks and Designs, Florida 142, 1337 Buenos Aires, Argentina.

sor or a logic circuit, or of a combination of said means, or when the means are being used for purposes of automation or in order to use their operative capacity;

The present norm does not deviate from what has been laid down by Rule No. 10/64;

Now, therefore:

THE FEDERAL DIRECTOR OF INDUSTRIAL PROPERTY RULES:

(1) In application whose main object is a computer, a data processor or a logic circuit, or a combination of means at least one of which is a computer, a data processor or a logic circuit, said object shall be defined by an enunciation of, and the relationship between, the means integrating the same; the definition of the interaction between the means will be accepted by way of complement when the interaction is univocal.

(2) In the applications mentioned in point (1) there shall not be accepted a claim if characterizing as the main object the program, understood as a set of instructions to control the operative sequence of said computer, data processor or logic circuit, or said combination of means at least one of which is a computer, a data processor or a logic circuit.

(3) In applications whose main object is a process, including one that comprises one or several steps of control, and which is carried out in part or wholly by means of a computer, data processor or logic circuit-control being understood in this case as the fact inducing a variation in a parameter or an operative condition dependent on the same or another parameter or operative condition—the definition of the process will be accepted only if the novel characteristic is not exclusively the processing of said parameter or operative condition by said computer, data processor or logic circuit. When the arrangement carrying out said process is also claimed, it shall be claimed with deviating from the provision of points 1 and 2.

(4) Let this Rule be registered, communicated, published etc. . . .

(signed) Ing. Gerardo Rodolfo Seeber
Federal Director of Industrial Property

LAW OF SEED AND PHYTOGENETIC CREATIONS

Chapter I: Generalities

Article 1. The present law has the objective to promote efficient activity in production and commercialization of seed to assure to the agricultural producers the identify and quality of purchased seed and to protect the owner of phytogenetic creations.

Article 2. The effect of this law is intended for:
(a) "Seed:" For vegetative structures destined for planting or propagation.
(b) "Phytogenetic Creations:" The cultivars obtained by discovery or application of scientific knowledge for inheritable improvement of the plants.

Article 3. The Ministry of Agriculture and Livestock, with the advisement of the National Commision of Seeds, will administer the present law and establish requirements, norms and general tolerances for class, category, and species of seed.

Chapter III: Of The Seed

Article 9: The seed available to the public or delivered to consumers (farmers) under whatever contract, ought to be positively identified, specified in the marking of the container for, at least, the following ways:

(a) Name and address of the identifier of the seed and his registration number.
(b) Name and address of the commercial seller of the seed and his registration number, where there is no identifier.

(c) Common name of the species, and the botanical name for those species that are regulated; in the case of a mixture of two or more species, it must be specified "mixture" and show the names and percentages of each one of the components which, individually or together, exceed the total percentage established by regulations.
(d) Name of cultivar and varietal purity of the same if appropriate; in cases to the contrary it should indicate the term "common."
(e) Percentage of physical-botanical purity, in weight when this is inferior to the values established by regulation.
(f) Percentage of germination, in number, and date of analysis (mouth and year), when this is inferior to the values established by regulation.
(g) Percentage of weeds, for those species established (as such) by regulation.
(h) Net content.
(i) Year of harvest.
(j) Origin, for imported seed.
(k) "Category" of seed, if applicable.
(l) "Treated seed — poison," in red letters, if the seed has been treated with toxic substance.

Article 10. Establishes the following "classes" of seed:
(a) "Identified" — Those that comply with the requirements of Article 9.
(b) "Farm" — Those that, besides complying with requirements demanded for "identified" seed and demonstrating a good behavior in officially approved trials, is amenable to official control during the stages of its cycle of production. Within this class is recognized the "categories:" "original" (basic or foundation) and "certified" in distinct grades.

The regulation enables the establishment of other categories within the above mentioned classes.

The Ministry of Agriculture and Livestock, with the guidance of the National Commission of Seed, will maintain under the system of farm production all the species that will be encountered in such a situation at the date of enactment of the present law and ought to be obligatorily incorporated into the regimen of "farm" seed, the production of these species that are considered proper for agronomical reasons or of general interest.

Chapter V: National Register of the Ownership of Cultivars

Article 19. To create, under the jurisdiction of the Ministry of Agriculture and Livestock, the National Register of the Ownership of Cultivars, with the object to protect the right of the owner of the creations or describers of new cultivars.

Article 20. To allow entries in the register created by Article 19 which will be considered "property" of those conforming to the present law, the phytogenetic creations or cultivars that will be distinguishable from others known at the date of presentation of the application of ownership, and whose individuals possess inheritable characters sufficiently homogeneous and stable. The pertinent step should be carried out for the creation or description under sponsors of agronomic engineering with the title of a national or licensee, and must record individually the new cultivar with a name that agrees with that established in the respective part of Article 17.

Article 21. The application of ownership of the new cultivar will detail the required characteristics in Article 20 and will be accompanied with seed and specimens of the same, if so required by the Ministry of Agriculture and Livestock. Said Ministry will be able to submit the new cultivar to tests, laboratory analysis and grow-out tests to verify attributed characters, (or) is able to accept as evidence the information of previous tests performed for the use of the owner and of official services.

With such elements of judgment and the guidance of the National Commission of Seeds, the Ministry of Agriculture and Livestock will decide the granting of corresponding titles of ownership. If this (title) is not granted, the respective cultivar will not be able to be sold nor offered for sale. The owner will maintain one live specimen of the cultivar at the disposal of the Ministry of Agriculture and Livestock while the respective title is in effect.

Article 22. The title of ownership over a cultivar will be granted for a period of no less than ten (10) nor more than twenty (20) years, depending on the species or group of species, and according to the established regulations. In the Title of Ownership will be shown the dates of issuance and of expiration.

BRAZIL

Code of Industrial Property and Normative Acts Nos. 15 and 17*
Non-Patentable Inventions. Chapter II:

Article 9. The following are not patentable:
(a) Inventions the purposes of which are contrary to the laws, to morality, to health, to public safety, to religious cults and to sentiments which are worthy of respect and veneration.
(b) Substances, matter or products obtained by chemical means or processes, but the respective processes of obtaining or modifying such substances, matter or products, shall be patentable.
(c) Food and chemical-pharmaceutical substances, and admixtures or products and medicaments of any kind, as well as the respective processes for obtaining or modifying them.
(d) Metallic admixtures and alloys in general, but those which, not being included in the category of the preceeding subsection, have specific intrinsic qualities (and are) precisely characterized by the nature and proportions of their ingredients or by special treatment to which they have been submitted, shall be patentable.
(e) Aggregations of known processes, means or organs, mere alterations in form, proportions, dimensions or materials, but they shall be patentable if, on the whole, the result thereof is a new or different technical effect and is not otherwise unpatentable in view of the provisions of this article.
(f) Uses or employment of means related to discoveries, including the discovery of varieties or species of microorganisms, for a specific purpose.
(g) Operating, surgical or therapeutical techniques, but the devices, apparatus or machines (for such purposes) shall be patentable.
(h) Systems and programming, plans or schemes for commercial bookkeeping, for calculation, for financing, for credit purposes, for raffles, for speculation or for advertising.
(i) Purely theoretical concepts.
(j) Substances, matter, admixtures, elements or products of any nature, and also the modification of their physicochemical properties and their respective methods of being obtained or modified, when they are the result of transformation of the atomic nucleus *("núcleo atómico").*

Chapter III: Utility Models and Industrial Models and Designs, Section I:
Patentable Models and Designs

Article 10. For the purpose of this Code a utility model is considered to be any new arrangement or new form given to or introduced in a known object, provided such new arrangement or new form serves for practical work or use.
Paragraph 1. The word "object" *("objeto")* comprises tools, working instruments or utensils.
Paragraph 2. Protection is granted solely in respect of the new form or new arrangement which imparts greater usefulness to the function to be performed by the object or by the part of the machine.

---

*The English translations of the Brazilian Code of Industrial Property and Normative Act Nos. 15 and 17 were prepared and made available for republication through the courtesy of Denis Allen Daniel, partner in the firm of Daniel & Cía; Patent and Trademark Attorneys, Rua da Alfandega, 108–7° andar, Rio de Janeiro, Brazil.

CHILE

*Law for Regulating the Service of Industrial Property: Degree Law No. 588**

Article 1. The services of Industrial Property treated in this law include patents for invention, trade marks and industrial models, and will be transacted by a public department to be called the Industrial Property Office, under the superintendence of the Minister for Agriculture and Industries.

This law defines the attributes conferred on the Chief of this office for the management of the department.

Patents for Inventions

Article 2. Any person whatsoever, natural or juridicial, national or foreign, wishing to obtain the monopoly guaranteed to any inventor by Article 10, number 11 of the Political Constitution of the State, must make application personally or by his legal representative, for a patent for his invention or production, for the time he considers fit, within the terms determined by this law.

Application for patents must be made to the Government through the Industrial Property Office in the form prescribed by the respective regulations.

Article 3. Patents for inventions are reserved for real inventions, that is to say, for the creation of something previously nonexistent, and that must possess some definite industrial application.

Every application for a patent must also contain a formal declaration of novelty and originality of the invention.

The Industrial Property Office will verify these conditions by examination, either directly or through the medium of technical experts whom it may be considered convenient to appoint.

The costs of this examination as well as all other fees incurred in the application for patents, will be at the charge of the applicants and fixed by the Office.

Article 4. The following are patentable:
(a) Any definite product possessing novelty and utility.
(b) Any new machine of tool, and any new instrument or apparatus of industrial utility or of medicinal, technical or scientific application.
(c) The invention of parts or elements or accessories of machines, mechanism or apparatus, by means of which a greater economy or perfection is obtained in the products or results.
(d) New combinations or groups of machines or apparatus by means of which greater economy or perfection is proved to be obtainable in the products or final results.
(e) The invention of new processes for the preparation of materials or objects of commercial or industrial utility.
(f) New processes for the preparation of chemical products and new methods of elaboration, extraction, and separation of natural substances.
(g) Reforms, improvements or modifications introduced into known contrivances, subject to proof of their novelty and their advantage over the contrivance already in use, in producing a resulting or product superior to that already existing.

Article 5. The following are not patentable:
(a) Drinks and food-stuffs whether for man or beast; all kinds of medicaments; pharmaceutical medicinal preparations and chemical preparations; reactions and combinations.
(b) Financial, speculative, commercial or business systems, combinations or plans, as well as those of simple fiscalization or control.

---

*This English translation of Chile's industrial property law was made available for republication in this series through the courtesy of Patricio Claro T., Esq. of Claro y Cía; Moneda 1025 – Casilla 1867, Santiago, Chile.

(c) The simple application or benefit of natural forces of substances, even though they be of recent discovery.

(d) Modes of work or secrets of manufacture *(tours de main)*.

(e) New applications of articles, objects or elements already known and applied to determine ends and the simple changes or variations in their form, dimensions or material.

(f) Inventions already sufficiently known in the country through the medium of printed works or in any other ostensible manner, and those that are known to the public from their execution, sale or publicity in this or in other countries previous to be respective application for the patent. This disposition, however, does not apply to foreign inventions that in accordance with their respective laws have been made public after being patented but have not been made known commercially in Chile previous to the application for the patent, and for which the foreign patent is still in force.

(g) Inventions from abroad that be publicly known in any country whatsoever, even though they be totally unkown in Chile.

(h) Theoretical or speculative inventions whose application to a practical and useful end have not been proved.

(i) Inventions contrary to the national laws, to the public health and order, to morality or good habits and to the safety of the State.

MEXICO

New Patent and Trademark Law in Mexico, 1976*
Statutes, Regulations & Treaties: Preliminary Dispositions

Article 1. This Law regulates the granting of patents of invention and improvement patents, of certificates of invention, the registration of industrial models and designs, the registration of trademarks, denominations of origin, and trade names, and advertisements, as well as the repression of unfair competition in relation to the rights granted in this Law.

Article 2. The dispositions of the present Law are of the public order and of social interest. Their application is encharged to the Federal Executive through the Ministry of Industry and Commerce.

The National Council of Science and Technology will be the consultative organ in the terms of the Law which established it.

Title One: Patents of Invention
Chapter I: General Rules

Article 3. The physical person who creates an invention, or his assignee, has the exclusive right to exploit it to his advantage, either by himself or by others with his permission, in accordance with the dispositions contained in this Law and its Regulations. This right is acquired by means of the privilege of patent which the State grants, and its exercise will be subject to the conditions which are dictated by the public interest. The applicant may, however, elect to receive a certificate of invention, in the terms of Article 80 of this Law.

Article 4. That invention is patentable which is new, the result of an inventive activity, and susceptible of industrial application, in the terms of this Law.

---

*This English translation of the new Mexican Patent and Trade Mark Law was obtained through the kind efforts of John J. McGlew, Esq. of McGlew and Tuttle, P. C., 28 West 44th Street, New York, N.Y. 10036. The translation was prepared by Dumont-Bergman, Bider & Co., International Patent Lawyers, Apdo. (P.O. Box) 6-1012, Mexico 6, D.F.

That invention is also patentable which constitutes an improvement upon another and which complies with the requirements of the foregoing paragraph.

Article 5. An invention will not be considered novel if it is comprehended within the state of the art, that is, if it has been made available to the public, in this country or abroad by means of an oral or written description, through use, or by any other means sufficient to permit its accomplishment, prior to the date of filing of the application for patent or the date of priority validly claimed.

Article 6. Divulgation of the invention prior to filing of the application does not constitute loss of its novelty if such divulgation results from the fact that the applicant or his assignee has exhibited the invention in an official or officially recognized international exposition, provided that before such exhibition the documents anticipated in the Regulations are deposited with the Ministry of Industry and Commerce and that the corresponding application for patent is filed in the same office within the four months following the closing of the exposition.

Article 7. An invention is considered to imply inventive activity if, at the date to which Article 5 refers and taking into account the state of the art, it is not evident to a technician in the subject matter.

Article 8. An invention is susceptible of industrial application if it can be manufactured or utilized in industry.

Article 9. For the effects of the present Law the following are not inventions:
(a) Theoretical or scientific principles and mathematical method.
(b) The discovery which consists merely in making known, evident, or ostensible that which already existed in Nature, even if it was previously unknown to man.
(c) Commercial, accounting, financial, educational and advertising systems and plans; typographical characters; rules of games; presentation of information and programs of computation.
(d) Artistic or literary creations.
(e) Surgical or therapeutic methods of treatment of the human body and those related to animals or vegetables, as well as diagnostic methods in these fields.

Article 10. The following are unpatentable:
(a) Vegetable varieties and breeds of animals, as well as the biological processes for obtaining them.
(b) Alloys.
(c) Chemical products, excepting new industrial processes for obtaining them and their new uses of an industrial character.
(d) Chemical-pharmaceutical products and their mixtures, medicaments, drink and foods for human or animal use, fertilizers, insecticides, herbicides, fungicides.
(e) Procedures for obtaining mixtures of chemical products, industrial procedures for obtaining alloys, and industrial procedures for obtaining, modifying or applying products and mixtures referred to in the preceding section.
(f) Inventions related to nuclear energy and nuclear safety.
(g) Anticontamination apparatus and equipment, and procedures for manufacture, modification, or application thereof.
(h) The juxtapositions of known inventions, their variation in form, dimensions, or materials, excepting where those inventions are in reality so combined or fused that they cannot function separately, or that the qualities or functions which are characteristic thereof are so modified as to obtain a novel industrial result.
(i) The application or the use, in an industry, of an invention already known or used in another industry, and the procedures which consist simply in the employment or use of a device, machine, or apparatus functioning according to principles already known beforehand, even where such employment may be new.

(j) Inventions which, publishes or exploited, would be contrary to law, to the public order, health, public safety, morals or good habits.

Article 11. The holders of patents may be physical or juridical persons.

Article 12. The person who claims to be the inventor in the patent application is presumed to be the inventor.
The inventor has the right to be mentioned in the patent or to oppose his being mentioned.

Article 13. Inventions created by persons who lend their services by virtue of a contract or a labor relationship will be governed, in terms of Article 163 of the Federal Labor Law, by the disposition of that Law.

Chapter III: Rights Conferred by the Patent

Article 37. Subject to the limitations provided in this Law, the patent confers upon its holder the right to exploit the invention exclusively, either by himself or by others with his consent.
The patent does not confer the right to import the patented product or one manufactured with the process patented.

Article 38. The scope of the privilege conferred by the patent will be determined by the tenor of the claims. The description and the drawings or plans will serve to interpret them.

Article 39. The rights conferred by a patent will have no effect:
(a) Against a third party who for the purpose of study, scientific or technological research, or experimental or recreational purposes manufactures a product or uses a process identical to that patented.
(b) Against any person who, prior to the date of filing of the patent application in the country or prior to the date priority validly claimed, has manufactured the product or used the process covered by the invention or who has made the necessary preparations for carrying out such manufacture or use.
(c) Against the use, aboard ships of other countries, of the means which are the subject of the patents in the hull of the ship, in its machinery, equipment or other accessories, when such ships temporarily enter the waters of the country, provided that such means are employed exclusively for the needs of the ship.
(d) Against the employment of the means which are the subject of the patent in the construction or functioning of devices of aerial or terrestrial locomotion or the accessories of such devices, when the latter temporarily enter the country.

Article 40: The period of duration of the patents will be ten years, not subject to extension, counting from the date of issue of the title, but the legal date of the patent will be held to be the day and hour of filing of its application.

Chapter V: Exploitation of the Patents

Article 41. The granting of the patent implies the obligation of exploiting it in national territory.
The exploitation must be initiated within a period of three years counting from the date of issue of the patent.

Article 50. Compulsory Licenses and those of Public Interest. When the period to which Article 41 refers has expired, any person may request of the Ministry of Industry and Commerce the concession of a compulsory license for exploiting a patent in the following cases:
(a) When the invention patented has not been exploited.
(b) If the exploitation of the patent has been suspended for more than six consecutive months.

(c) When exploitation of the patent does not satisfy the national market.
(d) When there exist export markets which are not being covered by the exploitation of the patent and some person manifests his interest in utilizing the patent for purposes of exportation.

In the cases of Fractions (c) and (d), before granting the license, the opportunity will be given to the holder of the patent to correct the insufficient exploitation thereof, conceding him the preferential right to enlarge his exploitation, to cover adequately the national consumption or the international demand. To this end the Ministry of Industry and Commerce will make known to him the application for a compulsory license, in order that in a period of two months he may present a program for manufacture under conditions at least similar to those programs presented by the person applying for the license, and give a bond to guarantee performance. The Ministry of Industry and Commerce may grant for a single time and for a period of up to an additional two months an extension to the period for presenting the program of manufacture, if the interested party so requests prior to the expiration of the term first granted.

PERU

New Law of Industrial Property No. 22532 Enforcing Decision 85 of the Cartagena Agreement (Andean Pact) as From May 17, 1979 *

Chapter 1: Patents of Invention

1. Patentability Requirements

Article 1. A patent of invention shall be granted to new creations susceptible of industrial application and to those which improve such creations.

Article 2. An invention shall not be considered new if it is included within the state of the art, that is, if it has been made available to the public in any place through an oral or written description, or through use of exploitation, or through any other means sufficient to enable its execution, prior to the date of filing of the patent application. Notwithstanding the provisions contained in this article, a release made during the year previous to the filing of the application shall not constitute loss of the novelty of the invention, when such release is the result of:
(a) An evident abuse in detriment of the applicant or its assignee, such as theft of plans or documents, misfeasance or infidelity of the agent or of the collaborators of workers of the inventor, industrial espionage or similar acts.
(b) The fact that the applicant or this assignee have exhibited the invention during an exhibition officially made and recognized in one of the Member Countries, or when he has accomplished experiments to test its industrial application.

Article 3. An invention shall be susceptible of industrial application when its object can be manufactured or used in any kind of industry.

Article 4. The following shall not be considered as inventions:
(a) Principles and discoveries of a scientific character.
(b) The mere discovery of materials existing in nature.
(c) Commercial, financial, accounting or other similar plans, game regulations or other systems in the measure in which they have a strictly abstract nature.

---

*The English translation of the New Law of Industrial Property No. 22532 was prepared and made available for republication by J. M. Colmenares & Co., International Patent & Trade Mark Agents, Carabaya Street No. 933, P.O. Box 777, Lima, Peru, Telex: 20338 CP HBOLI, Lima.

(d) Therapeutical or surgical methods for treatment of humans or animals and methods for diagnosis.
(e) Creations purely aesthetical.

Article 5. Patents shall not be granted for:
(a) Inventions contrary to public order or to morals.
(b) Vegetable varieties or animal breeds, the essentially biological procedures for obtaining vegetables or animals.
(c) Pharmaceutical products, medicines, active therapeutical substances, beverages and foods for human, animal or vegetal use.
(d) Foreign inventions whose patent is applied for one year after the date of filing of the patent application in the country of the first filing. Upon expiration of this period, no right derived from such application may be sustained.
(e) Inventions that affect the development of the respective Member Country or processes, products or groups of products whose patentability is excluded by the Governments.

7. Licenses

Article 39. When concerning patents of interest for public health or national development requirements, the Government of the respective Member Country may submit the patent to a compulsory license at any time and in such case the competent national office may grant the licenses that will be requested.

VENEZUELA

Venezuelan Industrial Property Law*

Chapter I: General Rules

Article 1. The present Law shall govern the rights of inventors, discoverers and introducers on creations, inventions or discoveries connected with industry, and those of producers, manufacturers or merchants on the special phrases or signs which they may adopt to distinguish the results of their work or activity from others that are similar.

Chapter II: Patents

Article 9: The patents of invention, improvement, or industrial model or drawing shall be issued for a term of five or ten years, at the will of the petitioner; and only the patent of introduction for only five years.

Article 14. A patent will cover:
(a) All new, definite and useful products.
(b) All new machines or tools, and all new instruments or an apparatus for industrial use or of medicinal, technical or scientific application.
(c) Parts or elements of machines, mechanisms, an apparatus, accessories through which a great saving or perfection in products or results may be achieved.

---

*The English translation of the Venezuelan Industrial Property Law was made available for republication through the courtesy of Dr. Roberto Picon-Parra of Escritorio Maury, Attorneys at Law, Este 3 No. 12 (Jesuitas A. Maturin), Apartado de Correos: 573, Caracas 101, Venezuela.

(d) New processes for the preparation of substances or objects for industrial or commercial use.
(e) New processes for the preparation of chemical products and the new methods of preparation, extraction and separation of natural substances.
(f) Reforms, improvements or modifications introduced in already known objects.
(g) All new models or drawings for industrial use.
(h) Any other invention or discovery suitable for an industrial application.
(i) An invention, improvement, industrial model or drawing which, having been patented abroad, has not been revealed, patented or placed in use in Venezuela.

The recapitulation contained in this article is merely enunciative and not restrictive, once the purpose of a Patent, in general, is the result of the inventive effort of the human talent, with the exceptions established by this Law.

Article 15. The following are not patentable:
(a) Drinks and food products, whether they be for man or animals; medicines of all kinds; pharmaceutical medicinal preparations and chemical preparations, reactions and compounds.
(b) Financial, speculative, commercial, publicity systems, schemes or plans, or those of simple control or inspection.
(c) The plain use or utilization of natural substances or forces, even though they be of recent discovery.
(d) The new use of articles, objects, substances or elements already known or employed for fixed purposes, and the mere change or variation in form, dimensions or material of which they may be composed.
(e) The working methods or manufacturing secrets.
(f) Inventions merely theoretical or speculative, whose industrial feasibility and application could not possibly by described or demonstrated, nor their well-defined industrial application.
(g) Inventions contrary to national laws, to health or public order, to moral or proper habits, and to the security of the State.
(h) The juxtaposition of elements already patented, or which may be of public knowledge, provided they are not united in such a way as not to function independently, losing their characteristic action.
(i) Inventions which may be made known to the public in the country, through publications or revealed in printed papers or otherwise, and those which may be of public knowledge by reason of their execution, sale or publicity within or outside of the country, prior to the petition for patent.

Article 16. When an invention or discovery may interest the State, or may be considered basically of public interest, the National Government may, by reason of social or public interest, decree the expropriation of the right of the inventor or discoverer, adhering to the requisites which the Law on the matter provides for the expropriation of property.

In the publications which have to be issued for this purpose, the subject matter of the invention or discovery shall be omitted, and it shall solely mention that same is covered by the terms of this article.

Article 17. Patents lose their validity:
(a) When the judgment of the competent Courts cancels them, because they are declared as detrimental to the better right of third parties.
(b) When they may be cancelled in accordance with Articles 12 and 21 of this Law.
(c) When the beneficiary of the patent may have allowed a lapse of two years from the date of the grant, without working out the invention in Venezuela causing the same, or when the working out of same may have been interrupted for an equal period of time, except in case of hazard or *force majeure,* duly proved before the Industrial Property Registry Office.
(d) For lack of payment of any of the annuities stipulated in Article 49.

(e) For expiration of the term.
(f) For express waiving of the inventor.

UNITED STATES PLANT VARIETY PROTECTION ACT

December 24, 1970, (84 Stat. 1542) (7 U.S.C. 2321 Et Seq.)

Chapter IV: Protectability of Plant Varieties

Section 41. Definitions and Rules of Construction.

The definitions and rules of construction set forth in this section apply for the purposes of this Act:
  (a) The term "novel variety" may be represented by, without limitation, seed, transplants, and plants, and is satisfied if there is:
    (1) distinctness in the sense that the variety clearly differs by one or more indentifiable morphological, physiological or other characteristics (which may include those evidenced by processing or product characteristics, for example, milling and baking characteristics in the case of wheat) as to which a difference in genealogy may contribute evidence, from all prior varieties of public knowledge at the date of determination within the provisions of section 42; and
    (2) uniformity in the sense that any variations are describable, predictable and commercially acceptable; and
    (3) stability in the sense that the variety, when sexually reproduced or reconstituted, will remain unchanged with regard to its essential and distinctive characteristics with a reasonable degree of reliability commensurate with that of varieties of the same category in which the same breeding method is employed.
  (b) The terms "United States" and "this country" means the United States of America, its territories and possessions, and the Commonwealth of Puerto Rico.
  (c) The term "kind" means one or more related species or subspecies singly or collectively known by one common name, for example, soybean, flax, or radish.
  (d) The term "date of determination" means the date when there has been at least tentative determination that the variety has been sexually reproduced with recognized characteristics, whether or not the novelty of those characteristics has been determined.
  (e) The term "breeder" shall mean the person who:
    (1) directs the final breeding creating the novel variety, or
    (2) discovers the novel variety, and
  makes the tentative determination described in subsection (d). Where such actions are conducted by an agent on behalf of his principal, the principal, rather than the agent, shall be considered the breeder. The terms "breed," "develop," "originate," and "discover," and derivatives thereof shall each include the other.
  (f) The term "sexually reproduced" shall include any production of a variety by seed.
  (g) The term "basic seed" means the seed planted to produce certified or commercial seed.
  (h) The term "testing" means testing or experimental use of a variety before any sale thereof. Sale for other than seed purposes of seed or other plant material produced as the result of testing shall not constitute a sale for the purpose of the preceding sentence or for the purpose of the following subsection.
  (i) The term "public variety" means a variety sold or used in this country, or existing in and publicly known in this country; but use for the purpose of testing, or sale or use as individual plants not known to be sexually reproducible, shall not make the variety a public variety.
  (j) A variety described in a publication as specified in Section 42(a) (1) (b) is "effectively available to workers in this country" if a source from which it can be purchased is indicated in such publication or readily determinable or if such publication teaches how to produce the variety from source-material effectively available to workers in this country.

Section 42. Right to Plant Variety Protection; Plant Varieties Protectable.
(a) The breeder of any novel variety of sexually reproduced plant (other than fungi, bacteria, or first generation hybrids) who has so reproduced the variety, or his successor in interest, shall be entitled to plant variety protection therefor, subject to the conditions and requirements of this title unless one of the following bars exists:
 (1) before the date of determination thereof by the breeder, or more than one year before the effective date of the application therefor, the variety was (a) a public variety in this country, or (b) effectively available to workers in this country and adequately described by a publication reasonably deemed a part of the public technical knowledge in this country, which description must include a disclosure of the principal characteristics by which the variety is distinguished;
 (2) an application for protection of the variety based on the same breeder's act, was filed in a foreign country by the owner or his privies more than one year before the effective filing data of the application filed in the United States;
 (3) another is entitled to an earlier date of determination for the same variety and such other (a) has a certificate of plant variety protection hereunder or (b) has been engaged in a continuing program of development and testing to commercialization, or (c) has within six months after such earlier date of determination adequately described the variety by a publication reasonably deemed a part of the public technical knowledge in this country, which description must include a disclosure of the principal characteristics by which the variety is distinguished.
(b) The Secretary may, by regulation, extend for a reasonable period of time the one year time period provided in subsection (a) for filing applications, and may in that event provide for at least commensurate reduction of the term of protection.

Chapter 5: Application; Form, Who May File, Relating Back, Confidentiality

Section 52. Contents of Application.
An application for a certificate recognizing plant variety rights shall contain:
(a) The name of the variety except that a temporary designation will suffice until the certificate is to be issued.
(b) A description of the variety setting forth its novelty and a description of the genealogy and breeding procedure, when known. The Secretary may require amplification, including the submission of adequate photographs or drawings or plant specimens, if the description is not adequate or as complete as is reasonably possible, and submission of records or proof of ownership or of allegations made in the application. An applicant may add to or correct the description at any time, before the certificate is issued, upon a showing acceptable to the Secretary that the required description is retroactively accurate. Courts shall protect others from any injustice which would result. The Secretary may accept records of the breeder and of any official seed certifying agency in the country as evidence of stability where applicable.
(c) A declaration that a viable sample of basic seed necessary for propagation of the variety will be deposited and replenished periodically in a public repository in accordance with regulations to be established hereunder. This declaration may be added by amendment.
(d) A statement of the basis of applicant's ownership.

Chapter 11: Infringement of Plant Variety Protection

Section 111: Infringement of Plant Variety Protection.
Except as otherwise provided in this title, it shall be an infringement of the rights of the owner of a novel variety to perform without authority, any of the following acts in the United States, or in commerce which can be regulated by Congress or affecting such commerce, prior to expiration of the right to plant variety protection but after either the issue of the certificate or the distribution of a novel plant variety with the notice under section 127:
(a) Sell the novel variety, or offer it or expose it for sale, deliver it, ship it, consign it, exchange it, or solicit an offer to buy it, or any other transfer of title or possession of it.

(b) Import the novel variety into, or export it from the United States.
(c) Sexually multiply the novel variety as a step in marketing (for growing purposes) the variety.
(d) Use the novel variety in producing (as distinguished from developing) a hybrid or different variety therefrom
(e) Use seed which had been marked "propagation prohibited" or progeny thereof to propagate the novel variety.
(f) Dispense the novel variety to another, in a form which can be propagated, without notice as to being a protected variety under which it was received.
(g) Perform any of the foregoing acts even in instances in which the novel variety is multiplied other than sexually, except in pursuance of a valid United States plant patent.
(h) Instigate or actively induce performance of any of the foregoing acts.

Section 112. Grandfather Clause.
Nothing in this Act shall abridge the right of any person, or his successor in interest, to reproduce or sell a variety developed and produced by such person more than one year prior to the effective filing date of an adverse application for a certificate of plant variety protection.

Section 113. Right to Save Seed; Crop Exemption.
Except to the extent that such action may constitute an infringement under subsections (c) and (d) of Section 111, it shall not infringe any right hereunder for a person to save seed produced by him from seed obtained, or descended from seed obtained, by authority of the owner of the variety for seeding purposes and use such saved seed in the production of a crop for use on his farm, or for sale as provided in this section: Provided, That without regard to the provisions of section 111 (c) it shall not infringe any right hereunder for a person, whose primary farming occupation is the growing of crops for sale for other than reproductive purposes, to sell such saved seed to other persons so engaged, for reproductive purposes, provided such sale is in compliance with such State laws governing the sale of seed as may be applicable. A bona fide sale for other than reproductive purposes, made in channels usual for such other purposes, of seed produced on a farm either from seed obtained by authority of the owner for seeding purposes or from seed produced by descent on such farm from seed obtained by authority of the owner for seeding purposes shall not constitute an infringement. A purchaser who diverts seed from such channels to seeding purposes shall be deemed to have notice under Section 127 that his actions constitute an infringement.

Section 114. Research Exemption.
The use and reproduction of a protected variety for plant breeding or other bona fide research shall not constitute an infringement of the protection provided under this Act.

NOTES

[1] To the extent that a manufacturing process could be applied to a wide range of agricultural machinery (steel making/welding techniques), it would have a wide adaptability. However, such processes are not considered as primarily agricultural, and will not be further considered herein.

[2] For example, most fertilizers the world over have the same basic chemical ingredients, present inventive activity being primarily in improving processes of production.

[3] Trade secrecy is becoming a central policy issue in the contemporary boom in biogenetic research in the private sector. Virtually all biological scientists with expertise in this field have currently entered into some consulting arrangement or more direct involvement with a private firm. Given this and current trade secrecy protection laws, we may see the virtual cessation of scientific communication in this field.

[4] For example, the inventor of the first tractor-mounted corn picker could exclude others from making, using or selling tractor-mounted corn pickers but may have to obtain the license or consent of an earlier inventor of the tractor.

[5] Most invention patents granted to national inventors in developing countries are not patented in other countries (See Table 8).

[6] The United States system also has a "buy-out" provision whereby the Secretary of Agriculture can limit patent protection on unusually valuable inventions.

[7] Appendix 1 includes a brief synopsis and excerpts from the statutes of the various countries cited in Table 1.

[8] The rationale for excluding certain types of invention from protection appears to be related to a basic needs concept and to an interest in preventing transnational firms from capturing monopoly rents. The elimination of national protection is a serious hindrance to local inventions, however.

[9] A comprehensive discussion of the development of biogenetic technology is contained in the April, 1981 report titled "Impact of Applied Genetics" (microorganisms, plants, and animals), published by the Office of Technology Assessment of the Congress of the United States.

[10] The hearings for this amendment brought out well-intentioned but misplaced anti-technology arguments. The central theme of the arguments was that improved crop varieties would replace current varieties, causing current varieties to be lost.

[11] Diamond v. Chakrabarty 447 U.S. 303 (1980).

[12] In this connection, see Appendix 1, which contains the express provisions of many countries' statutes that preclude pharmaceutical, medical, and food-related inventions, or those related to human and/or animal health.

[13] These data are discussed more fully in R. E. Evenson, "International Invention: Implications for Technology Market Analysis," Economic Growth Center Discussion Paper No. 419, Yale University, New Haven, 1982.

[14] Data on R & D spending and scientists and engineers are from annual issues of the *Statistical Yearbook* (1968, 1972, 1977, 1980), The United Nations Educational, Scientific, and Cultural Organization, Paris, France.

[15] Of course, many inventors will invent without patent protection. Nonetheless, it appears that patent protection does induce invention.

[16] These data are from Boyce and Evenson (1975). A recent effort to update them produced no new data.

[17] India is an exception. The Council for Scientific and Industrial Research engages in a broad range of inventive activity.

[18] This is more easily seen for process inventions than for product inventions. However, a product invention is an improved product which can be described as a larger quantity of the existing products. Costs expressed per unit of the old product will fall.

CONCLUSIONS:
TOWARDS AN INTERPRETATION OF TECHNICAL CHANGE IN LATIN AMERICAN AGRICULTURE

Martín Piñeiro and Eduardo Trigo

THE ORIGIN OF DIFFERENT PRODUCTION SITUATIONS AND CONSEQUENT FRAGMENTATION OF THE PROCESS OF MODERNIZATION AND TECHNICAL CHANGE

One of the consequences of the development of industrial capitalism in Europe during the second half of the 19th century was the integration of Latin American nations into the world market as suppliers of raw materials. This process affected all of Latin America, but underlying it were specific situations which led to profound differences in the later development of each of the countries and regions. Among the factors contributing to these differences were: (a) the forms in which production was organized for each of the commodities offered on the international market by each country or region; (b) the diversity of resources originally available in these countries or regions; (c) the presence of adequate labor; (d) the abundance and relative fertility of agricultural lands. The combination of these factors, especially the relative availability of land and labor, gave rise to significantly different situations, ranging from plantation economies to the settlement of desert plains, that either reinforced or destroyed previous social and economic relations, thus affecting the rest of the economy and society as a whole.

After 1930, Latin American involvement in the world market developed new characteristics, and most countries experienced further growth. The traditional role of the agricultural sector was altered by rapid population growth and by the accelerated urbanization process experienced in most countries. This created a new performance pattern in many economies.

Adjustment that took place in agriculture in response to the new conditions reinforced the operations of capitalism. The State, in general, helped create the conditions required for integrating agriculture into the new process of accumulation, fitting agricultural production into the operations of capitalism. With this in mind, programs and policies were designed for lifting the restrictions inherent in pre-capitalist property relations. As a result, the more rudimentary forms of production were increasingly rejected in favor of a reassignment of the reserve labor force that had sustained traditional Latin American agriculture. In addition, the State implemented policies that encouraged capital formation in the sector and considerable efforts were made to develop public institutions for generating and transfering technology. Technological innovation played an important role in this process, encouraging the investment of capital and, in many cases, the replacement of the agricultural labor force, thus accelerating certain changes in production relations caused by modernization.

These processes did not occur everywhere. Instead, they were concentrated in areas where a certain degree of capitalist development had already taken place, or which could be easily transformed and absorbed into markets expanding rapidly as a consequence of industrialization. In turn, modernization and technical change heralded a new process of differentiation between the types of production that had been generated during the first stages of development. This led to the fragmentation of the primary sector into sub-sectors which displayed different degrees of development in their productive forces. As a result, the agricultural sector in Latin America reflects considerable diversity in terms of types of production relations. Most countries contain, in addition to the typically capitalist forms of production, situations comparable to plantations or haciendas in decay, as well as the different types of campesino units.

These forms of production reflect different degrees of decline and of integration into the capitalist mode of production, and are of considerable importance. In addition, the various conditions of production differ among themselves, depending on their role in the economy as a whole. These are the elements that produced the different technological processes that have developed in recent decades.

The production situations included in the PROTAAL studies illustrate technological processes that, despite their variety, are representative of the changes that occurred in the agricultural sector as a consequence of the modernization processes affecting Latin American economies in the post-World War II period. They also illustrate the State's role in bringing about these changes, especially in defining certain qualitative aspects of technical change.

AGRICULTURAL RESEARCH IN THE PUBLIC SECTOR

The Creation of National Research Institutes

State participation in agricultural research up to the mid-1950s was concentrated mainly at the level of state universities and the research departments of the Ministries of Agriculture. Impact from the work undertaken by these institutions was mostly modest. At that time, however, a new organizational structure started to take shape, in an effort to emulate the successes in agricultural research conducted by the public institutions of industrialized countries. Accordingly, during the 1960s and 70s, most countries in Latin America created national research institutes, financed by growing support from international sources and increased national budgetary allocations. This new institutional model was based on a decentralized administrative system, integration of functions, and broad coverage of products, regions and types of farmers; all in all, a clear expression of the dominant role of public institutions in agricultural technology.

Social Forces Related to the Creation of National Agricultural Research Institutes

The simultaneous creation of research institutions in a number of countries with different socioeconomic and political characteristics was a phenomenon which overrode specific national social processes, taking on dimensions of continental scope. This process was linked, on the one hand, to the ECLA proposal which sought to transform and modernize national economies through public sectoral actions, and, on the other, to the international ambiance resulting from the Punta del Este meeting that gave birth to a number of international assistance programs. The process also raises the following queries: Why does technology surface as an important social issue only in the 1960s? Where did the adopted institutional models originate? Why are all the agencies so similar? Why didn't the individual conditions of each country influence the organizational format taken by each institution?

A possible answer to the first question emerges from an analysis of public policies and institutions with regard to technology, which suggests that, until 1960, debates on the organization of research took a back seat to other public policies of greater urgency. These policies received more attention since they threatened the very existence of certain sectors of the economy (as in the case of agrarian reform), or had a very clear economic impact (as in price policies). It was during these years that the slow growth of the agricultural sector began to restrict economic development, making the modernization of the agricultural sector an essential policy objective. The answers to the other queries concern the special relationships existing between technological policy and social context.

The most important policy instruments for defining an appropriate economic context (like the price for products and official credit lines) are relatively specific for each agricultural product. This means that the government's negotiation process could be limited to the social sector directly involved with a specific product, focusing on limited and concrete points. In contrast, the definition of a technological policy, especially for creating, organizing, and funding technological agencies, required by its very nature the definition of a general strategy on agricultural development. Discussions and negotiations related to State decisions were more general and abstract, and involved a broader range of social sectors. This not only slowed down and complicated the statement of policy by the different sectors, but also increased the complexity of the State's role in mediating negotiated solutions. In addition, the State's bureaucratic apparatus also derived greater relative autonomy in the process of defining public policies. This greater autonomy, and the classic acceptance of international ideas on science and technology, explain why it was relatively easy to have an impact on the selection of research priorities; moreover, the institutional models adopted for generating and disseminating technology were often developed abroad. This gave rise in some cases to institutional models and research priorities which had not been discussed with the social sectors concerned with the specific conditions of production. Thus, policies were often inconsistent with the economic and political conditions of a given situation.

THE MODERNIZATION PROCESS

The creation of national institutes undoubtedly had a great impact on technological innovation in Latin America, stimulating renewed efforts by the public sector in the generation and transfer of technology, and mobilizing public opinion on the need for effective national effort in this field.

Technological improvement in some products in turn furthered the process of adoption of newer technology. As a result, and contrary to general opinion, yield and production increased considerably in several commodities in a number of countries of the continent. This reflects the increasing influence of technologies adopted from abroad, and the greater capability at national levels to take advantage of them, even generating, in some cases, solutions to local production problems.

The research institutes were created as part of a general strategy directed to the modernization of the public sector, which was to become the main agent of change in Latin American economies. In this sense, the strategy was successful. The success, however, was partial, occurring only in those production situations based on an economic structure and social organization which guaranteed the conditions needed for articulating and furthering technological change. Evidently, these social processes had a bearing on the types of technologies provided in response to the needs expressed by different productive sectors.

Social factors also affected the way in which the national institutions operated, the allocation of resources by these institutes, and the activities conducted by the private sector. They influenced economic policy and, through it, the demand for technology. This last point is significant, since it indicates that the modernization strategy followed was defective, in that it was based on the premise that the very existence of available technology was sufficient for initiating the process of technical change. However, empirical evidence clearly demonstrates that these processes were possible only when certain economic and social conditions were present which ensured that politically important segments of society benefitted from the change.

Two principal types of technical change processes have been observed. One resembles the paradigm of agricultural development in developed countries and especially the mid-western United States, as extensively studied by Owen (1966), Cochrane (1958), and Hayami and Ruttan (1971), among others. In this case, the innovation process appears to have been stimulated by non-agricultural social sectors attempting to solve problems related to non-agrarian capital accumulation. This process can be considered as part of agricultural development subordinate to the interests and needs of urban-industrial development. They are a function of a capitalist market economy, which stimulates economic development through a political system able to mediate the interests of diverse social sectors.

The second type of technical change process arose from agrarian initiatives, where specific productive sectors have the political power to affect sectoral public policies, allowing them to define the economic context which favors their economic expansion. These processes, although developed in other parts of the world, especially in the United States, during the past two decades (tomatoes in California), also seem to prevail in Latin American agricultural development.

Such situations implied an increase in the capital-intensive nature of technical change, notably accelerating the trend toward economic concentration and the vertical integration of the productive processes. Their very presence indicates that the agricultural development process underway is fragmented and differentiated. These technological processes, focused geographically and by commodity, are strongly controlled by agrarian capital, and are one important cause of the generalized proletarianization of campesino units and the failure to generate employment, characteristic of the modernization of agriculture in Latin America.

THE QUALITATIVE NATURE OF TECHNICAL CHANGE: IMPACT ON RURAL EMPLOYMENT

The processes of technical change had a lasting effect on the use of capital and on rural employment. Although all processes analyzed were capital intensive in nature, thereby using less labor per unit of output, labor displacement did not always occur because of the increase in total production.

In the cases identified as agrarian initiatives, those controlling capital had to ensure an adequate supply of labor and to maintain adequate control over it. This was hindered by labor unionization, stimulated by the large number of workers employed in one site. Case studies clearly indicate that in these cases dominant social sectors made use of their capability to mobilize public policies in favor of their own needs, with technology playing a major role in the strategy employed. An example of this is provided by the three types of actions taken by the sugar processing industry to reduce their dependence on the increasingly unionized paid labor: (a) the mechanization of labor-intensive work; (b) the use of the job-lot crews, organized, controlled and hired independently; and (c) purchasing sugar cane from independent growers, on a contract basis.[1]

In those cases characterized by intersectoral negotiation, technical change had two main effects:[2] (a) reducing overall labor requirements, where cultivating practices were mechanized before harvesting; and (b) the virtual elimination of migrant salaried workers for harvesting. In summary, the overall effect of technical change was a reduction of labor requirements in general, particularly in those areas not dominated by family labor or permanent wage labor.

It is interesting to note that capital intensive technologies were used even in those situations where the cost of labor decreased in relation to the price of capital goods. This would indicate a strong technological bias toward a more intensive use of capital. It also indicates the effect of technologies generated in the developed world on the supply of technology at national levels, and the absence of corrective measures within the international process of technology transfer.

The qualitative nature of technical change and a natural tendency to concentrate on commercial agriculture has, indirectly, had a detrimental effect on **campesino** economies. These, because of poor access to benefits from technical change, have progressively lost their ability to compete with commercial agriculture, and have been displaced from the production of crops with dynamic international markets and greater profitability (De Janvry and Crouch, 1981).

ECONOMIC DETERMINANTS OF THE QUALITATIVE NATURE OF TECHNOLOGY

The apparent relationship between the type of social process that generates the adoption of technology and the qualitative nature of the technology incorporated, suggests the existence of inducement mechanisms of the type described by Hayami and Ruttan (1971). It is important to specify the nature of these mechanisms and the way they operate. Research findings indicate that the technological processes studied were capital intensive, despite the fact that in most situations the price of capital increased more rapidly than wages. This capital bias, which showed up with different degrees of intensity in case studies, was in general unrelated to the relative market prices of the factors of production.

This can be explained by three interrelated facts. The first is the existence of factor market imperfections, especially where market prices of the factors do not reflect their real cost to the production unit. The most noteworthy example is the case of highly concentrated enterprises where the cost involved in administering large bodies of labor as a result of labor disputes and the risk of strikes are not reflected in the market price of the labor force.

The second refers to restrictions that are implicit in the type of knowledge available throughout the world, and the limited capacity of Latin American institutions to create original solutions to local production problems. As a result, know-how available at international levels has served as a frame of reference for national agencies in adapting new technologies. However, this knowledge generally responds to the capital-intensive technological requirements of industrialized countries. This situation illustrates some of the deficiencies of the institutional model used, one which has been fairly successful in adapting existing technologies to local conditions, but not very good at developing local responses to the production problems encountered.

This bias in the available technology has affected the entire process of technology adoption by the production unit. Given a universe of technologies available throughout the world—the meta production function (Ahmand, 1966; de Janvry and Martínez, 1972)—the productive unit is limited to selecting one of the available technologies. If this universe includes efficient but capital-intensive techniques, then the production unit is forced to adopt those which increase overall profits (reduction of average cost), even though they may result in a higher use of the factor with faster price increases, in this case capital (Salter, 1960; Hicks, 1964; Ahmad, 1966).

The third explanatory element is related to the qualitative nature of the inducement mechanisms. Data from the case studies undertaken by PROTAAL suggest that some inducement mechanisms were developed under agrarian initiatives, where the productive sectors, by co-opting public institutions, had considerable impact on the qualitative nature of research endeavours conducted by them (sugar cane, tomatoes, milk). However, this influence and the corresponding result were not necessarily in line with the relative availability, and therefore social value, of productive factors at the national level. Technology produced was directed to conform with the economic conditions of given production situations, including the specific characteristics of the productive units themselves, and their access to, availability of, and control over needed resources.

MODERNIZATION AND INSTITUTIONAL CHANGE

The modernization process over the past 20 years has brought about a set of institutional changes that have substantially modified the framework within which Latin American agricultural research systems function, at both national and international levels (Trigo and Piñeiro, 1981). Two forms of development of major importance over the past few years were: (a) the progressive internationalization of the technological process; and (b) the growing importance of the private sector in the innovative process, including the direct participation of farmers' associations in technology transfer and development activities. The transnational nature of the technological process is clearly linked to the world-wide creation of new knowledge and the increasingly important role played by the international centers, transnational corporations, and international trade of technological inputs, all of which define the type of technology to be supplied at the national level.

The participation of transnational firms in the production of technological inputs has reinforced the participation of the private sector in the innovative process. As they operate simultaneously in various markets, these firms can meet research investment requirements for developing new products not necessarily justified by the level of demand in a given country, but made feasible from an overall business viewpoint. Thus, the participation of private industry in the generation of agricultural technologies is increasing faster than the conditions in each country would seem to merit. This participation, together with the growing importance of technological inputs (as a result of the faster growth of commercial as compared to traditional agriculture), means that technology increasingly depends on international trade patterns and the capital investment programs of these firms.

The growing participation of the private sector and its relationship with transnational firms bring forward a fundamental issue: the appropriate balance between control and incentives to the private sector and the role of patents and other legal instruments.

Evidence presented by Evenson and Evenson suggests that developing countries have tended to develop legal systems that do not foster creativeness in the private sector, creating instead excessive reliance on the inventive capacity of the developed world. Petty patents systems, as developed in Brazil, Mexico and a few other countries, along with a greater reluctance to adhere to international conventions regarding patents and other systems of protection that work in favor of large, monopolistic firms, could foster more autonomous private research in Latin America.

The Fragmentation of Public Institutions

Recently, farmers' organizations have made substantial progress in managing their own research programs. In most cases, this has been possible where homogeneity of products and structure of production facilitated a joint effort with regard to the orientation of technology generation and transfer activities. Moreover, public funds are very frequently redirected to help finance these activities.

These processes have had a significant impact on institutional organization and the operation of State institutions. In Chapter 7 it was argued that when processes of technological articulation were promoted by agrarian initiatives, it meant the co-opting of certain state institutions by the agrarian sectors responsible for organizing the innovation process. This process of penetrating public organizations leads, in certain cases, to the fragmentation of the state's institutional apparatus. An analysis of case studies indicates that this process is particularly evident in research and extension institutions, although it shows up, to a lesser degree, in other areas of public policy. In these cases, the productive sectors carried out corporate actions in an attempt to influence or control the initiatives of the public sector in relation to technology. In some cases, it was possible to do this within the public institutions themselves, due to the political power of the productive sector (milk in Ecuador, livestock in Uruguay) or the nature of the research institution (tomatoes in California). In those cases where these actions could not be carried out, the productive sector had to dismantle the public system and develop private research extension activities under its own control (sugar in Colombia).

Apparently, the process of co-opting is less intensive in the areas of economic policy, due to the resistance of other competing social sectors. The case studies show, however, that when the agrarian sector took the initiative, it acquired considerable influence and participation in the institutional mechanisms through which price and credit policies were established.

NOTES

[1] The use of independent work crews allowed the sugar processors to pass labor responsibilities on to the contractors, thereby diminishing state control over compliance with labor laws, and reducing the possibilities for forming unions. Moreover, establishing commodity contracts diminished the need for hiring salaried workers on a permanent basis, while providing a guaranteed supply of raw material.

[2] Rice in Colombia and corn in the Argentine pampas clearly demonstrate this situation; see Balcázar and others (1980).

REFERENCES

Ahmad, S. (1966). On the theory of induced innovation. *Economic Journal 76*, 344–357.

Ahmad, S. and A. A. Kubursi (1977). *Induced Adjustment and the Role of Agriculture in Economic Development.* Working Paper 77–20, Department of Economics, McMaster University, Hamilton, Ontario.

Alonso, J. M. and C. Pérez (1980). Adopción de Tecnología en la Ganadería Vacuna Uruguaya. Study Series No. 14, Centro de Investigaciones Económicas, Uruguay.

Alves, M. and R. Fiorentino (1981). *La Modernización Agropecuaria en el Sertão de Pernambuco.* PROTAAL, Document No. 64. IICA, Miscelaneous Publication No. 282, San José, Costa Rica.

Ardila, J., E. Trigo and M. Piñeiro (1981). *Human Resources in Agricultural Research: Three Cases in Latin America.* PROTAAL, Document No. 50. IICA, San José, Costa Rica.

Arndt, T. M., D. G. Dalrymple and V. W. Ruttan (Eds.) (1977). *Resource Allocation and Productivity in National and International Research.* University of Minnesota Press, Minneapolis.

Balcázar, A. and others (1980). *Estudio del Proceso de Generación, Difusión y Adaptación de Tecnología en la Producción de Arroz en Colombia.* IICA, Bogotá.

Barbato, C. (1980). *El Proceso de Generación, Difusión y Adopción de Tecnología en la Ganadería Vacuna: Uruguay (1950–1977).* PROTAAL, Document No. 59. IICA, Miscelaneous Publication No. 263, San José, Costa Rica.

Barker, R. and R. Herdt (1978). Equity implications of technology changes. In IRRI (Ed.), *Interpretive Analysis of Selected Papers from Changes in Rice Farming in Selected Areas of Asia.* International Rice Research Institute, Philippines.

Barsky, O. and G. Cosse (1980). *Iniciativa Terrateniente, Cambio Técnico y Modelo Institucional: El Caso de la Producción Lechera en la Sierra Ecuatoriana.* PROTAAL, Document No. 60. IICA, Miscelaneous Publication No. 225, San José, Costa Rica.

Barsky, O. and others (1980). *El Proceso de Transformación de la Producción Lechera Serrana, y el Aparato de Generación y Transferencia en Ecuador.* PROTAAL, Document No. 40. Facultad Latinoamericana de Ciencias Sociales, Quito.

BCRA (various years). *Memorias Anuales.* Banco Central de la República Argentina.

Beckford, G. (1972). Strategies for agricultural development: comment. *Food Research Institute Studies, 11(2),* 149–154.

Ben-Zion, U. and V. W. Ruttan (1978). Aggregate demand and the rate of technical change. In H. Binswanger and V. W. Ruttan (Eds.), *Induced Innovation: Technology, Institutions and Development.* Johns Hopkins University Press, Baltimore.

Berndt, E. and M. Khaled (1979). Parametric productivity measurement and choice among flexible, functional farms. *Journal of Political Economy, 87,* 1220–1245.

Biggs, S. (1981). *Institutions and Decision-Making in Agricultural Research.* Overseas Development Institute Discussion Paper No. 5, London.

Binswanger, H. (1974a). A micro-economic approach to induced innovation. *Economic Journal, 84,* 940–958.

Binswanger, H. (1974b). The measurement of technical change biases with many factors of production. *American Economic Review, 64,* 964–976.

Binswanger, H. (1977). Measuring the impact of economic factors on the direction of technical change. In T. M. Arndt, D. G. Dalrymple, and V. W. Ruttan (Eds.), *Resource Allocation and Productivity in National and International Agricultural Research.* University of Minnesota Press, Minneapolis.

Binswanger, H. and V. W. Ruttan (1978). *Induced Innovation: Technology, Institutions and Development.* Johns Hopkins University Press, Baltimore.

Boulding, K. E. (1978). *Ecodynamics: A New Theory of Societal Evolution.* Sage View, London.

Boyce, J. and R. Evenson (1975). *National and International Agricultural Research and Extension Programs.* Agricultural Development Council, Inc., New York.

Braverman, H. (1974). *Labor and Monopoly Capital.* Monthly Review Press, New York.

Brenner, R. (1977). The origins of capitalist development: a critique of neo-Smithian Marxism. *New Left Review, 104,* 25–92.

Bronfenbrenner, M. (1965). Das Kapital for the modern man. *Science and Society,* Autumn, 205–206.

Brown, M. (1970). *On the Theory and Measurement of Technical Change*. Cambridge University Press, Cambridge.

Cain, L. and D. Peterson (1981). Factor biases and technical change in manufacturing: the American system, 1850–1919. *Journal of Economic History, 41*, 341–360.

Capdevilla, P. V. (1978). *La Estancia Argentina*. Plus Ultra, Buenos Aires.

Cardoso, F. H. and E. Faletto (1971). *Dependencia y Desarrollo en América Latina*, 3rd ed. Siglo XXI, México.

CGIAR (1981). *Second Review of the CGIAR*. Consultative Group on International Agricultural Research, Washington.

CGIAR Secretariat (1978). *1978 Report on the Consultative Group and the International Agricultural Research System: An Integrative Report*. World Bank, Washington, D.C.

CGIAR Secretariat (1979). *1979 Report on the Consultative Group and the International Agricultural Research System: An Integrative Report*. World Bank, Washington, D. C.

Chandler, Jr., R. (1979). *Rice in the Tropics: A Guide to Development of National Programs*. International Agricultural Development Service. Westview Press, Boulder.

CIAP (1973). *Evaluation de la Labor Institutional de la INTA*. Instituto Torcuato di Tella, Buenos Aires.

CIAT (1981). Trend highlights for CIAT's commodities. *CIAT Econ., 6(3)*.

CIDA (1966). *Tenencia de la Tierra y Desarrollo Socioeconómico en el Sector Agrícola Argentino*. Comité Interamericano de Desarrollo Agrícola. Pan American Union – OAS, Washington.

CIP (1980). *Profile: The International Potato Center, 1970–2000*. CIP, Lima.

Cleaver, H. (1972). The contradictions of the green revolution. *American Economic Review, 52*.

Cochrane, W. W. (1958). *Farm Prices: Myth and Reality*. University of Minnesota Press, Minneapolis.

Cohan, H. (1981). *El Escenario Agropecuario en América Latina y el Caribe en la Década de 1980*. IICA, San José, Costa Rica.

Coscia, A. and J. C. Torchelli (1968). *La Productividad en la Mano de Obra del Maíz*. INTA, Technical Report No. 79. Instituto Nacional de Tecnología Agropecuaria, Argentina.

Coto, R. (1967). El IICA y la OEA. In Instituto Interamericano de Ciencias Agrícolas and Asociación Latinoamericana de Fitotécnica (Eds.), *Las Ciencias Agrícolas en América Latina*. Editorial Trejos, San José, Costa Rica. pp. 465–506.

Coulter, J. K. (1981). *International Technology: The International Agricultural Research Centers*. IICA–UNDP, San José, Costa Rica.

da Silva, J. G. (1980). Progreso técnico e relacões de trabalho na agricultura paulista. Ph.D. Dissertation, Universidad Estadual de Campinas, Brazil.

Dagnino Pastore, J. M. (1965). *La Industria del Tractor en la Argentina*. Instituto Torcuato di Tella, Buenos Aires.

Dalrymple, D. G. (1978). *Development and Spread of High Yielding Varieties of Wheat and Rice in the Less Developed Nations*. USDA Foreign Agricultural Economic Report No. 95, Washington, D. C.

de Janvry, A. (1973). A socioeconomic model for induced innovation for Argentine agricultural development. *Quarterly Journal of Economics, 87*, 410–435.

de Janvry, A. (1975). The importance of small farmer technology for rural development. In *International Workshop on Economic Analysis in the Design of New Technology for the Small Farmer*. International Center of Tropical Agriculture, Cali.

de Janvry, A. (1977). Inducement of technological and institutional innovations: an interpretive framework. In T. M. Arndt, D. G. Dalrymple and V. W. Ruttan (Eds.), *Resource Allocation and Productivity in National and International Research*. University of Minnesota Press, Minneapolis.

de Janvry, A. (1978). Social structure and biased technical change in Argentine agriculture. In H. Binswanger and V. W. Ruttan (Eds.), *Induced Innovation: Technology, Institutions and Development*. Johns Hopkins University Press, Baltimore. pp. 297–323.

de Janvry, A. and J. C. Martínez (1972). Introducción de innovaciones y desarrollo agropecuario argentino. *Económica, 18(2)*, 63–95.

de Janvry, A. and L. Crouch (1981). *Technological Change and Peasants in Latin America*. PROTAAL, Document No. 56. IICA, Miscelaneous Publication No. 226, San José, Costa Rica.

de Janvry, A., P. LeVeen and D. Runsten (1981). *The Political Economy of Technological Change:*

Mechanization of Tomato Harvesting in California. PROTAAL, Document No. 63. IICA, San José, Costa Rica.

del Carril, B. (1892). Praderas de Alfalfa en la República Argentina. *Sociedad Rural Argentina, XXVI*, 273–274.

Díaz Alejandro, C. (1975). *Ensayo sobre la Historia Económica Argentina.* Amorrotu, Buenos Aires.

Dobb, M. (1963). *Studies in the Development of Capitalism.* International Publishers, New York.

Drandakis, E. and E. Phelps (1966). A model of induced innovation, growth and distribution. *The Economic Journal, 76*, 823–840.

Draper, A. and H. Draper (1968). *The Dirt on California: Agribusiness and the University.* Independent Socialist Clubs of America, Berkeley.

ECLA (1959). *El Desarrollo Económico para América Latina, Vol. I.* United Nations, Mexico.

Edquist, C. and O. Edquist (1979). *Social Carriers of Techniques for Development.* Sarec Report R3, Sweden.

Elgueta, M. (1967). Evolución de la investigación agrícola en América Latina. In Instituto Interamericano de Ciencias Agrícolas and Asociación Latinoamericana de Fitotécnica (Eds.), *Las Ciencias Agrícolas en América Latina.* Editorial Trejos, San José, Costa Rica. pp. 125–141.

Evenson, R. (1968). The contribution of agricultural research and extension to agricultural production. Ph.D. dissertation, University of Chicago, Chicago.

Evenson, R. (1973). Comparative evidence on returns to investment in national and international research institutions. In W. Fishel (Ed.), *Resource Allocation in Agricultural Research.* University of Minnesota Press, Minneapolis.

Evenson, R. (1977). Cycles of research productivity in sugar cane, wheat and rice. In T. M. Arndt, D. G. Dalrymple, and V. W. Ruttan (Eds.), *Resource Allocation and Productivity in National and International Research.* University of Minnesota Press, Minneapolis, pp. 209–236.

Evenson, R. and Y. Kislev (1975). *Agricultural Research and Productivity.* Yale University Press, New Haven.

Evenson, R., P. Waggoner and V. W. Ruttan (1979). Economic benefits from research: an example from agriculture. *Science, 205*, 1101–1107.

FAO (various years). *FAO Year Books.* Office of Statistics, Department of Economic and Social Policy, Rome.

FBB (1976). *Argentina, Evolución Económica 1915–1976.* Fundación Banco de Boston, Buenos Aires.

FEDESARROLLO (1976). *Las Industrias Panelera y Azucarera en Colombia.* Fundación para la Educación Superior y el Desarrollo, Bogotá.

FEDESARROLLO (1980). Coyuntura económica: análisis y perspectivas de la economía colombiana. *Presencia, 7(4)*, 204.

Feeny, D. (1976). Technical and institutional change in Thai agriculture, 1880–1940. Ph.D. dissertation, University of Wisconsin, Madison.

Feeny, D. (1978). Induced technical and institutional change: a Thai case study. In G. Mears (Ed.), *The Past in Southeast Asia's Present.* Canadian Society for Asian Studies, Ottawa. pp. 56–69.

Fellner, W. (1960). *Productivity and Technical Change.* Cambridge University Press, Cambridge.

Fellner, W. (1961). Two propositions in the theory of induced innovations. *Economic Journal, 71*, 305–308.

Fellner, W. (1971). Empirical support for the theory of induced innovation. *Quarterly Journal of Economics, 85*, 580–604.

Ferrer, A. (1963). *La Economía Argentina.* Fondo de Cultura Económica, México.

Fiorentino, R. (1977). Uma visão geral dos problemas de emprego e renda no sector rural do nordeste brasileiro. *Revista Económica do Nordeste, 22*, 30.

Fishel, W. (Ed.) (1971). *Resource Allocation in Agricultural Research.* University of Minnesota Press, Minneapolis.

Fisk, E. (1978). From communities to commodities: the private domination of agricultural research policy in California, 1942–1948. Paper presented at annual meeting of the Rural Sociological Society, San Francisco.

Flichman, G. (1977). *La Renta del Suelo y el Desarrollo Agrario Argentino.* Siglo XXI, México.

Flores, O. (1977). An historical analysis of Peru's agricultural export sector and the development of agricultural technology. Ph.D. dissertation, University of Wisconsin, Madison.

Fuller, V. (1939). The supply of agricultural labor as a factor in the evolution of farm organization in California. Ph.D. dissertation, University of California, Berkeley.
Gaitskell, A. (1959). *Gezira: A Story of Development in the Sudan.* Faber and Faber, London.
Germani, G. (1962). *Política y Sociedad en una Epoca de Transición.* Paidós, Buenos Aires.
Giberti, E. H., D. W. Norman, and F. E. Winch (1980). *Farming Systems Research: A Critical Appraisal.* Paper No. 6, MSU Rural Development, Department of Agricultural Economics, Michigan State University, Lansing.
Giberti, H. (1964). *El Desarrollo Argentino.* Editorial Universitaria de Buenos Aires, Buenos Aires.
Giberti, H. (1974). *Historia Económica de la Ganadería Argentina.* Solar/Hackette, Buenos Aires.
Gintis, H. (1978). The nature of labor exchange and the theory of capitalist production. *Review of Radical Political Economics, 8,* 36–54.
Gómez, G. and A. Pérez (1979). El proceso de modernización de la agricultura latinoamericana. *Revista de la CEPAL, 8.*
Graboski, R. (1979). Implications of an induced innovation model. *Economic Development and Cultural Change, 27,* 723–734.
Grela, P. (1958). *El Grito de Alcorta: Historia de la Rebelión Campesina de 1912.* Tierra Nuestra, Rosario, Argentina.
Griliches, Z. (1957). Hybrid corn: an exploration of the economics of technological change. *Econométrica, 25,* 501–522.
Griliches, Z. (1968). Agriculture: productivity and technology. In *International Encyclopedia of the Social Sciences,* Vol. 1. MacMillan and Free Press, New York. pp. 241–245.
Gutman, J. M. (1978). Interest groups and the demand for agricultural research. *Journal of Political Economy, 86(31).*
Haines, K. A. (1967). Los países amigos. In Instituto Interamericano de Ciencias Agrícolas y Asociación Latinoamericana de Fitotécnica (Eds.), *Las Ciencias Agrícolas en América Latina.* Editorial Trejos, San José, Costa Rica. pp. 453–464.
Hartmans, E. H. (1981). *Land Development and Management in Tropical Africa.* International Institute of Tropical Agriculture, Ibadan, Nigeria.
Hayami, Y. and V. W. Ruttan (1970). Factor prices and technical change in agricultural development: the United States and Japan, 1880–1960. *Journal of Political Economy, 78,* 1115–1141.
Hayami, Y. and V. W. Ruttan (1971). *Agricultural Development: An International Perspective.* Johns Hopkins University Press, Baltimore.
Hayami, Y. and others (1975). *A Century of Agricultural Growth in Japan.* University of Minnesota Press and University of Tokyo Press, Minneapolis and Tokyo.
Hewitt de Alcántara, C. (1976). *Modernizing Mexican Agriculture: Socio-economic Implications of Technological Change, 1940–1970.* United Nations Research Institute for Social Development, Geneva.
Hicks, J. (1964). *The Theory of Wages.* McMillan, London.
Hirschman, A. O. (1977). A generalized linkage approach to development with special reference to staples. *Economic Development and Cultural Change, 25,* 67–98.
Huret, J. (1913). *De La Plata á la Cordillere des Andes.* Eugéne Fasquelle, Paris.
ICA (1976). *Memoria.* Instituto Colombiano Agropecuario, Bogotá.
IFPRI (1980). *Annual Report for 1980.* International Food Policy Research Institute, Washington, D.C.
IITA (1981). *Tasks for the Eighties: A Long Range Plan.* International Institute of Tropical Agriculture, Ibadan, Nigeria.
INEC (1950–1978). *Series de Comercio Exterior.* Instituto Nacional de Estadísticas y Censos, Buenos Aires.
IRRI (1977). *Constraints on High Yields on Asian Rice Farms: An Interim Report.* International Rice Research Institute, Philippines.
Jorgenson, D. and L. Lau (1973). *Duality and Differentiability in Production.* Discussion Paper No. 308, Harvard Institute of Economic Research, Cambridge.
Kalmanovitz, S. (1978). *Desarrollo de la Agricultura en Colombia.* La Carreta, Bogotá.
Kapp, R. and V. K. Smith (1981). The measurement of non-neutral technological change. Paper presented at University of North Carolina, Department of Economics.

Katz, J. (1978). *Cambio Tecnológico, Desarrollo Económico y las Relaciones Intra y Extra–Regionales de la América Latina.* Programa BID/CEPAL de Investigaciones en Temas de Ciencia y Tecnología, Monografía de Trabajo No. 30, Buenos Aires.

Katz, T. and R. Cibotti (1976). *Marco de Referencia para un Programa de Investigación en Temas de Ciencia y Tecnología en América Latina.* Programa BID/CEPAL de Investigaciones en Temas de Ciencia y Tecnología, Buenos Aires.

Kay, C. (1974). Comparative development of the European manorial system and the Latin American hacienda system. *Journal of Peasant Studies, 2,* 69–98.

Kennedy, C. (1964). Induced bias in innovation and the theory of distribution. *The Economic Journal, 74,* 541–547.

Kislev, Y. and W. Peterson (1981). Induced innovation and farm mechanization. *American Journal of Agricultural Economics.*

Krug, C. A. (1967). La FAO y las Naciones Unidas. In Instituto Interamericano de Ciencias Agrícolas and Asociación Latinoamericana de Fitotécnica (Eds.), *Las Ciencias Agrícolas en América Latina.* Editorial Trejos, San José, Costa Rica. pp. 383–402.

Lahitte, E. (1912). Crédito agrícola. In Dirección General de Economía Rural y Estadística Agrícola (Ed.), *La Cooperación Rural.* Ministerio de Agricultura, Buenos Aires.

Lazo, J. (1980). *Situación Mundial de la Productividad de Maíz, Arroz, Papa, Caña de Azúcar y Leche.* IICA, San José, Costa Rica.

Leach, G. (1979). Energy report. Presented at the conference "Agricultural Production: Research and Development Strategies for the 1980s." GTZ, Bonn.

Lele, V. (1981). Rural Africa: modernization, equity and long-term development. *Science, 211,* 547–553.

Lipset, S. M. (1968). *Agrarian Socialism: The Cooperative Common Health Federation in Saskatchewan.* Doubleday, New York.

Lucas, Jr., R. (1967). Tests of a capital–theoretic model of technological change. *Review of Economic Studies, 34,* 175–180.

Mallon, R. and J. Sourrouille (1976). *La Política Económica en una Sociedad Conflictiva: El Caso Argentino.* Amorrotu, Buenos Aires.

Martínez, J. C. (1973). On the economics of technological change: induced innovations in Argentine agriculture. Ph.D. dissertation, Iowa State University, Ames.

Martínez, J. C., D. Fienup, and C. Chevallier (1976). *Aspectos Económicos y Tecnológicos de la Producción Cerealera Argentina: Trigo, Maíz, Sorgo.* Centro Internacional del Mejoramiento de Maíz y Trigo, México.

Martínez, J. C., M. Piñeiro, and C. Chevallier (1976). Nuevamente en torno al problema de asignación de recursos en el sector agropecuario pampeano. *Desarrollo Económico IDES, 16(61).*

Martínez de Hoz, J. A. (1961). *La Agricultura y la Ganadería Argentina en el Período 1930–1960.* Sudamericana, Buenos Aires.

Marx, K. (1966). *The Capital.* Fondo de Cultura Económica, México.

Marzocca, A. (1967). Los pioneros. In Instituto Interamericano de Ciencias Agrícolas and Asociación Latinoamericana de Fitotécnica (Eds.), *Las Ciencias Agrícolas en América Latina.* Editorial Trejos, San José, C.R. pp. 27–66.

McConnell, G. (1958). *The Decline of Agrarian Democracy.* University of California Press, Berkeley.

McLean, I. (1979). *The Analysis of Agricultural Productivity: Alternative Views and Victorian Evidence.* Working Paper 79–4, Economics Department, University of Adelaide, Adelaide.

Mellor, J. (1977). Relating research resource allocation to multiple goals. In T. M. Arndt, D. G. Dalrymple and V. W. Ruttan (Eds.), *Resource Allocation and Productivity in National and International Research.* University of Minnesota Press, Minneapolis.

Montes, M., R. Candelo, and A. Muñoz de Gaviria (1980). La economía del arroz en Colombia. *Revista de Planeación y Desarrollo, 12(1),* 73–131.

Murmis, M. and J. C. Portantiero (1968). *Crecimiento Industrial y Alianza de Clases en la Argentina (1930–1940).* Instituto Torcuato di Tella, Buenos Aires.

Nelson, R. and S. Winter (1973). Toward an evolutionary theory of economic capabilities. *American Economic Review, 63,* 440–449.

Nelson, R. and S. Winter (1974). Neoclassical vs. evolutionary theories of economic growth: critique and prospectus. *Economic Journal, 84,* 886–905.

Nelson, R. and S. Winter (1975). Factor price changes and substitution in an evolutionary model. *The Bell Journal of Economics, 6(2)*.

Nelson, R. and S. Winter (1977). In search of useful theory of innovation. *Research Policy, 6*, 36–76.

Nelson, R., S. Winter, and H. Schuette (1976). Technical change in an evolutionary model. *Quarterly Journal of Economics, 40*, 90–118.

Nordhaus, W. (1967). The optimal rate and direction of technical change. In K. Shell (Ed.), *Essays in the Theory of Optimal Economic Growth.* MIT Press, Cambridge.

Nordhaus, W. (1973). Some skeptical thoughts on the theory of induced innovation. *The Quarterly Journal of Economics, 87*, 209–219.

Norman, D. (1980). *El Método de Investigación de Sistemas Agropecuarios: Su Pertinencia para el Pequeño Productor.* Report No. 5, Rural Development Studies Series, Department of Agricultural Economics, Michigan State University, Lansing.

Norton, G. and others (1981). *Evaluation of Agricultural Research.* University of Minnesota Agricultural Experiment Station Miscelaneous Publication 8–1981, St. Paul.

O'Donnell, G. (1976). *Apuntes para una Teoría del Estado.* CEDES, Buenos Aires.

O'Donnell, G. (1977). Reflexiones sobre las tendencias de cambio del estado burocrático autoritario. *Revista Mexicana de Sociología, 39(1)*.

OECEI (1973). *Argentina Económica y Social.* FIAT – Oficina de Estudios para la Colaboración Económica Internacional, Buenos Aires.

Olcese, O. (1967). La moderna fundación y su ingreso al campo agrícola en la América Latina. In Instituto Interamericano de Ciencias Agrícolas y Asociación Latinoamericana de Fitotécnica (Eds.), *Las Ciencias Agrícolas en América Latina.* Editorial Trejos, San José, Costa Rica. pp. 507–522.

Oliveira, F. (1975). *Elegía para uma Regiao.* Paz e Tene, Rio de Janeiro. pp. 120–123.

Oszlak, O. (1978). Formación histórica del estado en América Latina: elementos teoricometodológicos para su estudio. *Estudios CEDES, 1*, 46.

Oszlak, O., J. E. Roulet, and J. F. Sábato (1971). *Determinación de Objetivos y Asiganación de Recursos en el INTA.* CIAP – Instituto Torcuato di Tella, Buenos Aires.

Owen, W. (1966). The double development squeeze on agriculture. *American Economic Review, 56*, 43–70.

Pastore, J. and E. Alves (1977). Reforming the Brazilian agricultural system. In T. M. Arndt, D. G. Dalrymple, and V. W. Ruttan (Eds.), *Resource Allocation and Productivity in National and International Agricultural Research.* University of Minnesota Press, Minneapolis. pp. 394–403.

Pereda, H. V. (1939). *La Ganadería Argentina es una Sola.* Printed by author, Buenos Aires.

Piñeiro, M. and E. Trigo (1977). *Un Marco General para el Análisis del Progreso Tecnológico.* PROTAAL, Document No. 3. IICA, Miscelaneous Publication No. 149, Bogotá.

Piñeiro, M., and E. Trigo (1983). *Procesos Sociales e Innovación Tecnológica en la Agricultura de América Latina.* IICA, San José, Costa Rica.

Piñeiro, M., E. Trigo and R. Fiorentino (1977). *La Generación y Transferencia de Tecnología Agropecuaria: Notas sobre la Funcionalidad de los Centros Nacionales de Investigación.* PROTAAL, Document No. 9 (Addendum to Document 6). IICA, Serie de Cursos, Informes y Reuniones No. 138, Bogotá.

Piñeiro, M., E. Trigo and R. Fiorentino (1979). Technical change in Latin American agriculture: a conceptual framework for its interpretation. *Food Policy, 4*, 169–177.

Piñeiro, M., J. C. Martínez and C. Armelin (1975). *Política Tecnológica para el Sector Agropecuario,* INTA–EPGCA, Departamento de Economía, Serie Investigación No. 18, Castelar, Argentina.

Piñeiro, M. and others (1979). *El Proceso de Generación, Difusión y Adopción de Tecnología en la Producción Azucarera de Colombia.* PROTAAL, Document No. 37. IICA, Bogotá.

Piñeiro, M. and others (1982). *Articulación Social y Cambio Técnico: El Caso del Azúcar en Colombia.* IICA, San José, Costa Rica.

Pizarro, J. and M. G. Carcciamani (1979). *Insumos de Mano de Obra en Cultivos Agrícolas.* INTA, Basic Information No. 5, Buenos Aires.

Portantiero, J. C. (1977). Economía y política en la crisis argentina, 1958–1973. *Revista Mexicana de Sociología, 39(2)*.

PREALC (1976). *El Problema del Empleo en América Latina: Situación, Perspectivas, y Políticas.* Programa Regional de Empleo para América Latina y el Caribe, Santiago de Chile.

President's Council of Economic Advisers (1978). *Economic Report of the President.* Washington, D. C.

Proyecto Cooperativo de Investigación sobre Tecnología Agropecuaria en América Latina (1978). *Asignación de Prioridades y Recursos a la Investigación Agropecuaria en Colombia.* PROTAAl, Document No. 13. IICA, Serie de Informes, Cursos y Reuniones No. 153, Bogotá.

Ras, N. and R. Levis (no date). *El Precio de la Tierra: Su Evolución entre los Años 1916 y 1978.* Sociedad Rural Argentina, Buenos Aires.

Reca, L. G. (1967). The price and production duality within Argentine agriculture, 1923–1965. Ph.D. dissertation, University of Chicago, Chicago.

Rosenberg, N. (1976). Marx as a student of technology. *Monthly Review, 28(3).*

Ruttan, V. W. (1971). Technology and the environment. *American Journal of Agricultural Economics, 53,* 707–717.

Ruttan, V. W. (1978). Induced innovation in socialist agriculture. Mimeograph, Department of Agricultural Economics, University of Minnesota, Minneapolis.

Ruttan, V. W. (1980). Bureaucratic productivity: the case of agricultural research. *Public Choice, 35,* 529–547.

Ruttan, V. W. (1982). *Agricultural Research Policy.* University of Minnesota Press, Minneapolis.

Ruttan, V. W., H. Binswanger and Y. Hayami (1980). Induced innovation in agriculture. In C. Bliss and M. Boserup (Eds.), *Economic Growth and Resources: Natural Resources,* Vol. 3. Proceedings of Fifth World Congress of the International Economic Association. MacMillan, London. pp. 162–189.

Sábato, J. F. (1980). *El Agro Pampeano Argentino y la Adopción de Tecnología entre 1950–1978: Un Análisis a través del Cultivo del Maíz.* PROTAAL, Document No. 58. IICA, Miscellaneous. Publication No. 262, San José, Costa Rica.

Sagasti, F. (no date). A review of schools of thought on science technology, development and technical change. Manuscript, STPI Project.

Salter, W. E. (1960). *Productivity and Technical Change.* Cambridge University Press, Cambridge.

Samper, A. (1977). National systems of agricultural research in Latin America. Paper presented at the preparatory meeting for the International Congress on Potential for Cooperation between National Agricultural Research Systems, Bellagio, Italy.

Samuelson, P. (1965). A theory of induced innovation along Kennedy – Weizacher lines. *Review of Economics and Statistics, 47.*

Sanders, J. and V. W. Ruttan (1978). Biased choice of technology in Brazilian agriculture. In H. Binswanger and V. W. Ruttan (Eds.), *Induced Innovation: Technology, Institutions, and Development.* Johns Hopkins University Press, Baltimore. pp. 276–296.

Schmidt, A. and D. Seckler (1970). Mechanized agriculture and social welfare: the case of the tomato harvester. *American Journal of Agricultural Economics, 52.*

Schmookler, J. (1966). *Invention and Economic Growth.* Harvard University Press, Cambridge.

Schmookler, J. (1972). *Patents, Invention and Economic Growth: Data and Selected Essays.* Harvard University Press, Cambridge.

Schultz, T. W. (1964). *Transforming Traditional Agriculture.* Yale University Press, New Haven.

Schultz, T. W. (1968). *Economic Growth and Agriculture.* McGraw Hill, New York.

Schultz, T. W. (1980). Economics and agricultural research. *Desarrollo Rural en las Américas, 12,* 171–180.

Schumacher, E. F. (1975). *Small Is Beautiful: Economics as if People Mattered.* Perennial, New York.

Scobie, G. (1968). *Revolución en las Pampas: Historia Social del Trigo Argentino, 1860–1910.* Solar/Hackette, Buenos Aires.

Scobie, G. and R. Posada (1976). *The Impact of High Yielding Rice Varieties in Latin America with Special Emphasis on Colombia.* International Center of Tropical Agriculture, Cali.

Scobie, G. and R. Posada (1978). The impact of technical change on income distribution: the case of rice in Colombia. *American Journal of Agricultural Economics, 60(1),* 85–92.

Sen, A. K. (1959). The choice of agricultural techniques in underdeveloped countries. *Economic Development and Cultural Change, 7,* 279–285.

Sen, A. K. (1962). *Choice of Techniques.* Basil Blackwell, Oxford.

Silva, J. F. (1981). Brazil's gasoline replacement program. *Foreign Agriculture, 19(5)*.

Smith, A. (1958). *The Wealth of Nations*. Fondo de Cultura Económica, México.

Smith, P. H. (1968). *Carne y Política en la Argentina*. Paidós, Buenos Aires.

Smith, V. K. (1974). *Technical Change, Relative Prices, and Environmental Resource Evaluation*. Johns Hopkins University Press, Baltimore.

Sraffa, P. (1960). *Production of Commodities by Means of Commodities*. Cambridge University Press, Cambridge.

Sweezy, P. (1978). A critique. In R. Hilton (Ed.), *The Transition from Feudalism to Capitalism*. Verso, London.

TAC Secretariat (1979). *TAC Review of Priorities for International Support to Agricultural Research*. FAO, Rome.

TAC Secretariat (1980). *Report on the Stripe Analysis of the Off-Campus Activities of the International Agricultural Research Centers*. FAO, Rome.

Takahashi, K. (1978). A contribution to the discussion. In R. Hilton (Ed.), *The Transition from Feudalism to Capitalism*. Verso, London.

Taylor, C. (1948). *Rural Life in Argentina*. The Louisiana University Press, Baton Rouge.

Trigo, E. and M. Piñeiro (1981a). Dynamics of agricultural research organization in Latin America. *Food Policy, 6,* 2–10.

Trigo, E. and M. Piñeiro (1981b). *La Investigación Agropecuaria a Nivel Nacional en América Latina: Problemas y Perspectivas en la Década de 1980*. PROTAAL, Document No. 77. IICA, San José, Costa Rica.

Trigo, E., M. Piñeiro and J. Ardila (1979). *Aspectos Institucionales de la Investigación Agropecuaria en América Latina: Problemas y Perspectivas*. CIGTAT, Document No. 1. IICA, San José, Costa Rica.

Trigo, E., M. Piñeiro and J. Ardila (1982). *Organización de la Investigación Agropecuaria en América Latina*. IICA, San José, Costa Rica.

Trigo, E., M. Piñeiro and J. Chapman (1981). *Assigning Priorities to Agricultural Research: A Critical Evaluation of the Use of Programs by Product–Line and Production Systems*. PROTAAL, Document No. 70. IICA, San José, Costa Rica.

Trigo, E., M. Piñeiro and J. F. Sábato (1981). *La Cuestión Tecnológica y la Organización de la Investigación Agropecuaria en América Latina*. PROTAAL, Document No. 71. IICA, San José, Costa Rica.

Trigo, E., R. Fiorentino and M. Piñeiro (1978). *Notas Comparativas Sobre la Evolución de la Producción y Productividad de Productos Agropecuarios en Colombia y en Países Seleccionados de América Latina y el Resto del Mundo*. PROTAAL, Document No. 23. IICA, Miscellaneous Publication No. 178, Bogotá.

Turner, W. (1965). No dice for braceros. *Ramparts, 4,* No. 5.

U. S. Congress Report (1981). USC 161 and 163. Washington, D. C.

Valdes, A., G. Scobie and J. Dillon (Eds.) (1979). *Economics and the Design of Small Farmer Technology*. Iowa State University Press, Ames.

Wade, W. (1973). Institutional determinants of technical change and productivity growth: Denmark, France and Great Britain, 1880–1965. Ph.D. dissertation, University of Minnesota, Minneapolis.

Wan, Jr., H. Y. (1971). *Economic Growth*. Harcourt, Brace, Johanovich, New York.

Weber, A. (1973). *Productivity of German Agriculture, 1850 to 1970*. Department of Agricultural and Applied Economics Staff Paper 73–1. University of Minnesota, St. Paul.

White, D. and others (1979). Análisis económico de la maquinaria agrícola. *Convenio, 1(6),* 15–29.

WIPO (1965–1972). *Reports*. World Intellectual Property Organization, Geneva.

Yamada, S. and V. W. Ruttan (1980). International comparisons of productivity in agriculture. In J. Kendrick and B. Vaccara (Eds.), *New Developments in Productivity Measurement and Analysis*. National Bureau of Economic Research Studies in Income and Wealth, Vol. 44. The University of Chicago Press, Chicago. pp. 509–594.